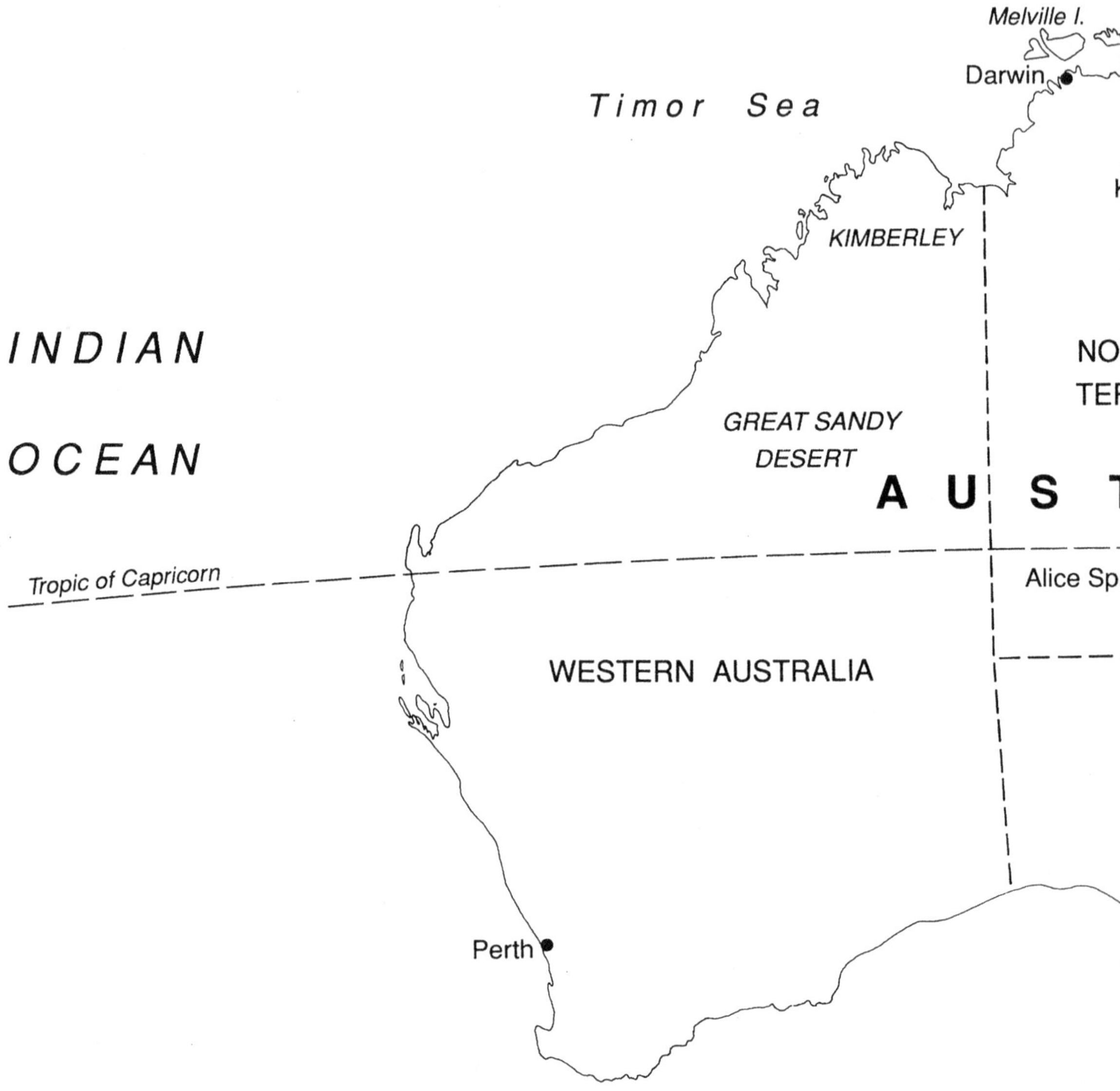
Melville I.
Darwin
Timor Sea
KIMBERLEY
INDIAN
OCEAN
GREAT SANDY
DESERT
NOR
TER
A U S T
Tropic of Capricorn
Alice Spr
WESTERN AUSTRALIA
Perth

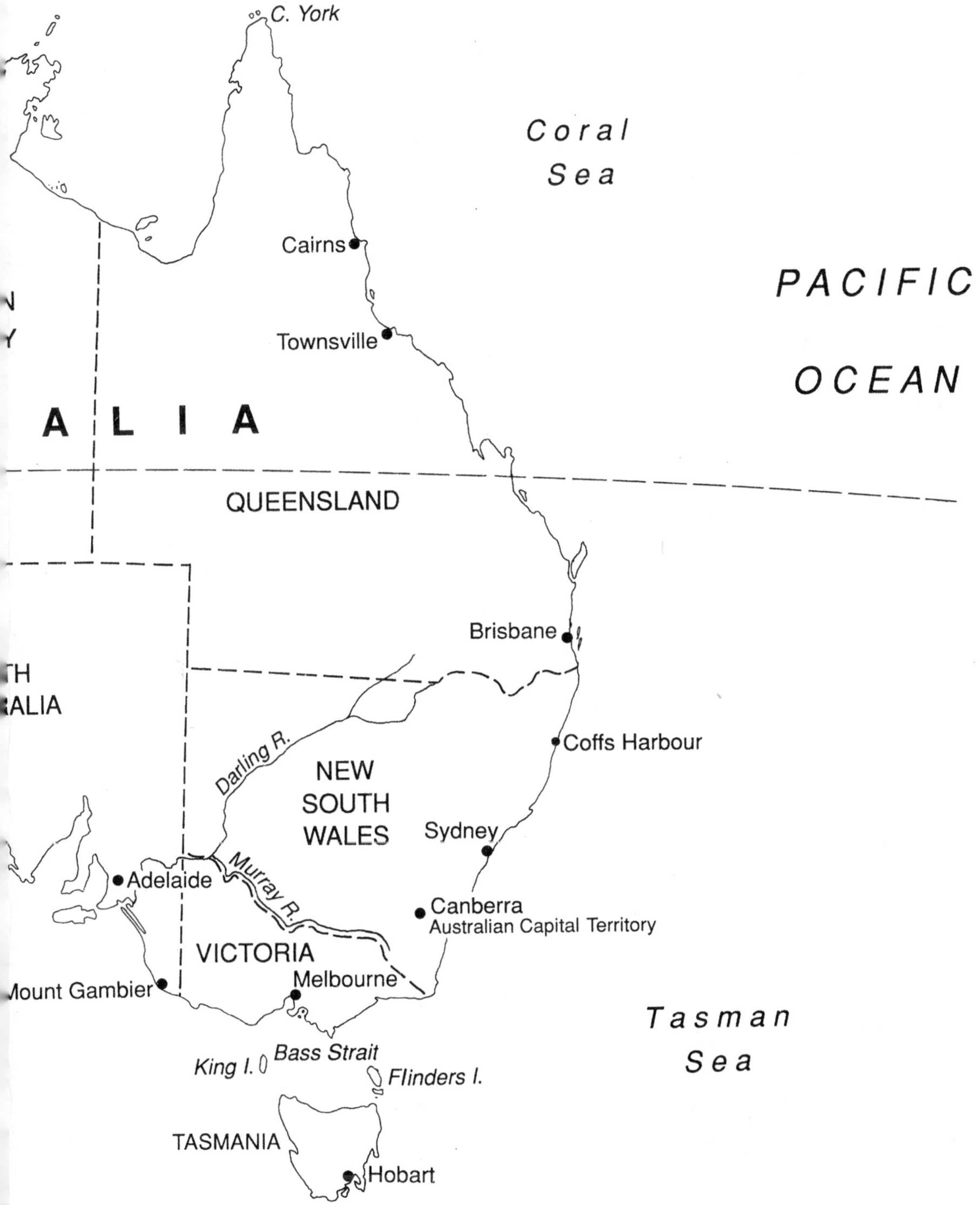
C. York
Coral Sea
Cairns
PACIFIC OCEAN
Townsville
ALIA
QUEENSLAND
Brisbane
TH
ALIA
Coffs Harbour
Darling R.
NEW SOUTH WALES
Sydney
Murray R.
Adelaide
Canberra
Australian Capital Territory
VICTORIA
Melbourne
Mount Gambier
Tasman Sea
Bass Strait
King I.
Flinders I.
TASMANIA
Hobart

Eucalypt Domestication
and
Breeding

Eucalypt Domestication and Breeding

KEN ELDRIDGE

CSIRO Forestry and Forest Products
Canberra, Australia

JOHN DAVIDSON

Eucalyptus and Forestry Services
Armidale, Australia

CHRIS HARWOOD

CSIRO Forestry and Forest Products
Canberra, Australia

and

GERRIT VAN WYK

Faculty of Forestry, University of Stellenbosch
Matieland, South Africa

CLARENDON PRESS • OXFORD

This book has been printed digitally in order to ensure its continuing availability

OXFORD
UNIVERSITY PRESS

Great Clarendon Street, Oxford OX2 6DP

Oxford University Press is a department of the University of Oxford.
It furthers the University's objective of excellence in research, scholarship,
and education by publishing worldwide in

Oxford New York

Auckland Bangkok Buenos Aires Cape Town Chennai
Dar es Salaam Delhi Hong Kong Istanbul Karachi Kolkata
Kuala Lumpur Madrid Melbourne Mexico City Mumbai Nairobi
São Paulo Shanghai Singapore Taipei Tokyo Toronto

with an associated company in Berlin

Oxford is a registered trade mark of Oxford University Press
in the UK and in certain other countries

Published in the United States
by Oxford University Press Inc., New York

First published 1993

First published in paperback 1994

Reprinted 1997, 2001

A catalogue record for this book is available from the British Library

Library of Congress Cataloging in Publication Data
Eucalypt domestication and breeding/Ken Eldridge ... [et al.].
Includes bibliographical references and index.
1. Eucalyptus. 2. Eucalyptus - Breeding. I. Eldridge, Ken.
SD397.E8E87 1993 634.9'7342—dc20 93-10412

ISBN 0-19-854866-4

To

Max Jacobs and Lindsay Pryor

whose infectious enthusiasm for the eucalypts
has influenced nearly every page

Preface

Forest plantations are costly to establish, and land available for planting trees is limited. It is therefore disappointing to see a poor stand of eucalypts growing on land that would support much better trees and a more valuable crop if planting stock of higher genetic quality had been used. The co-authors, having worked on the domestication and breeding of eucalypts in many countries, wrote this book as our contribution to increasing the benefits from growing eucalypts as a useful renewable resource. Our aim is to show how the imaginative use of biological knowledge leads to dramatic genetic improvement as wild eucalypts are brought into cultivation.

When the IUFRO Working Party on Breeding Eucalypts was formed in 1977 the four co-chairmen, Arno Brune, John Davidson, Ken Eldridge, and Gerrit van Wyk, decided that writing a book about eucalypt seed orchards would be a worthwhile project for the new group. We circulated a draft outline in 1977 but there was little progress in preparing drafts until 1981 when John Davidson came to Canberra on sabbatical leave as a Visiting Scientist at CSIRO. The co-authors met at a IUFRO meeting at Bordeaux in France in 1983 where we decided to approach Oxford University Press to publish a textbook with wide scope. Ken Eldridge was appointed co-ordinating editor. In 1984 he started editing and re-writing in Oxford where he was a Visiting Fellow at Wolfson College and the Commonwealth Forestry Institute. As well as drafting several chapters himself, his task was to edit each author's contributions so that the finished text would have a logical flow and a continuity of style. We decided not to attribute chapters to particular authors as two or three contributed to each chapter.

As the project developed there were changes in authorship and content. Arno Brune, who had contributed to planning the structure of the book and early drafts, withdrew in 1991 and was replaced by Chris Harwood who had drafted several of the species chapters. The scope of the book changed to concentrate on the first generation or two of domestication of 'wild' eucalypts for wood production, covering those aspects of variation, selection, and reproduction which are unique to eucalypts. We decided not to provide much detail on advanced generation breeding and genetic engineering which we considered are not unique to eucalypts and are well covered in other textbooks. We see a continuing need to use the sound genetic principles set out in this book for eucalypt planting around the world, from choosing the first handful of seed for a new project to reviewing an established project which has already achieved benefits from selection and breeding.

Many people provided their time and knowledge and skills for this book and the authors thank them all. They contributed with a generous spirit knowing their work would help to produce a better book for the readers. We are grateful for the encouragement and support of successive Chiefs of the CSIRO Division of Forestry, Max Day, Joe Landsberg, and Alan Brown. Indeed the sustained contribution of CSIRO was an essential element in the completion of the book. The staff of the CSIRO Division of Forestry Library in Canberra were unfailingly helpful in locating reference material. This Library houses almost all the publications and reports referred to in the Reference list.

For review of draft chapters we thank Peter Ades, Doug Boland, Trevor Booth, Ian Brooker, Alan Brown, Arno Brune, Peter Burgess, Trevor Butcher, Christian Cossalter, Paul Cotterill, John Doran, Rod Griffin, Julian Evans, Vic Hartney, Alan House, Max Hodgson, Neville Howcroft, Peter Kanowski, David Lea, Colin Matheson, Mike Moncur, Gavin Moran, Jock Morse, Garth Nikles, Leon Pederick, Brad Potts, Lindsay Pryor, Carolyn Raymond, Don Rockwood, Tony Shelbourne, David Spencer, David Stockfeld, Lex Thomson, Wayne Tibbits, John Turnbull, Peter Volker, Kevin White, Mike Wilcox, and Emlyn Williams. For typing, checking references, preparing Tables and Figures, and other assistance we thank Michael Cashmore, Gloria Davidson, Helen Desmond, Marian Eldridge, Gillian Foley, Anita Gracie, Kathy Hanrahan, Alan House, Susan House, Peter Martensz, Patrick Milimo, Eva Morrow, Vlad Mosmondor, Karin Munro, Helen Neave, Joe Oros, and John Owen.

It is our sincere hope that well-planned breeding programs developed from a study of this book will help to make better use of the land set aside for growing eucalypts.

Canberra
May 1993

K.G.E
J.D
C.E.H
G.v.W

Acknowledgements

The following have kindly granted permission to reproduce material: Edward Arnold (Publishing) Ltd and L. D. Pryor for Fig. 1.1. The Chief, CSIRO Division of Plant Industry for Fig. 1.2. The Manager, Commonwealth Information Services, Australian Government Publishing Service for Figs 4.1, 4.2, 4.3, 4.4, and 4.5 (Commonwealth of Australia copyright reproduced by permission), Elsevier Science Publishers and T. H. Booth for Fig. 4.6. The Chief, CSIRO Division of Forestry for Fig. 5.1 and for citing Hand *et al.* (1989). Food and Agriculture Organization of the United Nations for Fig. 6.8. The Manager, CSIRO Editorial Services for Figs 6.1, 8.2, 8.3, 9.1, 10.2, 11.1, 12.1, 12.2, 13.1, 13.2, 13.4, 14.1, 14.4, 14.6, 15.2, 17.1, 17.3. The Director, AFOCEL for Figs 8.4, 8.5. The Editor, New Zealand Journal of Forestry Science for Fig. 9.4. The Director-General, Australian National Library for Fig. 10.1. The Editor, Australian Forestry for Fig. 10.4. The Chief, Division of Education and Research, Department of Conservation and Environment, Victoria and L.A. Pederick for Figs 12.2 12.4, 13.6 and for citing Pederick (1974, 1977, 1984, 1988, 1990). New Phytologist Trust and J. W. Green for Fig. 13.4. The Editor, Heredity for Fig 13.5. The Editor, Appita Journal for Fig. 14.5. The Editor, Commonwealth Forestry Review for Fig. 15.3. The Editor, Bois et Forêts des Tropiques and C. Cossalter and B. V. Gunn for Fig. 16.1. The Editor, South African Forestry Journal for Figs 17.4, 17.5 and for quoting Van Wyk (1977). Oxford & IBH Publishing Co. Pvt. Ltd for Fig. 18.2. Springer-Verlag for Fig. 18.1. Kluwer Academic Publishers for Fig. 18.3. G. Chapman for Fig. 19.6. The Gippsland Manager, A. P. M. Forests Pty Ltd for citing Appleton (1988), Davis and O'Connor (1988), and Mann (1988).

Contents

Abbreviations

Common abbreviations used in the text and in the list of References.

AFOCEL	Association Forêt-Cellulose, France
APM	Australian Paper Manufacturers Ltd
CELBI	Celulose Beira Industrial S.A., Portugal
APPM	Associated Pulp and Paper Mills, Australia
CSIRO	Commonwealth Scientific and Industrial Research Organisation, Australia
CTFT	Centre Technique Forestier Tropical, France (now CIRAD Forêt)
dbhob	diameter at breast height (1.3 m) over bark
FAO	Food and Agriculture Organization of the United Nations
EMBRAPA	Empresa Brasileira de Pesquisa Agropecuaria
IBDF	Instituto Brasileiro de Desenvolvimento Florestal
INRA	Institut National de la Recherche Agronomique, France
IUCN	International Union for Conservation of Nature and Natural Resources (now World Conservation Union)
IUFRO	International Union of Forestry Research Organizations
OECD	Organisation for Economic Cooperation and Development
PICOP	Paper Industries Corporation of the Philippines
STBA	Southern Tree Breeding Association Inc., Australia
UNDP	United Nations Development Programme
USDA	United States Department of Agriculture

1

Introduction

1.1 Objectives of the book

Many people have asked us 'how do you breed eucalypts?' Although eucalypts are in great demand for planting because of their rapid growth of useful wood, many growers are not satisfied with the present yield and quality. They want improvements in such things as growth rate, stem straightness, wood quality, coppicing ability, and tolerance of salinity or frost. Therefore this book is directed to eucalypt breeders, plantation managers, students, and other people who will be involved in domesticating the eucalypts as crop plants for wood production.

Eucalypt plantations are expanding so rapidly worldwide that by the year 2000 they may exceed 10 million ha, mainly little improved compared with their wild ancestors. We have aimed to establish, from scientific knowledge and practical experience, an up-to-date basis for making wise decisions in eucalypt breeding, both for the short and the long term.

Domestication and breeding simply mean a sequence of repeated selection and mating which brings about a change in gene frequencies. One might think that to be successful at breeding eucalypts one must know a great deal about gene frequencies and apply the statistical theory of genetic selection. The facts are that gene frequencies of the traits eucalypt breeders want to improve are not known, and may not be for many years. Besides, current selection theory does not predict accurately what will happen in the mixed mating system of the eucalypts, with both selfing and outcrossing, though predominantly outcrossing.

Much improvement can be achieved by applying the knowledge now available of the genetics and biology of eucalypts. We have gathered together as much information as we can to put decision-making on the best scientific basis.

Inevitably, all students and practitioners of eucalypt breeding will look at the subject from their own perspective with some strong opinions based on their own experience. While we acknowledge that there will be several ways to approach eucalypt breeding now and in the future, we consider that the varied experience of the co-authors will give a broad enough perspective to reveal the common scientific principles of eucalypt genetics and breeding.

Since eucalypt breeding, like all forest tree breeding, is at an early stage, a large part of this book places special emphasis on genetic resources and their domestication from wild populations, and the best ways to choose and maintain a suitable base population for selection. As the natural populations of the major eucalypt species retain most of their genetic variability despite much clearing, particularly for farming, breeders have an immense range of material for selection in the eucalypts, far greater than for most agricultural crop plants. The rest of the book is concerned with methods of breeding which are similar to those used in genetic improvement of outcrossing crop plants.

The most important scientific discipline needed at present in eucalypt breeding is mathematical statistics for efficiently selecting among large numbers of varieties and families in field trials. Other sciences employed are the ecology of natural distribution, ecophysiology of adaptation and growth to facilitate early selection, biology of flowering, horticultural science in vegetative propagation, and, increasingly, quantitative genetics for predicting the consequences of selection.

Some decisions can greatly affect the profitability of a plantation, large or small. Sometimes the decision on choice of seed for one small plantation has an effect through succeeding generations of larger plantations because that first plantation itself becomes a source of seed. Examples are *Eucalyptus globulus* in Ethiopia, Spain, and Portugal and

E. tereticornis (Mysore gum or 'eucalyptus hybrid') in India where the extensive plantations are suspected to have originated from a restricted genetic base of too few Australian trees. Thus some apparently short-term decisions, such as what seed to order now to sow this year's nursery, can have an effect on wood production in a region for decades.

Other decisions are needed about long-term plans for continuous, cumulative genetic improvement of a species as a crop plant. Several different plans are in use by eucalypt breeders in different parts of the world. This book will help decide, for example, whether clonal propagation of inter-specific hybrids would be the best way to improve particular eucalypts or whether a simple seed production area and use of seedlings would be better.

Once a sound strategy has been chosen, genetic improvement of eucalypts will require ingenuity, good technical management, and the continued availability of sufficient funds. Often a simple solution will result from the most profound scientific understanding. Sound breeding strategy and operational plans will depend greatly on the scientific knowledge as well as practical common sense of the people making decisions. We give examples of successes, and some failures, in eucalypt breeding to stimulate practitioners to make their own decisions. We would be disappointed if these examples were used unimaginatively as prescriptions for breeding eucalypts without sufficient understanding of scientific principles.

Tree breeding is an institutional activity requiring continuity of good organization over many years. Key staff must be well trained, imaginative, and well motivated. Changes in staff can be accommodated without too much loss of momentum to the program, provided there is an operational plan revised periodically, and good records are kept. We consider that the quality of decisions in long-term planning, and their recording, are more important than continued employment of one outstanding individual to control the program over several decades.

Eucalypt growers are domesticating new crop plants for wood production. Many wild populations from among the 500 or more species of eucalypts have been brought under human control in plantations in the last century. When such immediately useful natural populations are brought into cultivation they are sometimes called 'found varieties'. These found varieties, though reasonably suitable for wood production in their semi-wild condition, are still far from ideal.

Fortunately eucalypt populations, both natural and planted, contain considerable genetic variation. There are large and small trees, straight and crooked trees, trees with small, low-angle branches which prune themselves, other trees with thick, steeply-angled, persistent branches, trees with dense wood, others with light wood, and so on.

Such variation gives the opportunity for selection of improved individuals and varieties. Planting improved stock would make the relatively fixed costs of establishment yield more profit through greater quantity and quality of wood becoming available at harvest. A small amount of genetic improvement takes place when eucalypt growers thin to remove smaller and poorer quality trees from plantations, and collect seed from the best of the final crop trees. One case of success of this simple procedure is the reduction in end-splitting in logs of *E. grandis* after two or three generations of mass selection in South Africa (Nel 1965). The more elaborate, and better documented, procedure of selection of *E. grandis* over four generations for growth and frost tolerance in southern Florida, U.S.A. (Meskimen 1983) is discussed in detail in Sections 18.2.3 and 21.2.6.

A small amount of genetic improvement by natural and silvicultural selection in plantations can lead to a 'land race' (Section 3.4) which is better adapted to a particular environment than the seed which was originally introduced.

More spectacular and obvious genetic improvement of eucalypts has been achieved in recent years by collecting seed from superior natural populations selected on the basis of provenance trials, and by mass vegetative propagation of outstanding individuals in plantations.

The word 'profit' as used in this book needs some discussion. We talk about profit in the sense of a greater return, through improved yield and quality, on the investment in the cost of establishing a plantation. The word 'profit' can be used whether the return is on a 30 000 ha plantation for a pulpmill, or 30 trees on some wasteland to provide firewood for a rural family.

The costs either as money or as opportunity costs include land, access, clearing, cultivation, obtaining seed or cuttings, raising planting stock, planting, fertilizing, weeding, protection from grazing animals and from fire, fungi and insects, and perhaps pruning. Among these items weed control is often of overriding importance, and very expensive. All these costs except seed or cuttings are usually the same whether unselected planting stock or an improved variety is used. (Sometimes an improved variety is planted at a wider spacing, thus reducing costs since fewer plants are required.) Usually the cost of eucalypt seed is a very small percentage of establishment costs because 1 kg costing say US$100 is usually enough to plant 50 ha at 1000 stems ha^{-1} (*E. camaldulensis, E. deglupta, E. grandis*, and *E. tereticornis* all frequently have more than 500 000 viable seeds per kilogram (Boland *et al.* 1980a)). Even if improved seed cost $500 kg^{-1}, only $8 ha^{-1} is added to establishment costs, which usually total several hundred dollars per hectare. With 4 or 5 coppice harvests, the additional seed cost at establishment becomes a minute part of the costs of growing each coppice crop. Since genetic potential of seed has such a large effect on yield, quality, and profitability of a plantation project, and seed is relatively cheap, it is difficult to understand the reluctance of many plantation managers to spend more on a good genetic foundation for their investment.

Usually the land is the most important item on the list of costs of establishment because of the opportunity forgone to use the land for food crops or for grazing or for maintenance of a natural ecosystem. Increased profit from the plantation, through higher yield and quality resulting from tree breeding and other improved silvicultural practices, helps to justify the decision to use the land for this purpose. Of course the decision to use land for a eucalypt plantation involves many other considerations, not least the effects of eucalypt cropping on this particular piece of soil and its groundwater (Florence 1986).

Not everyone is enthusiastic about the value of eucalypt plantations. Indeed there is much opposition to eucalypt plantations in India (Shiva and Bandyopadhyay 1983; Shah 1985) and in other countries. In Spain an organization promoting the use of native trees is called the Phoracantha Club, taking the generic name of a bark beetle which kills the exotic eucalypts.

In a review of the ecological effects of eucalypt plantations, Poore and Fries (1985) stressed 'that there can be no universal answer, either favourable or unfavourable, to the planting of eucalypts. Nor should there be any universal answer: each case should be examined on its individual merits'. Similar conclusions were reached in reviews on eucalypts in Australia, Bangladesh, and Portugal (Florence 1986; Davidson 1985a; Kardell *et al.* 1986). 'The problem arises' as Khoshoo (1986, p. 12) observed 'when there is a mis-match between species on the one hand and end use and land use on the other'. The fault is not with the trees but with the people whose combination of enthusiasm and ignorance led to planting eucalypts in unsuitable places. It is not appropriate to have large areas of close-spaced, fast-growing trees where water is a far more important commodity than wood.

Like any other plants, the eucalypts require large amounts of water to sustain their rapid growth. The rate of photosynthesis and hence growth is roughly proportional to the rate of transpiration. The question to be asked is 'how efficient are particular eucalypts in particular situations?' Preliminary studies of water use indicate that eucalypts are comparable with other agricultural crop plants and more frugal than many slow-growing native tree communities in the number of tonnes of water required per tonne of dry matter harvested (Tiwari and Mathur 1983; Poore and Fries 1985; Davidson 1985a; Florence 1986).

There are many ways to reduce or avoid the deleterious effects on the soil of densely planted, fast-growing tree crops, including the eucalypts. Mixtures of eucalypts with nitrogen-fixing trees are being attempted with increasing success in several places. For example, in Hawaii DeBell *et al.* (1985) found that the biomass production of mixtures of eucalypts and nitrogen-fixing trees was much greater than the production of pure eucalypt plantations. Alternating crops of eucalypts and leguminous pasture or underplanting eucalypts with legumes are other approaches aimed at proper care of the land (Davidson 1986a).

Often eucalypt plantations are established on so-called 'wasteland' for rehabilitation of degraded sites where almost nothing else will grow, and also beside railways, roads, and canals (Tiwari 1983). Eucalypts are growing extremely well in some places where the natural vegetation grows much more slowly as in part of the 'cerrado' area of Brazil and in savanna regions of Africa. In such places they are providing firewood, charcoal, poles, shade, and sometimes honey and eucalyptus oil, protecting the soil from erosion, recycling nutrients, sheltering wildlife in some places, stemming the migration of rural people to the cities, and improving the environment for people to live in. By making the land more productive the eucalypts are contributing to the national wealth in many countries. Also, the provision of wood from eucalypt plantations on previously cleared land contributes to some extent to protection of remaining natural forests from exploitation for wood production (Davidson 1985b).

Insufficient information on the eucalypts as crop plants means that neither the proponents nor the opponents of eucalypt plantations are at present well enough informed on the effects of the eucalypts to justify the vehemence of the debate. There is clearly a need for responsible land use planning to define which areas are suitable for eucalypt planting and which areas are better for agricultural crops. Further research on the effects of eucalypts on soil and water will assist this planning process. However, whether or not to plant eucalypts is often a social and economic question rather than a biological one. Eucalypt breeding helps to make the best use of land committed to wood production and is part of the implementation of responsible land use planning.

As the debate continues so does the planting of eucalypts by individuals, companies, and government authorities all attracted by their fast growth and their potential to produce a profitable crop. This book is a contribution to making well-planned eucalypt plantations better.

1.2 Importance of eucalypts in the world

1.2.1 The eucalypts

The genus *Eucalyptus* contains a remarkably wide range of tree species in regard to adaptation to sites, types of management systems, and variety of uses, both in natural forests and in plantations.

No tree genus has ever been so widely propagated throughout the world as the pungent-smelling, evergreen, emigrant eucalypts (Zacharin 1978). Besides the homelands of Australia and its northern neighbours (Fig. 1.1), eucalypts can be grown in most of the tropical and temperate climatic regions of the world between latitudes 45°S and 40°N (Figs 4.2, 4.3, 4.4, 4.5, pp. 41–43). Survival outside this range depends on unusually favourable local climates, while growth at economic rates in forest plantations

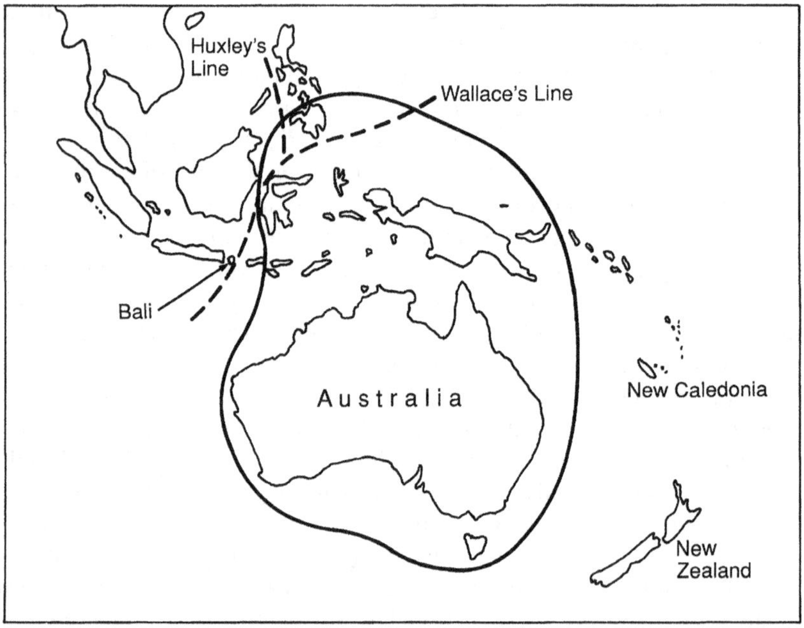

Fig. 1.1. The limits of natural distribution of *Eucalyptus* showing Wallace's line and Huxley's line (after Pryor 1976 and George 1981).

occurs only in a narrower latitudinal range which varies from continent to continent.

The genus *Eucalyptus* with more than 500 species (Chippendale 1988) is regarded by some taxonomists as several subgenera or perhaps even genera (Carr and Carr 1962; Pryor and Johnson 1971). If a proposal to split the genus into eight genera prevails (Johnson 1976; Johnson and Briggs 1983), it seems likely that within a decade or two the name *Eucalyptus*, in the strict sense, would be confined to the 100 or so species closely related to the type species *E. obliqua*. It was named by a French magistrate and botanist, Charles Louis L'Héritier de Brutelle, in 1788 (L'Héritier 1788). His Latin description and the fine illustration by Pierre Joseph Redouté (Fig. 1.2), then unknown but later famous for painting roses, were based on a specimen collected on Bruny Island, Tasmania, probably in 1777 on James Cook's third expedition.

The type specimen was in Joseph Banks' herbarium when L'Héritier and Redouté visited London and it can still be seen in the British Museum, Natural History. The majority of species of the original genus, including nine out of ten of the species most commonly planted as exotics, are in the group of about 300 species currently known as the informal subgenus *Symphyomyrtus* (Pryor and Johnson 1971) and in future proposed as the genus *Symphyomyrtus* (Johnson and Briggs 1983). It is proposed that the remaining 100 species be classified into six new genera. However, one can confidently predict that the common names 'eucalypt' and 'Eucalyptus' will remain in general use.

The names of the eucalypts in this book follow the nomenclature of the *Flora of Australia* (Chippendale 1988). However, we have not followed the *Flora* exactly for *E. globulus* (Section 10.3). The reader is referred to the *Flora* for a detailed botanical description of each species. A list of all *Eucalyptus* species mentioned in this book is given in Table 2.1 (p. 18) with the author of each botanical name and a common name.

Our guess at ranking the ten most important eucalypts around the world, in terms of current annual increment of wood, would include: *E. grandis, E. camaldulensis, E. tereticornis, E. globulus, E. urophylla, E. viminalis, E. saligna, E. deglupta, E. exserta*, and then either *E. citriodora, E. paniculata*, or *E. robusta*. Of these the first four are by far the most important. All these species are in the subgenus *Symphyomyrtus*, except *E. citriodora* which is in the subgenus *Corymbia*. *E. citriodora* is being planted on a substantial scale in southern China (Turnbull 1981).

Several publications in recent years have provided detailed information on the biogeography, ecology, silviculture, and utilization of the more commonly planted eucalypts. With so many useful species in such diverse habitats these digests are encyclopaedic. The reader is referred to papers by Pryor and Johnson (1981), Ovington and Pryor (1983) and Florence (1981) as an introduction, and for more detail to books by Jacobs (1955), Penfold and Willis (1961), Pryor (1976), Hillis and Brown (1984), Jacobs (1981), Hall *et al.* (1981), Boland *et al.* (1980a), Chippendale and Wolf (1981), Boland *et al.* (1984), and numerous papers from a IUFRO conference on eucalypts in 1991 (Schönau 1991). In addition there are large treatises on the silviculture and identification of eucalypts in a particular region or country including Brazil (Andrade 1961), Portugal (Goes 1985), southern Africa (Poynton 1979), Spain (Guttierrez 1977), and Chile (Prado and Barros 1989).

1.2.2 Plantation area and products

Brazil has the largest plantation area of eucalypts in the world, and has been growing them for wood production since 1910 (Andrade 1961). Recent estimates put the plantation area in Brazil at nearly three million hectares with a planting rate of more than 200 000 ha per year from 1972 to 1982 (Galvão and Couto 1984), mainly *E. grandis*. India appears to be the second eucalypt nation with more than 400 000 ha in 1972 (Jacobs 1981) and a rapid increase of planting since then, especially of *E. tereticornis* ('eucalyptus hybrid'). Spain was estimated to have 390 000 ha of eucalypt plantations in 1973 mainly *E. globulus* and *E. camaldulensis* (Jacobs 1981), and Portugal about 300 000 ha of *E. globulus* in 1983 (Kardell *et al.* 1986). Spain and Portugal greatly increased their areas of *E. globulus* plantations in the 1980s. South Africa had 470 000 ha of eucalypt plantations in 1989, of which 300 000 ha were *E. grandis* (Anon. 1990b). Angola had a similar area by 1980 according to unpublished FAO figures, mainly *E. tereticornis*. China also has large and rapidly increasing areas of eucalypt plantations

Fig. 1.2. P. J. Redouté's drawing of *E. obliqua*, the type specimen of the genus *Eucalyptus* (L'Héritier 1788).

estimated at over 400 000 ha in 1989, mainly *E. citriodora*, *E. exserta*, and *E. globulus* (J.W. Turnbull, personal communication). Ethiopia is estimated to have over 250 000 ha, mainly *E. globulus* and *E. camaldulensis* (Davidson 1989).

Several other countries have eucalypt plantations estimated to be more than about 100 000 ha in 1985, according to various sources listed in Table 1.1. About 20 other countries now have large eucalypt plantations between about 10 000 and 90 000 ha. Probably another 50 countries grow eucalypts in small plantations, woodlots, windbreaks, and ornamental plantings.

The total area of eucalypt plantations in the world may well have exceeded 6 million ha by 1985 (Table 1.1). However, it must be emphasized that statistical information on plantation areas from many countries is so incomplete that the figure of 6 million ha in Table 1.1 must be regarded as only a rough approximation.

It is difficult to obtain accurate and up-to-date information on areas of eucalypts planted for wood production. The statistical information on eucalypt plantation resources in Brazil provides an example. The net area of eucalypt plantation established under the fiscal incentive scheme from 1966 to 1981 was estimated to be 1.96 million ha in one report (Anon. 1983) and about 2.5 million ha in another (Galvão and Couto 1984). These areas included both new plantations and replanting of harvested or failed plantations. Some plantations of very low production were included and some older and privately planted areas were excluded. As another example, in Australia the area of 55 000 ha of eucalypts planted up to 1989 (ABARE 1990) does not include the replanting of seedlings in harvested native forests, perhaps another 30 000 ha.

The 41 million ha of natural Australian eucalypt forest (defining 'forest' as having mature stand height greater than 20 m) produces a greater variety of wood products than do the fast-growing plantations in other countries. The large logs from trees more than 100 years old produce high-quality sawn timber for furniture and joinery, construction timber particularly

Table 1.1

Estimates of approximate area of eucalypt plantations in the countries with the largest plantation areas. Published sources of information are noted. 'FAO' is from 1985 summaries of Inventory and Global Environmental Monitoring Projects by FAO, Rome. The estimate for more than '60 other countries' is less reliable than the estimates for the 13 countries listed.

	Area (ha)	Principal species	Source
Angola	390 000	*E. tereticornis*	FAO
Argentina	240 000	*E. grandis*	Golfari (1985)
Brazil	2 500 000	*E. grandis*	Galvão and Couto (1984)
Chile	60 000	*E. globulus*	Prado and Barros (1989)
China	300 000	*E. exserta*	Turnbull (1981)
Ethiopia	250 000	*E. globulus*	Davidson (1989)
India	550 000	*E. tereticornis*	FAO
Morocco	180 000	*E. camaldulensis*	Jacobs (1981)
Peru	90 000	*E. globulus*	Jacobs (1981)
Portugal	430 000	*E. globulus*	Pereira and Pereira (1988)
South Africa	470 000	*E. grandis*	Anon. (1990)
Spain	390 000	*E. globulus*	San Miguel (1988)
Uruguay	110 000	*E. globulus*	Jacobs (1981)
More than 60 other countries	500 000		
Approximate world total in 1985	6 000 000 ha		

for frames of houses, and railway sleepers. These forests also produce about half the country's requirements of pulpwood for the domestic paper industry, and some 5 million green tonnes per year of woodchips for export to Japan. The total annual harvest of eucalypt wood in Australia, excluding fuelwood, is of the order of 13 million green tonnes. Of this harvest, much less than one per cent comes from Australia's 55 000 ha of eucalypt plantations (ABARE 1990). Fuelwood amounts to more than 4 million tonnes air-dry, nearly all eucalypt (Fortech 1989).

Throughout the world eucalypts are cultivated to meet a wide range of needs, including amenity plantings for shade and shelter. Young trees are a source of paper pulp, charcoal and fuelwood, poles, mining timber, and fibreboard; mature trees provide strong and durable timber, and all sizes are capable of use for other forest products such as volatile oils for pharmaceutical and industrial uses, and honey (Penfold and Willis 1961; Jacobs 1981; Hillis and Brown 1984; Boland *et al.* 1991). Decorative eucalypt foliage is a notable part of the cut-flower trade in western Europe.

Of the many uses for eucalypt wood there is probably more used for fuel than for any other purpose. The economic importance of eucalypt firewood is overlooked in most countries as it is generally not bought and sold and therefore exact statistics are not available. However, eucalypt charcoal is of major economic importance for the iron and steel industry of Brazil and for domestic use in many other countries. The greatest production of industrial eucalypt wood is for the pulp and paper industry, mainly bleached kraft pulp which is in demand because of the excellent properties which it imparts to printing, writing, and tissue papers, as well as its availability and price (Sidaway 1988). There has been a rapid increase in the world production of chemical eucalypt pulp, from about 40 000 tonnes per year in the early 1960s to nearly 6 million tonnes per year in 1985 (Molleda 1984). The worldwide demand and production continue to increase with large new industrial eucalypt plantations and new pulp mills starting every few years.

1.2.3 Plantation yield

Performance of a eucalypt species on a particular site is the result of the interaction of that species' genetic make-up with climatic, edaphic, and other physical and biological factors. A mean annual increment of 10–15 m^3 ha^{-1} yr^{-1} of stem wood is commonly obtained from the already large areas of eucalypt plantations in temperate and tropical regions (Jacobs 1981). Assuming wood density of about 400 kg m^{-3} and total area of about 6 million ha, the potential yield of the world's eucalypt plantations is of the order of 30 million t of dry wood per year.

There is now ample evidence that much faster growth rates of eucalypts in plantations are obtainable following careful site selection, intensive cultural practices, selection of the best species and provenances, and genetic improvement. Small sample plots of eucalypts aged 6–8 years in East and West Africa, Brazil, and Papua New Guinea have grown at the rate of 70–90 m^3 ha^{-1} yr^{-1} of stemwood. If stemwood basic density is about 400 kg m^{-3}, some of these stands are approaching the practical upper limit of 50 t ha^{-1} yr^{-1} for fibre yield suggested by Bevege (1976). These high growth rates combined with increasing rates of planting mean that eucalypt pulp may challenge conifer pulps as the dominant pulp in future world markets and eucalypts are likely to dominate bioenergy plantations.

1.3 Base populations

The importance of the origin of seed used in establishment of base populations for future selection cannot be emphasized too strongly. Since natural variability of most eucalypts is enormous, provenance research is the first step in breeding eucalypts. The task is to assess and describe the variability which is present and capable of being used immediately for rapid genetic advance at relatively low cost in terms of energy and finance.

Continuing botanical exploration in Australia and in neighbouring countries to the north, especially of stands of limited extent and only recently accessible, has disclosed a wide range of variation within some well-known species and also disclosed some additional species not previously known. It must be expected that still more eucalypt species will be recognized and described in the future. Despite this, it is quite unlikely that any species with an extensive distribution and comprising large stands remains to be discovered, although some redefinition of species represented by extensive populations may still be expected. Identification of natural stands whose progeny will have desirable silvicultural characteristics in plantations is greatly facilitated by recognition of patterns of variation through genecological investigations.

Genecology is concerned mainly with botanical attributes of the trees in the natural forests, usually morphology of leaves, flower buds, and fruits, but sometimes also chemical and physiological characteristics of naturally growing trees and their seedling progeny. A list of eucalypt species which have been subject to genecological studies is given in Table 3.1 (p. 28). Provenance studies (Chapters 3 and 6–17) are different in that they include field trials with the silvicultural objectives of choosing the best seed sources for plantations of high yield and quality. A provenance trial brings together populations from different natural environments for comparison of their yield and quality in a single environment. Provenance studies usually require many more years than genecological studies.

Introduction of exotic eucalypts outside Australia has tended to follow two patterns. First and older is a pattern typified by the import of small seedlots of unknown provenance and limited number of parents. Often these have hybridized with other species or formed new land races over a few generations before becoming the source of seed for large-scale plantations. The second pattern is the more scientific method of importing seed of several provenances and providing a good base population for future genetic improvement.

The first pattern applies to most early overseas plantations grown on a large scale which are now known or suspected to be based on seed from very few parent trees. The first trees established in the importing country became a seed stand, and its progeny in turn were used as seed stands for seed supplies to industrial plantations. Although the first seed stand itself may exhibit apparent superior growth and form, the quality of seed collected from it might be poor because of a high degree of inbreeding. Here we give five cases of inappropriate genetic history of introduction of eucalypts (the first pattern). More details of these species (*E. globulus, E. tereticornis,* and *E. urophylla*) and several other species appear in the digests of provenance information, Chapters 6–17.

1. The case of 'Brazil alba' (*E. urophylla*) has become well known and is related in Chapter 16. Seed of a Timor eucalypt was collected and introduced into Java, where the species was not indigenous. In 1919 Edmundo Navarro de Andrade introduced the species to the Rio Claro Arboretum in Brazil also under the name *E.* 'alba', probably using seed from trees in the Bogor Botanic Gardens. Seed from the three plots in the Rio Claro Arboretum was used to plant extensive areas in Brazil. Much of this seed has since proved to be of hybrid origin, because of the ease with which the Rio Claro trees, now definitely identified as *E. urophylla*, hybridize with certain other species of eucalypt growing in the same arboretum. In most cases the performance of the F_1 generation was excellent, but F_2 and subsequent generations have shown poor growth and form and extreme variation.
2. It appears that most Indian plantations of *E. tereticornis* are derived from one of the first introductions planted as a double row of some 20 trees in the garden of a palace in the Nandi Hills, Mysore (Pryor 1966). Some hybridization had taken place and the complex of land races and hybrids are now known variously as 'Mysore hybrid', 'Eucalyptus hybrid', and 'Mysore gum' (Boland 1981). Performance in plantations has left much to be desired due to great variability within stands and slow growth compared with pure species of appropriate provenance.
3. A variety of *E. tereticornis* called *E.* '12ABL' (Section 22.2.4) has been planted widely in West Africa, particularly in the Republic of Congo, but it appears it may be descended from a single tree in Madagascar (Martin 1971) (Section 15.2).
4. The introduction of *E. globulus* to southern Africa is well documented (Poynton 1979, p.328). Nine seedlings imported from Mauritius were planted in the garden of the Governor's residence in Cape Town from whence the species was soon

distributed further afield. These nine plants were the origin of all *E. globulus* planted in southern Africa over the next 80 years.

5. In Ecuador there is a record of the planting of the country's first two eucalypts in 1865. Extensive plantings of *E. globulus* in Ecuador by 1880 were mainly from seed of these two trees (Acosta-Solis 1963).

The second and later pattern of introduction has now become common. This has involved introduction of several known provenances (though sometimes still with inadequate numbers of parent trees represented), conduct of well-designed provenance trials, selection of the best individuals within the best provenances, then creating seed production areas by heavy thinning of good quality stands or establishing seed orchards of the best individuals with or without progeny testing.

Selection carried out in populations with a narrow genetic base (unsuitable land races or single source Australian provenances) leads to erosion of genetic variability in plantations, thus reducing the opportunity of making further genetic improvement in future generations.

Lately there is also the added dimension of planting vegetatively propagated eucalypts on a large scale using selected superior individuals or hybrids as ortets. If large stands of a single clone are used for seed production for plantations the seed will be selfed and the result will be a reversal of progress and a further narrowing of the genetic base in an exotic environment (Brune 1983).

In order to ensure sufficient genetic variability for immediate and future use, genetic base populations should be set up in two distinct ways, one by mixing different provenances and one by maintaining separate isolated provenances (Libby 1973), provided there are sufficient resources to do both. If not, work should be concentrated on the separate isolated provenances first. The mixed provenances will, over a number of generations, recombine to form a single land race while the separate isolated provenances recombine within themselves to form several land races. Since many of the main species used commercially, such as *E. citriodora*, *E. deglupta*, *E. grandis*, *E. saligna*, or *E. urophylla*, flower outside their native habitat within 1 to 3 years of planting, and the flowering season of different provenances and individuals may overlap considerably, generation turnover can be very rapid.

The difference between the two types of base population is that mixed provenances will provide a larger genetic base (analagous to the 'recombination plantation' of Libby (1973)), and it is likely that inter-provenance hybrids will be produced. Some of these new gene combinations which did not exist in native stands might be more useful than those which did. This scheme cannot be used by itself since there is a disadvantage that once recombination has taken place, pure sources are no longer available should they be required (this has been one of the main lessons learnt from food plant domestication). If several provenances are also kept separate and genetically isolated as they exist in nature, they can be continued separately generation after generation (analagous to the 'random variability archive' of Libby (1973)) to form several adapted land races. Such specially managed plantations of particular provenances are also known as 'gene resource conservation stands' (Section 1.4.3) and 'provenance resource stands' (Nikles and Newton 1983). Any two or more of these land races may be brought together at any time in a new 'recombination plantation' with the likelihood that an even more desirable end product will be obtained than was the case with the recombination plantation containing all available provenances. Maintenance of separate populations and bringing them together is the basis of the 'multiple population' breeding strategy discussed in Section 18.2.3.

'Random variability archives' and 'recombination plantations' can be set up in a representative sample of environmental situations in recipient countries or groups of countries. Seed from these populations should ensure relative independence from wild stands in Australia and islands to the north in future, and in most cases could provide material better adapted for local conditions. These populations would supply material for breeding populations at any desired stage in the future.

Certain characteristics of eucalypts are in need of improvement and are capable of being improved by selection and breeding. For most plantations the selection criteria are faster growth, resistance to certain pests and diseases, straighter stems, and greater uniformity. In some cases better wood properties are required for particular uses, for example for higher yield of pulp. More information on selection criteria for breeding plantation eucalypts is given in Section 20.1.

It is important to make a clear distinction between ***base populations***, which are broadly based genetic reserves and serve genetic conservation goals, ***breeding populations*** which serve the purpose of genetic improvement, and ***propagation populations*** for mass production of improved seed or cuttings for use in the ***wood-producing populations*** which produce the commercial product and do not provide further reproductive material. This concept of four major populations in domestication strategies of forest trees was first set out by Libby (1973). The conceptual model called 'the breeding cycle', described by White (1987), concentrates on the components and activities of the breeding population. Eucalypt breeding strategy is explained in more detail in Chapter 18.

1.4 Sources of eucalypt seed

Whenever eucalypt seedlings are planted it is possible to make a choice of seed source. This selection of seed source is critical to success. Within eucalypt species there is clear evidence of genetic variation between populations in traits of importance in plantation silviculture, many examples of which are given in Chapters 6 to 17. Because this variation is frequently the result of adaptation and is inherited, the collection of seed from certain localities (provenances, Section 3.1) and not from others may confer a distinct advantage in subsequent silviculture.

1.4.1 Natural stands

The natural eucalypt forests will continue for many years to be the best initial sources of material for selection to improve eucalypt plantations. The genus *Eucalyptus* dominates the tree vegetation (mainly open woodland) of more than half the Australian continent (Ovington and Pryor 1983). It also occurs to a very limited extent outside Australia in Papua New Guinea, the island of Mindanao in the Philippines, in Irian Jaya and in the islands of Sulawesi (Celebes), Ceram, Timor, and some of the Lesser Sunda Isles of Indonesia. Each year many thousands of small seedlots collected in the natural forests are exported for research trials (Fig. 1.3), and in addition several tonnes of seed are exported for direct use in commercial plantations. Also, the natural forests currently supply several tonnes of eucalypt seed each year for aerial re-seeding of harvested forests in Australia.

Fig. 1.3. Seed collectors' camp in a small stand of *E. camaldulensis* at the edge of the Gibson Desert, Carnegie, Western Australia (photo CSIRO).

It must be assumed that the processes of natural selection which have been going on for a very long time in the natural habitat of the eucalypts have led to many cases of very precise site adaptation so that separate populations within species generally differ one from another throughout the geoclimatic range of the species. Notable cases of genetic variation detected in growth and other traits between adjacent populations are variation with altitude of seed source in *E. pauciflora*, *E. regnans*, and *E. urophylla*, and variation with soil type in *E. obliqua* and *E. viminalis*. Genetic variation between widely separated populations is obvious in provenance trials of *E. camaldulensis*, *E. cloeziana*, *E. deglupta*, *E. delegatensis*, *E. grandis*, *E. nitens*, and *E. obliqua*. This capacity for precise site adaptation is quite general in eucalypts and other forest trees and is a condition of basic value in any improvement program.

1.4.2 Plantations and other planted trees

Eucalypt seed for plantations has often been collected with a most casual disregard for genetic history of the seed trees and the genetic potential of the resulting seedlings. Collections in small quantities are made from individuals and groups of trees in gardens, streets, and arboreta and in larger quantities from routine plantations, despite the fact that collection costs are often high and both the genetic qualities of the seed and their capacity for germination are frequently poor, especially when collected from isolated trees.

Collection from unselected trees in routine plantations is often an undesirable practice because many eucalypt plantations have been made with an unsuitable provenance and a narrow genetic base. If the plantation is derived from too few trees (perhaps from a single street tree) the seed collected will be inbred and will not grow as well as the original tree or stand, even though that tree or stand itself may have had satisfactory growth and form. Collection from a stand of uncertain origin may be tolerated in the early stages of a seed supply program provided no alternatives are available. On no account should the resulting stands themselves be used for seed production.

However, if the genetic base of a successful plantation is known to be wide (say from 50 or more trees scattered through a large natural stand of appropriate provenance) then the resulting plantation will be an excellent source of seed for planting in that environment. New stands from plantation seed will grow somewhat faster than the original plantation. The reason is that when a plantation flowers the mating is more likely to be between unrelated trees than is the case in a natural stand. In the natural forest mating between close relatives prevails and much of the seed produced may be somewhat inbred, resulting in slower growing trees. In the plantation this neighbourhood inbreeding effect is broken (Section 19.6.2).

1.4.3 Gene resource conservation stands

Adaptation of eucalypts to a vast range of environments has provided foresters with a heritage of eucalypt gene resources which have been only partly explored. Prudence demands that sufficient samples of this natural heritage be retained in their original state or as near as possible to serve needs of the present and possible demands of the future. Gene resource conservation can be *in situ* for those stands which are accessible. *Ex situ* conservation stands are desirable for species which are wide spread and difficult and expensive to collect (e.g. *E. camaldulensis*, *E. deglupta*, *E. pellita*, and *E. urophylla*) or under pressure of extensive exploitation or destruction (Brune 1983). Losses due to dysgenic effects of less controlled seed collection and distribution also are avoided.

A seedlot collected for conservation from a sufficiently large number of trees (say over 50 unrelated parents) would contain most of the genes existing in the mother stand (Namkoong *et al.* 1980, p. 14). Plantations derived from these seedlots, with suitable isolation from neighbouring gene pools, can be used for seed production for gene conservation. These plantations can be commercially utilized when they become mature and, provided the seeds are collected again from trees representing the whole stand and not just from a few intensively selected trees, the gene resource can be replanted and maintained. Selective, heavy thinning of the whole conservation stand to form a seed production area would negate the gene conservation function by narrowing the genetic base. However, if the area is sufficiently large a small portion might be managed in this way.

1.4.4 Seed production areas

Selection and management of seed production areas is a commonly adopted strategy representing the first stage of eucalypt tree improvement, after selection of the best species and provenances. Such seed production areas could be stands specially planted for seed production or they could be existing stands specially managed for seed production, provided their genetic origin is appropriate.

The cautionary note on the origin of the stand again must be stressed. In addition to a possible narrow genetic base, existing eucalypt plantations often have a high incidence of hybrids, these being more common in stands which have been established using seed from exotic plantations. Another important question is isolation of the seed production area from potentially contaminating genetic sources.

Collection of seed from better-than-average stands which have been thinned early to remove inferior trees and to promote development of large crowns capable of heavy seed production is often the quickest and best interim measure to meet the need for large quantities of seed of known origin, and with at least some genetic improvement. We recommend that in all eucalypt breeding programs seed production areas should be made in suitable stands to provide an inexpensive supplement to more elaborate plans. Seed production areas are discussed in more detail in Section 21.1.

1.4.5 Seed orchards

Seed orchards are a means of obtaining large quantities of genetically improved seed relatively cheaply by allowing selected outstanding trees ('plus' trees) or their progeny to cross, usually under natural conditions (Faulkner 1975; Zobel and Talbert 1984). Grafts or cuttings from the plus trees are used to make clonal seed orchards, or seedlings raised from seed collected from the plus trees are used to make seedling seed orchards. In both cases the orchards are isolated and managed for seed production, as described in Section 21.4.

Clonal seed orchards are preferred if finance and technical skills are available for vegetative propagation, particularly for a species like *E. dunnii* which takes many years to reach flowering age. Grafts are used unless incompatibility of stock and scion is a problem, in which case air layers or cuttings are sometimes used. If vegetative propagation is difficult it is better to make a seedling seed orchard composed of open-pollinated families with family identity retained. Species with early flowering and managed on a short rotation are most suitable for seedling seed orchards.

For eucalypts, most seed orchards have been established using open-pollinated seed from outstanding trees in plantations (plus trees). These seedling seed orchards have been found suitable for many eucalypt species and have been established relatively easily. Inferior trees in the orchards have been culled early, sometimes using information from separate progeny tests, to further increase the genetic gain provided by selection of the plus trees. Open-pollinated seedling seed orchards have been established for *E. citriodora, E. cloeziana, E. deglupta, E. diversicolor, E. globulus, E. grandis, E. nitens, E. regnans, E. robusta, E. tereticornis*, and *E. urophylla.*

A eucalypt seed orchard program in Florida which shows great promise without requiring very difficult techniques has been developed by Franklin and Meskimen (1973, 1974, 1984). A system of combined (between and within family) recurrent selection was used with successive generations of open-pollinated families in *E. grandis* and other species. Seedling seed orchards were derived from each successive genetic base population by a system called 'two-stage selection'. The first stage, in year 3 for *E. grandis*, involved selection of from 1 to 3 trees in each family which had acceptable trees, and a very heavy thinning to remove the unselected trees. One year later seed was collected from all trees to produce progeny for the next generation genetic base population. The second stage selection and thinning, in year 6, left the best trees of the best families as a wide-spaced seed orchard. This process was repeated for four generations in *E. grandis* in Florida.

Grafted seed orchards of several widely-grown eucalypt species have been established in several countries. Incompatibility of stock and scion has controlled the degree of success of these orchards. In South Africa, in highly productive orchards of *E. grandis*, two or three grafts are planted in each planting position to alleviate later losses due to incompatibility (Van Wyk and Hodgson 1983). Detailed accounts of grafted eucalypt orchards are given in Section 21.3.

1.5 Interspecific hybrids

Interspecific hybridization is an important part of successful improvement programs for many crop plants, usually by the introduction of a desired trait (commonly disease resistance) from wild parents of another species into a well adapted stock by back-crossing (Simmonds 1979). By contrast, eucalypt hybrids which are now in use in large plantations are generally F_1 crosses between two wild, or almost wild, parents of different species. The hybrids are usually not propagated by seed, as with grain crops, but by rooted cuttings.

Many species of eucalypts hybridize in nature or can be crossed readily, within subgeneric groups (Section 19.6.3). In recent years interspecific hybrids of eucalypts have come into commercial use through the successful mass propagation by cuttings of outstanding individuals of hybrid origin (Section 22.1). For the most part these outstanding individuals arose as chance crosses in arboreta and plantations, and not necessarily from the best-adapted populations or individuals of the parent species. These undoubted current successes have encouraged a prominent role for hybrids in some plans for future genetic improvement of eucalypts (for example Martin 1987, Nikles 1987). Improvement of growth rate and quality started with interspecific hybrids, so why not make hybrids the basis of future strategy? Current eucalypt breeding at Pointe Noire in the Congo by Centre Technique Forestier Tropical (CTFT) and in Brazil by Aracruz Florestal S.A. emphasizes further interspecific hybrids between selected individuals of compatible species (Martin 1989; Campinhos and Ikemori 1989).

Hybrids can bring together desirable characteristics of both parents. For example, the hybrid *E. grandis* × *urophylla* combines the fast growth of *E. grandis* with the greater resistance to fungal diseases and greater coppicing ability of *E. urophylla*. Fast growth is an attractive feature of most eucalypt hybrids, with claims of double the growth rate of previous plantations being common (for example Delwaulle (1985a)). When planning to repeat such performance in other circumstances it is as well to ask what this apparent hybrid vigour (faster growth than the mid value of the parents) really represents (Section 20.5.2). At Pointe Noire the hybrid *E. grandis* × *tereticornis* is well adapted to a 'hybrid habitat' (Anderson 1949) and both the pure species grow slowly. No amount of selection and crossing within pure lines of these poorly adapted parent species offers any promise of growth rate approaching that of the hybrids (Martin 1989).

In other cases it is not so clear that true hybrid vigour prevails, and the superior growth of some interspecific eucalypt hybrids may simply be due to outcrossing of fortuitously inbred lines; similar vigour may have been achieved by crossing selected unrelated individuals within a pure species. Until the outcrossed (non-inbred) progeny of the parent species are used as comparisons with interspecific hybrids the question of hybrid vigour will not be resolved. A research project with this objective was started with *E. nitens* and *E. globulus* in 1986 in Tasmania (A.R. Griffin, personal communication).

Hybrids are in vogue because many of the outstanding trees in existing plantations are certainly of hybrid origin. However, much better individuals of pure species, and subsequently even better hybrids, are likely to result from careful attention to species and provenance selection.

1.6 Vegetative propagation

Vegetative propagation is one means of capturing exceptional gene combinations and it is in those eucalypt species which can be easily propagated vegetatively that interspecific hybridization has had and will continue to have its greatest impact. Also, for many eucalypt breeding strategies with pure species, advanced material can be put into wood-producing plantations much more quickly through vegetative propagation than through seed orchards. The spectacular success of mass production of cuttings of selected eucalypts at Aracruz in Brazil (Fig. 1.4) and at Pointe Noire in Congo has brought a new era to tree breeding for tropical plantations.

Fig. 1.4. Spectacular growth and uniformity of a five-year-old plantation of cuttings of *E. grandis* × *urophylla* at Aracruz, Brazil (photo E. Campinhos).

1.6.1 Cuttings

Advances made in the techniques for mass production of rooted cuttings of eucalypts since 1975 have enabled the cloning of desirable individuals on a commercial scale. In addition to the two large programs at Aracruz in Brazil and at Pointe Noire in the Congo, there are several smaller programs in operation in other countries. Outstandingly large straight trees, usually hybrids, are felled and the physiologically juvenile coppice shoots are used to make rooted cuttings. The resulting clones are then selected for high rooting ability. Several thousand hectares of cuttings are being established each year at both Pointe Noire and Aracruz using several clones, planted in blocks of up to 50 hectares of a single clone (Delwaulle 1985a).

At Pointe Noire the plantation area established by cuttings was 5 ha in 1975 and 2750 ha by 1979. An equally rapid expansion was achieved at Aracruz from the first experimental planting of cuttings in 1976 to 17 000 ha in 1983 (Wallenberg 1984). Both these programs are in the tropics: Pointe Noire is at latitude 5°S and Aracruz 20°S. Other programs in cooler climates and with other species have found the cuttings technology more expensive to apply and their expansion of planted area has been less rapid.

The eucalypts propagated in the largest programs are mainly the hybrids *E. tereticornis* × *grandis* at Pointe Noire (Delwaulle 1983, 1985a), *E. grandis* × *urophylla* at Aracruz (Brandao *et al.* 1984), *E. gunnii* × *dalrympleana* in southern France (Chaperon 1983), and the species *E. grandis* in South Africa (Denison and Quaile 1987), *E. globulus* in Portugal (CELBI 1982), and *E. camaldulensis* in Nepal (White 1986). Further details of techniques for growing eucalypts from cuttings are given in Chapter 22.

The appropriate number of clones to include in clonal plantations is presently much discussed (Heybroek 1984; Lindgren *et al.* 1989). Although several hundred clones have been propagated and tested in each program, the number constituting the bulk of each year's planting is usually small, often fewer than ten. The reason is that a few clones prove to be much easier to root than others and are therefore produced in largest numbers.

Planting single-clone blocks is easier to manage than planting mixtures of clones. Also, single-clone blocks permit eradication of any clone with extreme susceptibility to disease or pest attack. However, a clonal plantation made with a mixture of many clones in each block, similar to the genetic structure of a natural forest, may be less susceptible to the build up of intense outbreaks of pests and diseases (Libby 1982).

The large-scale use of cloning does not in any way negate the need for continuing a sexual breeding program. A well-managed breeding population is required, supported by broadly based genetic reserves. Breeding strategy in relation to mass vegetative propagation is discussed further in Chapter 18.

1.6.2 Micropropagation

There has been a period of great advance in tissue culture of eucalypts since the mid-1970s. The clonal culture in sterile media of meristems, hypocotyls, leaf discs, and other tissues has opened up new avenues for conservation of genotypes, rejuvenation of physiologically old shoots, rapid multiplication of desirable individuals, and mass propagation on a commercial scale. Each of these activities is still in the research and development stage (Boulay 1983; McComb and Bennett 1985; Hartney and Svensson 1992). However, micropropagation of seedlings of *E. camaldulensis* selected for tolerance of salinity in the soil is close to commercial scale operation (Hartney and Kabay 1984). In 1987 20 000 plantlets of salt-tolerant *E. camaldulensis* clones were propagated for planting to rehabilitate bauxite mines near Perth, Western Australia (E.D. Kabay, personal communication).

In southern China the age of eucalypt tissue culture has certainly arrived. In 1991 more than one million tissue culture plantlets were produced by micropropagation of selected *E. urophylla* and hybrid eucalypts. They were used as juvenile foundation stock for nurseries producing rooted cuttings and for some direct planting in the field (Yan Muqin, personal communication).

Tissue culture techniques have been used for the transport of sterile plantlets over long distances without causing quarantine problems. In future these techniques may be used to select *in vitro* for tolerance of diseases, salinity, or other environmental factors. Cell cultures offer an additional source of variation (somaclonal variation), and the production of haploid plants and somatic hybrids are other possibilities for the future.

At present the mass propagation of eucalypt clones by tissue culture techniques is more expensive than propagation by rooted cuttings. The most expensive operation is manual subculturing, a restriction which could be overcome by automated machine systems which would revolutionize clonal forestry. Until such machines become available it seems that the principal role of micropropagation will be to rapidly multiply selected seedlings. The seedlings would either be from scarce seed of a cross of known high performance, or individual seedlings selected, for example, for unusual tolerance of frost or of high salinity in hydroponic culture. Sterilized bud tissue would be propagated *in vitro* to produce sufficient hedged mother plants for the mass production of planting stock by rooted cuttings.

2

Eucalypts natural and planted

2.1 Taxonomy

2.1.1 Names of eucalypt species

Species of the genus *Eucalyptus* are commonly known as 'eucalypts' throughout the world, although in Australia they are often called 'gum trees' because of the gum (kino) that exudes from the trunk of older trees (Zacharin 1978). The use of correct botanical names is important to growers of eucalypts when they order seed or seek information on appropriate practices in breeding and plantation silviculture. As the common name is often used for more than one species we avoid common names, except where there is no ambiguity. In this book we use botanical names.

While most eucalypts can be identified fairly easily in the field as mature trees with the help of books such as those by Chippendale (1988), Boland *et al.* (1984), Brooker and Kleinig (1983, 1990), Gutierrez (1977), or Goes (1985), there are some problems. It is certainly convenient to consider eucalypt species and subspecies as discrete taxa but it is important to remember that within most species, and between some pairs of species, there is a continuum of variation in many characteristics.

There are several examples of such taxonomic difficulties in well-known plantation species due to what appears to be continuous variation between the taxa in certain intermediate populations. These include *E. citriodora* and *E. maculata*, *E. globulus* and *E. pseudoglobulus* (Section 10.3), *E. grandis* and *E. saligna* (Section 11.3), *E. regnans* and *E. obliqua* (Section 13.5), *E. tereticornis* and *E. camaldulensis* (Section 15.1), and *E. dalrympleana* and *E. viminalis* (Section 17.3).

Although the nomenclature of Australian timbers is the subject of Australian Standard 2543 (SAA 1983) the recommended standard trade common names are not used consistently in Australia or other countries. In addition to the standard trade common name there are two or three other common names for some of the eucalypts included in *Forest trees of Australia* (Boland *et al.* 1984). Hence there is merit in the increasingly common practice for plantation foresters to refer to a eucalypt only by its specific name, for example as 'grandis' or 'globulus'.

The common names used in Australia can be both useful and confusing. They are usually descriptive and make sense in the particular natural forest where the species occurs and they are useful in referring to groups of related species such as the red gums (*E. camaldulensis*, *E. tereticornis*, and others) or the ashes (*E. fastigata*, *E. regnans*, and others). But it is confusing that *E. tereticornis* is known locally as red gum in the south of its range and blue gum in the north, and that swamp gum is *E. ovata* in Victoria and it is *E. regnans* in Tasmania. There is a story that foresters who admired some fine young stands of *E. regnans* in New Zealand ordered substantial quantities of seed from Australia of swamp gum and received *E. ovata*. A considerable area of this totally unrelated species of slower growth and inferior wood quality was planted in New Zealand (Weston 1957). Similar problems with the supply of the wrong seed have arisen in other countries due to mistakes with records and labels in arboreta and other trials.

It is essential for eucalypt growers, and particularly eucalypt breeders, to know the botanical identity and the exact geographic origin of a seedlot under test so that more seed of the same species and provenance can be obtained in future.

2.1.2 Taxonomic relationships

It is important to recognize that eucalypts belong to discrete groups. The hierarchy of groups used by Pryor and Johnson (1971) is subgenus, section, series, subseries, superspecies, species, subspecies. The groupings are based on shared diagnostic morphological characteristics of the constituent species. Two major groups (informal subgenera) were first shown by Carr and Carr (1959, 1962) to comprise species either with one operculum, *Monocalyptus,* or with two, *Symphyomyrtus*. *Monocalyptus* is also known as *Eucalyptus sensu stricto* because this group (about 100 species) is based on the type or first-named species, *E. obliqua*. *Symphyomyrtus* includes more than 300 species. The remaining species (about 100) are classified into six other subgenera (Table 2.1). The subgeneric grouping was substantiated by the successful manipulation of hybrids within the groups and failure of attempts to hybridize between them (Pryor and Johnson 1981; Griffin *et al.* 1988). Johnson (1976) proposed the groups should become nine genera of the 'Eucalyptus alliance', including the previously separate genus *Angophora* which has seven species and which is also a member of the family *Myrtaceae*. However, this suggestion has not been formalized. Table 2.1 lists eight subgeneric groups of eucalypts with the names of the species referred to in this book. Table 2.1 also includes the approximate number of species in each subgenus according to Johnson and Briggs (1983) and the standard trade common name (SAA 1983).

Table 2.1

Taxonomic subdivision of the genus Eucalyptus *into informal subgenera, after Pryor and Johnson (1971) and Johnson and Briggs (1983). Numbers of species in each subgenus after Johnson and Briggs (1983). The species and authors of botanical names of eucalypts mentioned in this book are listed here. Botanical names of species and authors of botanical names are from Chippendale (1988). Standard trade common names are from Australian Standard 2543 (SAA 1983). Common names in brackets are from Boland et al. (1984).*

Subgenera and species	Common name
Subgenus *Blakella* (approx. 9 species)	
E. confertiflora F. Muell.	broad-leaved carbeen
E. papuana F. Muell.	ghost gum
E. tessellaris F. Muell.	carbeen
Subgenus *Corymbia* (approx. 35 species)	
E. centralis D. Carr & S. Carr	–
E. citriodora Hook.	spotted gum (or lemon-scented gum)
E. ficifolia F. Muell.	red-flowering gum
E. intermedia R. Baker	red bloodwood
E. latifolia F. Muell.	round-leaved bloodwood
E. maculata Hook.	spotted gum
E. polycarpa F. Muell.	red bloodwood
E. torelliana F. Muell.	cadaga
Subgenus *Eudesmia* (approx. 15 species)	
Subgenus *Gaubaea* (2 species)	
Subgenus *Idiogenes* (1 species)	
E. cloeziana F. Muell.	Gympie messmate

Table 2.1
(Continued)

Subgenera and species	Common name
Subgenus *Monocalyptus* (= *Eucalyptus sensu stricto*) (approx. 100 species)	
E. amygdalina Labill.	black peppermint
E. coccifera J.D. Hook.	(Tasmanian snow gum)
E. delegatensis R. Baker	alpine ash
E. dives Schauer	broad-leaved peppermint
E. elata Dehnh.	river peppermint
E. fastigata Deane & Maiden	brownbarrel
E. macrorhyncha F. Muell. ex Benth.	red stringybark
E. obliqua L'Hér.	messmate
E. olsenii L. Johnson & Blaxell	–
E. pauciflora Sieber ex Sprengel	snow gum
E. pilularis Smith	blackbutt
E. pulchella Desf.	white peppermint
E. pyrocarpa L. Johnson & Blaxell	large-fruited blackbutt
E. radiata Sieber ex DC.	narrow-leaved peppermint
E. regnans F. Muell.	mountain ash
E. risdonii J. D. Hook.	(Risdon peppermint)
E. robertsonii Blakely	narrow-leaved peppermint
E. sieberi L. Johnson	silvertop ash
E. stellulata Sieber ex DC.	black sally
Subgenus *Telocalyptus* (4 species)	
E. deglupta Blume	–
Subgenus *Symphyomyrtus* (approx. 300 species)	
E. alba Reinw. ex Blume	poplar gum
E. amplifolia Naudin	cabbage gum
E. benthamii Maiden & Cambage	Camden white gum
E. bicostata Maiden, Blakely & J. Simm.	southern blue gum (or eurabbie)
E. blakelyi Maiden	forest red gum (or Blakely's red gum)
E. botryoides Smith	southern mahogany
E. brassiana S.T. Blake	brassiana gum
E. caesia Benth.	–
E. camaldulensis Dehnh.	river red gum
E. cinerea F. Muell. ex Benth.	Argyle apple
E. cladocalyx F. Muell.	sugar gum
E. crebra F. Muell.	narrow-leaved red ironbark
E. cypellocarpa L. Johnson	mountain grey gum
E. dalrympleana Maiden	mountain gum
E. deanei Maiden	round-leaved gum
E. diversicolor F. Muell.	karri

Table 2.1
(Continued)

Subgenera and species	Common name
E. dunnii Maiden	Dunn's white gum
E. exserta F. Muell.	Queensland peppermint
E. globulus Labill.	southern blue gum (or Tasmanian blue gum)
E. gomphocephala DC.	tuart
E. grandis W. Hill ex Maiden	rose gum (or flooded gum)
E. gunnii J.D. Hook.	cider gum
E. kitsoniana Maiden	Gippsland mallee
E. leptophleba F. Muell.	Molloy red box
E. leucoxylon F. Muell.	yellow gum
E. macarthurii Deane & Maiden	Camden woollybutt
E. maidenii F. Muell.	southern blue gum (or Maiden's gum)
E. melanophloia F. Muell.	silver-leaved ironbark
E. melliodora Cunn. ex Schauer	yellow box
E. microcorys F. Muell.	tallowwood
E. microtheca F. Muell.	coolibah
E. moluccana Roxb.	grey box
E. nitens (Deane & Maiden) Maiden	shining gum
E. nova-anglica Deane & Maiden	New England peppermint
E. occidentalis Endl.	flat-topped yate
E. ovata Labill.	swamp gum
E. paniculata Smith	grey ironbark
E. pellita F. Muell.	red mahogany
E. polyanthemos Schauer	red box
E. populnea F. Muell.	(poplar box)
E. pryoriana L. Johnson	(Gippsland manna gum)
E. pseudoglobulus Naudin ex Maiden	southern blue gum (or Gippsland blue gum)
E. pulverulenta Sims	(silver leaved mountain gum)
E. resinifera Smith	red mahogany
E. rhodantha Blakely & Steedman	–
E. robusta Smith	swamp mahogany
E. rubida Deane & Maiden	candlebark
E. rudis Endl.	flooded gum
E. saligna Smith	Sydney blue gum
E. sideroxylon Cunn. ex Woolls	red ironbark
E. stoatei C. Gardner	–
E. tereticornis Smith	forest red gum
E. urnigera J.D. Hook.	urn gum
E. urophylla S.T. Blake	(Timor mountain gum)
E. viminalis Labill.	manna gum
E. woodwardii Maiden	–

Further taxonomic subdivision, for example into 110 series by Chippendale (1988), is of interest to tree breeders because natural hybrids are more frequent and manipulated hybrids are more successful between species within series than between the obviously less related species in different series or sections (groups of series). This was clearly demonstrated by Griffin *et al.* (1988) from their search of herbarium records and the literature.

The informal subgenus *Symphyomyrtus* is the most commonly planted group of eucalypt species throughout the world. Among the nine sections in *Symphyomyrtus* (Pryor and Johnson 1971) three sections contain nearly all the species which are most widely planted as exotics:

Section *Transversaria: E. grandis, E. saligna, E. urophylla;*
Section *Exsertaria* (red gums): *E. camaldulensis, E. exserta, E. tereticornis;*
Section *Maidenaria: E. dalrympleana, E. globulus, E. gunnii, E. maidenii, E. nitens, E. viminalis.*

Many interspecific hybrids between species within these three sections have been made in recent years to provide foundation stock for clonal forestry. Pairs of species to be crossed are often chosen for their close-relatedness as indicated by their taxonomic affinities (Martin 1987) (Section 20.5.2).

2.2 Eucalypts in their natural environment

The genus *Eucalyptus* is almost entirely Australian with only two species not found there: *E. urophylla* in Timor and some adjacent islands of Indonesia and *E. deglupta* in Papua New Guinea, Irian Jaya and the Moluccas of Indonesia, and Mindanao in the southern Philippines. Ten eucalypts, *E. alba, E. brassiana, E. confertiflora, E. latifolia, E. leptophleba, E. papuana, E. pellita, E. polycarpa, E. tessellaris,* and *E. tereticornis,* occur in southern New Guinea (Irian Jaya and Papua New Guinea) and also in northern Australia (Carr 1972; CSIRO records). One of these species, *E. alba*, extends to Timor and nearby islands in Indonesia. Among large genera of woody plants, the genus *Eucalyptus* is almost unique in being confined to a single compact region of the globe and essentially to one continent (Pryor 1976). The natural range of latitude is between 9°N (*E. deglupta*) and 44°S (*E. obliqua* and other species). While there is increasing interest in tropical eucalypts for planting, the majority of species and the best natural stands of most of the widely planted species occur south of the Tropic of Capricorn.

This section provides the reader with a background on the main features of the Australian environment within which the eucalypts have evolved. Accounts of the Australian environment are numerous and include those by Leeper (1970), Beadle (1981), Keast (1981, several chapters in Volume 1), and Brown and Turnbull (1986).

2.2.1 Natural history of the Australian continent

Fossil evidence indicates that 200 million years ago Australia and southern New Guinea, Antarctica, India, Arabia, Africa, Madagascar, and South America were joined in one vast southern continent called Gondwana (Barlow 1981; Whitmore 1981). Beginning about 130 million years ago this great landmass began to break up into the modern continents and the pieces started to move as separate tectonic plates into their present positions. Australia was isolated from Antarctica last of all to form the world's smallest continent and began to move northwards in the early Tertiary period, some 60–65 million years ago. Australia and New Guinea drifted north as part of the Indo-Australian Plate, advancing about seven centimetres per year.

Minimal volcanic activity (mainly near the east coast) and incessant erosion has produced the lowest and flattest of all the continents, regular both in surface and outline. Rain and wind over millions of years have levelled ancient ranges in the west into vast plains and reduced many of the younger uplifted mountains in the east to large plateaux. Few areas are higher than 1300 m, with the highest summit, Mount Kosciusko, only 2228 m above sea level at latitude 36°S in the south-east of the continent, while further south in Tasmania the highest altitude is 1617 m at 42°S.

It is generally agreed that the early part of the Tertiary period was uniformly humid over the Australian continent which was covered with vegetation adapted to abundant water and nutrients (see for example Florence 1981). Later in the Tertiary this vegetation was greatly changed due to increasing drought and declining soil fertility (Truswell 1990).

An important element in the evolution of the eucalypts is that the Australian soils on which they grow now and on which they have evolved over millions of years are generally highly leached, deficient in phosphorus and other plant nutrients, have an excess of iron and aluminium, and are mostly acidic. Soils in other continents are generally younger and richer in minerals than Australian soils which have developed on the weathered and depleted bedrock of ancient land surfaces, often on sedimentary rocks of low nutrient status.

Present-day physiography has been described in detail by Jennings and Mabbutt (1977). Their nomenclature is followed in this book and a simplified version of their map is presented on a different basemap projection in Fig. 2.1. The continent is divided east–west into three major landform zones: the eastern uplands, the interior lowlands, and the western plateau. The eastern uplands include all the narrow eastern coastal plain as well as the adjacent and parallel Great Dividing Range, 150–400 km wide, made up of tablelands, low mountains, and higher mountain ranges. The interior lowlands comprise the arid central basin, including the Great Artesian Basin and the Carpentaria Basin, and have low and gently sloping topography. The western half of Australia is largely one great ancient plateau between 300 and 600m in altitude broken only by the low but frequently rugged ranges which reach 1000–1500 m.

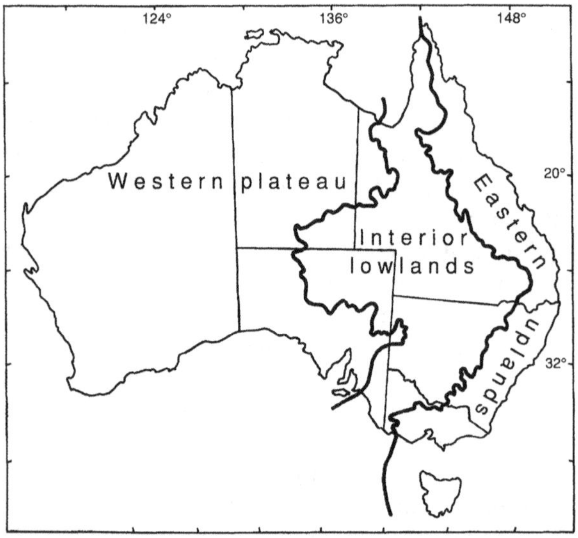

Fig. 2.1. Major physiographic zones of Australia (after Jennings and Mabbutt 1977).

2.2.2 Evolution of the eucalypts

There are five processes of organic evolution (Stebbins 1971, p.17):

(i) gene mutation,

(ii) changes in chromosome structure and number,

(iii) genetic recombination which results from intercrossing between individuals of the population as well as between them and occasional new genotypes which enter by migration,

(iv) natural selection (the 'fittest' individuals produce the most surviving offspring),

(v) reproductive isolation.

The first three processes provide the genetic variability without which change cannot take place, while natural selection and reproductive isolation produce progressive change in the population which keeps it adapted to the changing environment (Stebbins 1971).

These processes are acting in the natural populations of the genus *Eucalyptus* and also in the plantation populations manipulated by tree breeders. However, no variation in chromosome number has been found in the fast-growing plantation species of the genus *Eucalyptus*. Of 108 species studied 95 had the basic number $2n = 22$ common to most genera in the family *Myrtaceae* (see reviews by Eldridge 1969 and Rye 1980). Among the 13 species with $2n = 24$, only *E. cladocalyx* has been planted for wood production. Eucalypt chromosomes are small and difficult to count (J. E. Grant personal communication) and there has been no repetition to confirm the reports of $2n = 24$, most of which were by Ruggeri (for example Ruggeri 1961). Although there are no reports of natural polyploids in the genus *Eucalyptus*, there are reports of artifically induced polyploids. Colchicine has been used to induce autopolyploidy in *E. citriodora* (Janaki Ammal and Khosla 1969) and *E.*

tereticornis (Kapoor and Khedharnath 1974). Allotetraploidy has been reported in the hybrid *E. tereticornis* × *grandis* (Kapoor and Sharma 1985).

Very little fossil evidence has been found to describe the form of the precursors of the eucalypts. In the rainforest flora that once dominated the Australian continent in the early- and pre-Tertiary period, genera of the family *Myrtaceae* were probably numerous, as they still are in the rainforests of northern Australia and southern New Guinea.

With the onset of the changes in climate, topography, and soils over millions of years of the Tertiary period, new environments were formed which caused the contraction of rainforests. Mountain ranges were uplifted in the east and there were climatic changes featuring periods of high aridity. Concurrently there were soil changes such as lowering the fertility through leaching, laterization, accumulation of salt, and mass movement of sand and sediments by wind and water over much of the continent. These changes provided numerous new and different niches which plants could colonize, and to which they would become adapted by the processes of evolution.

Of the numerous genera and species available for new colonization, the ancestral eucalypt must have had the adaptive and competitive traits, either latent or already developed, which were necessary to spread from the retreating rainforests to reach every part of Australia, and islands immediately to the north.

According to Brooker (1986) the first 'eucalypts' were probably trees of the tropical rainforest or rainforest margins. Several of the 'primitive' morphological characters which he considered the ancestral eucalypts may have possessed (smooth bark, leaves of different colour on each side, leaves arranged in opposite pairs, leaf veins at a large angle to the midrib, a dense network of veinlets, terminal inflorescences, etc.) are found now in some present-day eucalypts at the margins of tropical rainforest, for example *E. intermedia* among the bloodwoods. *E. deglupta*, the only eucalypt now confined to tropical rainforest, has nearly all the features of Brooker's concept of a 'Proto-eucalyptus'.

Red gums (for example *E. camaldulensis*), boxes (for example *E. microtheca*), ironbarks (for example *E. sideroxylon*), and bloodwoods (for example *E. centralis*) are nowadays hardy and vigorous species inhabiting the arid and semi-arid zones of Australia. Two large groups, the ashes (for example *E. regnans* and *E. delegatensis*) and the mountain gums (for example *E. viminalis* and *E. dalrympleana*) have adapted to the cooler, moister, elevated parts in the southeast of the continent.

Eucalyptus may not be derived from one single ancestral species but from more than one. The floral operculum, which is the critical character separating this genus from others, is of such diverse nature that it may have evolved independently in different groups of plants and not entirely by variation and divergence from one ancestral proto-eucalyptus type (Johnson 1976).

Taxonomic relationships of groups of species should reflect evolutionary history. As that history becomes clearer through continued research, we must expect rearrangements in the theoretical evolutionary relationships of the eucalypts and further changes in names and groupings.

Eucalypt species maintain their distinctiveness today by restriction of gene exchange through the operation of many factors including geographical isolation, differing flowering times, and pollen-stigma incompatibility (Pryor 1976). Nevertheless, many pairs are cross-fertile and some hybrid swarms are found in nature when compatible species overlap in distribution and are able to interbreed (Section 19.6.3).

Consideration of the evolutionary history of the eucalypts helps to explain their success as exotics and why they very often grow faster than native species when cultivated together on the same site in the exotic location. Part of the answer may lie in a great difference in selection pressures which operated on the progenitors of the eucalypts and on the progenitors of the native trees now growing in places where eucalypts are planted.

2.2.3 Success as exotics

One might ask why so many eucalypts have the intrinsic capacity for fast growth and the ability to withstand and recover from harsh environmental conditions. Having developed from rainforest progenitors and having evolved over millions of years under alternating extreme environmental conditions (first wet and fertile, then dry and infertile), the present-day eucalypts have inherited many adaptations which allow them to take advantage of periods favourable for rapid growth between unfavourable periods. These adaptations, not all of them present in a single species, include an indeterminate shoot pattern, epicormic shoots, coppice, lignotubers, fire-resistant bark, and mechanisms to withstand drought and insect attack (Jacobs 1955). Furthermore, the mixed breeding system of both selfing and outcrossing and the abundant production of small seed would help adapt the eucalypts to survive the repeated cycles of near extermination and recolonization. It seems likely that it is these inherited characteristics which have given the eucalypts a competitive edge in other parts of the world over native species which have evolved under conditions which were not so extremely variable.

Another reason for success of eucalypts in exotic situations is that many of the restricting environmental factors affecting growth in native forest are greatly different in the new environment. It is frequently stated (for example Pryor 1976) that eucalypts perform better as exotics because they are separated from the many insect pests and diseases which co-evolved with them in their native habitat. However, Ohmart (1984) considers that 'there are insufficient data to claim that insects are a major factor responsible for the growth differential between indigenous and overseas plantation-grown eucalypts'. The superior growth of eucalypts in exotic plantations compared with their growth rate in native forests could be attributed to many factors in addition to the absence of pests and diseases, including planting on land with greater fertility and higher or more reliable rainfall, the planting of uniformly-spaced seedlings on well prepared sites, weed control, and protection from fire. These advantages are not available to eucalypts regenerating from seed in the native habitat.

Also many of the sites on which eucalypt planting is attempted are recently degraded and in effect, they have become exotic sites for the native species. In contrast with the eucalypts, it seems that few native trees have the adaptability to suit them to the new conditions which accompany degradation, such as soil acidity and low fertility following leaching. In such circumstances the eucalypts commonly perform better than the native species.

2.2.4 Present natural distribution of the eucalypts

A number of publications in the past have featured small maps of the distribution of eucalypts, for example, an early edition of *Forest trees of Australia* (Hall *et al.* 1975). However, more authoritative mapping has been made possible with the computerization of the locations of all eucalypt collections made in natural forests in Australia. The program and database is called EUCALIST (Chippendale and Wolf 1984). Maps based on EUCALIST have been published by Chippendale and Wolf (1981), Boland *et al.* (1984), Chippendale (1988), and are used in this book. More importantly, location information has been presented in the form of $1° \times 1.5°$ (115×135 km) cells of the global geographic grid of which the 1:250 000 Australian topographic survey is part. EUCALIST maps are wholly based on herbarium data and may be deficient because some localities have not yet been visited by botanical collectors. In using EUCALIST it is helpful to have an alphabetic index to locations of 541 Australian grid cells available from AUSLIG (PO Box 2, Belconnen, ACT, 2616, Australia).

The number of species varies greatly within grid cells of $1° \times 1.5°$. There are no eucalypts in some of the flat, arid, inland areas and large numbers (97 in a grid cell near Sydney) in areas of very diverse habitats including mountains and coast with fairly high rainfall (Chippendale 1981). Disjunctions are a feature of the distribution of eucalypt species. No eucalypt, apart from those few consisting of a single population, is actually distributed continuously (Pryor and Johnson 1971).

2.3 Eucalypts in plantations, advantages and disadvantages

The obvious success of many eucalypt plantations in quickly producing large amounts of useful wood has led to some subsequent plantings based more on enthusiasm than wisdom. Adverse public reaction has resulted (FAO 1988a). We wish to emphasize here that, in our opinion, eucalypts should only be planted on a large scale in places that are suitable for eucalypts, and only after appropriate land use planning (Section 4.2).

Much of the critical comment has failed to include suggestions for management practices or use of suitable alternative species. Only recently have these criticisms started to receive balanced scientific review (Davidson 1985a; Poore and Fries 1985; Florence 1986; Eldridge and Cromer 1987).

New plantations of eucalypts may be 'unsuitable' because they are:

(i) composed of species or provenances ill-adapted to the site;

(ii) poor genotypes of a species and provenance which normally would be expected to be successful on the site;

(iii) the wrong species for the end use desired; or

(iv) socially unacceptable in a particular locality for one or more reasons.

By 'unsuitable' here is meant death or poor growth of trees, or inability to provide the products or services expected even though the trees may have grown well, or unexpected adverse effects such as reduced stream flow or lowered water table.

These pitfalls can be minimized or avoided by undertaking:

(i) site assessment and classification followed by better matching of species to site (Chapter 4);

(ii) installation of well-designed species and provenance trials (Chapter 5);

(iii) better consideration of wood properties and end use (Section 20.1); and

(iv) consultation with local people and end users on the kind of plantation or tree planting scheme desired and products needed (Section 2.3.1).

2.3.1 Social aspects

Eucalypts have often been promoted as miracle trees that will bring immediate relief to local wood shortages and soil deterioration problems without qualifying which species is meant. When plantings are unsuccessful because they are of the wrong species or have been planted on the wrong sites, there is a wave of disappointed local expectations often leading to criticism of 'adverse ecological effects' of all 'eucalypts'. Much of the blame should be attributed to the poor land use planning and silvicultural practices which failed to consider proper matching of species to site and adequate species and provenance testing before large-scale planting took place, quite aside from assessment of the social benefits and costs of creating man-made plantations in a place where there were none before.

In some localities and under some social conditions eucalypts have been too successful for the wrong reasons. For example, eucalypts in some parts of India intended for fuelwood to be used locally by poor communities have grown so well that paper pulp and rayon companies have caused social disruption by paying very high prices for the wood. Scaffold poles for large building projects are another profitable product from commercial eucalypt crops in India. Fuelwood supplies to local villagers have not improved because the poor could not afford to buy wood at the inflated market prices. Lucrative tree farming displaced local foodcrops such as millet, causing more people to go hungry. A decrease in labour requirements of the eucalypt tree crops compared with agricultural crops deprived many landless poor of their subsistance living as farm labourers, leading them to retaliate by uprooting eucalypt seedlings planted by larger landowners, and no wonder! (This backlash probably would have occurred with any fast-growing tree with a high commercial value for poles and pulpwood grown in a similar situation.)

Farmers in northern India planted eucalypts as cash crops on a massive scale in the early 1980s but planting almost stopped in the late 1980s. This was despite an overall shortage of wood in the country. Low yields, low prices, and extreme difficulties with marketing eucalypt wood and poles were largely to blame, according to Saxena (1991).

Social effects of growing eucalypts by farmers, for example in Gujarat State in India, are often even more complex than stated in the previous paragraphs, as can be learned from a number of case studies described by FAO (1988b) based on the work of Shobhita Jain. She has analysed some of the questions raised by previous reports through studies of various social groups in different communities and involved in contrasting forestry schemes. Her findings and insights shed light on the complexities of successful farm forestry using eucalypts and on the danger of generalizing, especially on such issues as trees replacing food crops or conflicts of goals between the forest service and farmers.

In contrast, for Africa, three successful small-scale eucalypt projects established to fulfil local needs in Lesotho, Botswana, and Sudan and managed by local populations have been described by Pryor (1978).

The widespread acceptance of *E. globulus* and *E. camaldulensis* by Ethiopian farmers is well known (NAS 1980).

Since the Eighth World Forestry Congress was held in Jakarta in 1978 with its theme 'Forests for People', there has been some change from imposing tree planting on people towards first asking them what they want.

2.3.2 Ecological aspects

There is no shortage of publications which give detailed information on the biogeography, ecology, silviculture, and utilization of the more commonly planted eucalypts. Florence (1986) listed more than twenty such publications and goes on to examine the physiological and genetic attributes of fast-growing eucalypts, pointing out characteristics which have made them well-adapted as exotics.

These characteristics include five distinctive mechanisms for producing new leafy shoots and rapid recovery of a crown after defoliation by fire, fungi, insects, grazing animals, or drought. They are naked buds, accessory buds, axillary buds, lignotubers (Jacobs 1955; Cremer 1972), and rhizomes (Lacey 1974). Some eucalypts can survive drought by stomatal control or by growing roots deep enough to reach water unavailable to other plants. Many eucalypts have mechanisms which recycle nutrients and permit good growth on soils of low fertility (Florence 1986).

Unfortunately, for a large proportion of the area planted to eucalypts in the world, much of this literature on ecological characteristics has been ignored. Perhaps this is because it has appeared in English which is a minority language amongst local professionals and technicians in many developing countries where eucalypts do have a part to play in improving rural living standards. Other reasons may be shortage of experienced staff and too much pressure to achieve positive results quickly on virtually non-existent budgets. One still observes eucalypt planting starting off on unsuitable sites and with seed collected from single trees in gardens, in arboreta, or along roadsides despite the severe lessons of the past such as those from Brazil, Africa and India, described in Sections 1.3 and 3.4. A frequent problem encountered in forestry and land use planning outside Australia is the enthusiasm for planting 'Eucalyptus' without appreciating that the genus *Eucalyptus* has more than 500 species, adapted to survive and grow in an enormous range of environmental conditions.

3

Genetic resources of eucalypts

3.1 Concepts of species and provenance in the eucalypts

It has been recognized by foresters for a long time that different geographic origins of seed used for planting a particular species of tree can lead to substantial differences in growth, form, and health of the resulting plantations. The differences within a species can be so large that one lot of seed will result in successful, profitable plantations and another in complete failure. Thus it is important for foresters to know the place of origin of seed and to obtain seed from the best places.

In museums the place of origin of a specimen, for example a bone or a pot from an archaeological site, is described as its 'provenance'. In forestry the term provenance is also used for 'place of origin', as defined in dictionaries, for example the Concise Oxford Dictionary. Also it is used to refer to the population of trees growing at the place of origin, the seed collected from them, and their progeny growing in plantations. The concept is clear enough when used to refer to the original geographic origin in a natural forest. However, it is not so clear or consistently used when the place of origin is a plantation, perhaps two or more generations removed from the original natural forest and possibly of mixed origin.

In a comprehensive review of the concept of provenance and its importance in forestry, Turnbull and Griffin (1986) considered the definitions and explanations of current usage of the term provenance by such organizations as the Society of American Foresters and the Organization for Economic Co-operation and Development (OECD). They noted that the several published definitions were somewhat contradictory and were loosely used in forestry practice, but nevertheless 'this concept is of major practical and economic significance'. They concluded that 'it is incumbent upon the user to provide the appropriate definition'. In a review of provenance nomenclature in relation to seed certification Jones and Burley (1973) emphazised the need for adequate information on genetic history of a planted stand and its antecedents. This is a matter of increasing importance in the eucalypts with the expanding trade in seed from plantations.

In this book we use the term 'provenance' only when the place of origin of seed is a natural forest. When seed is collected from planted trees we prefer the general term 'seed source' or in some cases 'land race' (Section 3.4).

The use of genetically-based geographic variation is the first step in tree breeding when assembling a suitable base population for future generations of selection, crossing, and genetic improvement. The concept of provenance implies that genetic patterns of variation are associated closely with the ecological conditions in which the species evolved (Turnbull and Griffin 1986). The recognition of such patterns of variation is made through genecological studies, mainly of phenotypic variation in natural forests.

Genecology concerns the patterns of adaptation within species, genetic, physiological, ecological and taxonomic, sometimes emphasizing one aspect more than another (Turesson 1923; Heslop-Harrison 1964). When the emphasis is on taxonomy the terms 'biosystematics' or 'experimental taxonomy' are more appropriate. Biosystematic studies which reveal patterns of clinal (continuous) variation facilitate prediction of the location of stands of desirable provenance, but if the variation is discontinuous, as it often is, a very thorough sampling is required of populations throughout the range of the species (Pryor 1963). Genecology is therefore a significant element in eucalypt breeding and will have substantial impact on future developments in the improvement of the genus because of the need to understand the seed source.

Several genecological studies of eucalypts (Table 3.1) have indicated the importance of

Table 3.1
Eucalypt species which have been the subject of genecological studies.

Species	Author(s)
E. camaldulensis	Pryor and Byrne (1969), Banks and Hillis (1969), Awe (1973), Grunwald and Karschon (1983)
E. cloeziana	Turnbull (1979, 1983)
E. coccifera	Shaw *et al.* (1984)
E. dalrympleana	Phillips and Reid (1980); Thibout (1984)
E. deglupta	Davidson (1972, 1983a)
E. delegatensis	Boland and Dunn (1985)
E. diversicolor	Coates and Sokolowski (1989)
E. globulus	Kirkpatrick (1973, 1975)
E. grandis and *E. saligna*	Burgess and Bell (1983)
E. gunnii	Potts (1983, 1985), Potts and Reid (1985a, b)
E. leucoxylon	Boland (1978)
E. maculata and *E. citriodora*	Larsen (1965)
E. nitens	Pederick (1976a, 1979), Shepherd *et al.* (1976), Cook and Ladiges (1991)
E. obliqua	Green (1971), Brown *et al.* (1976), Anderson and Ladiges (1982)
E. pauciflora	Pryor (1956), Green (1969), Harwood (1980, 1981, 1984)
E. pilularis	Florence (1969)
E. regnans	Eldridge (1968,1972)
E. urnigera	Barber and Jackson (1957), Thomas and Barber (1974)
E. urophylla and *E. alba*	Martin and Cossalter (1975-1976)
E. viminalis	Banks (1972), Ladiges and Ashton (1974), Phillips and Reid (1980)

sampling marginal (boundary) sites to determine the full range of diversity within a species. Patterns of variation recognized from these studies can improve the efficiency of sampling to detect provenances with desirable characteristics but are no substitute for thorough provenance trials under field conditions. Both continuous and discontinuous patterns of variation have been revealed in genecological studies and provenance trials of eucalypt species. There are many cases of continuous variation across pairs of eucalypt species and subspecies (Section 2.1.1) where it is difficult to assign a name.

Economically important characteristics which can vary among provenances include growth rate, stem straightness, resistance to frost, drought, insects, or fungi, and the ability to tolerate salinity, alkalinity, acidity, or waterlogging of the soil. Best provenances for planting in a particular area may be local, for example local *E. regnans* in the Strzelecki Ranges in south-eastern Victoria (Griffin *et al.* 1982a) or a number of provenances may be equally suitable, for example several northern Australian provenances of *E. camaldulensis* in Bangladesh (Davidson and Das 1985). Provenance differences in the eucalypts may be slight or may be so large as to determine the success or failure of a new planting program.

Up to the present, provenances have been labelled and referred to by locality, as in 'Petford camaldulensis'. In many species, however, several different locality names have been applied to the same provenance-collecting area over time. For example, collections of *E. camaldulensis*, in north Queensland from an area of the Walsh River drainage centred on Petford, are all referred to as 'Petford' provenance, though to be precise they may be from Emu Creek, Gibbs Creek, Eccles Creek, etc., 10–20 km from Petford (Doran 1984). Many workers have used the CSIRO Australian Tree Seed Centre seedlot identification numbers as synonymous with provenance, but this number changes with each subsequent collection from the same stand of trees. What is needed in every case is latitude and longitude in degrees and minutes and altitude for the centre of the provenance collecting area.

3.2 Sampling provenances within a species

Present-day natural distribution of the eucalypts is not static in terms of a long time frame. Stands may be expanding or contracting. Some have been eliminated or have been left to survive in isolated but sheltered locations following disturbance by extreme environmental events and man. Distributions are often discontinuous (see Section 2.2.4) and many small populations have evolved which are adapted to a wide range of climatic and edaphic conditions.

As will be seen in Chapters 6 to 17, patterns of provenance variation found today range from regional in *E. camaldulensis* and *E. deglupta* to localized in *E. viminalis*, or with an altitudinal gradient in *E. pauciflora*, *E. regnans*, and *E. urophylla*. Geographic variation is more pronounced in species in which populations are separated by ocean barriers as in *E. deglupta*, or which stretch across severe climatic gradients such as the varying length of the dry season between the coast and inland of northern Australia as in *E. camaldulensis*.

When sampled for the first time nothing is known of the pattern of variation of a species. The pattern of sampling will often be irregular and determined by a cursory study of stand differences which appear to have resulted from differences in terrain, soil, or climate. In preparation for sampling the opportunity should be taken to include study of available aerial photographs and documentation such as forestry maps for State Forests. As the resolution of satellite imagery improves this may also prove valuable for planning future collections.

There are two separate and distinct requirements for sampling provenances within a species:

(i) sampling a limited number of provenances, often only three, to represent the range of provenance variation in each species in species elimination trials, and

(ii) wide-ranging sampling of many provenances to represent part or the whole of the distribution for use in provenance trials.

For the first requirement, one suggestion is that seed should come from that part of the natural range where the species appears to be growing best. This is termed the optimal site and normally could be expected to be found somewhere in the middle of the species' range. For eucalypts this judgement of performance would be measured simply in terms of height and diameter growth, straightness and self-pruning ability. The theory behind this reasoning is that populations at the edge of the distribution would have evolved conservative survival strategies at the expense of fast growth while those populations on optimal sites have evolved under more intense competition for light and nutrients, thus favouring rapid growth rather than a survival strategy (Namkoong 1969). According to Edwards (1963) at least three sources should be used to represent a species for testing: the optimal site, a marginal site, and the closest climatic match. This approach is recommended. Climatic matching is discussed in detail in Chapter 4.

In order to choose a marginal site from that part of the boundary of the distribution that extends in the right direction (for example towards tolerance of a long dry season in a monsoonal climate) and to choose the closest climatic match, prior information on the proposed trial site is required.

For the second requirement (wide-ranging sampling for provenance trials) the number of sources sampled will depend on the extent of the natural distribution, the diversity of the species, ease of access, and the time, money, human, and other resources available to mount a collecting expedition.

A coarse grid system of sampling localities was described by Green (1971) in a study designed to provide basic information on genetic variation in *E. obliqua*. He superimposed a one degree (approx. 110 km) square grid on a map of the known distribution of the species. With some adjustments 22 approximate locations were selected and marked on the map. The locations were visited and the final collecting sites further adjusted according to seed crop abundance, lack of human disturbance to the stand, and convenience of access. Usually, collections will not be as objective and systematic as this example. One reason is that, given a finite package of available resources, there will be a trade-off between numbers of provenances collected and numbers of trees sampled per provenance (few provenances with a large number of trees per provenance as against a larger number of provenances with few trees per provenance) depending on total distance required to be travelled in a set period of time.

If there is feedback available from prior species trials in the proposed planting region which included more than one provenance, it may be possible to adjust the sampling program to concentrate on a certain part of the range of a wide-ranging species. Published reports of provenance trials of the same species on similar planting sites in other parts of the world will also be helpful.

The choice of seed sources to represent species for introduction and testing in future should involve careful and detailed study of the climatic, edaphic, and other factors within the natural distribution in order that limited resources available for collection can be directed to the most strategic locations. Homoclime analyses and ecological zoning described in Sections 4.3 and 4.5 will assist greatly.

3.3 Sampling to represent a single provenance

Experience over a number of decades has resulted in a set of rules for sampling within a provenance (FAO 1969, p.37):

(i) collect seed only from dominant and co-dominant trees of 'average' quality within 'normal' stands rather than plus (outstanding) trees in superior stands; if collections are made from plus trees they should be kept separate from the representative collection;

(ii) collect from a minimum of about 10 trees; record the number of trees sampled as an approximate percentage of the stand; stands exhibiting large variation from tree to tree should be represented by a higher number of samples;

(iii) choose sample trees which are at least seed-fall distance apart from each other; for eucalypts in Australia this means about twice the average height of the trees, although an arbitrary distance of 100 m is often used in practice;

(iv) mark on a large-scale map the positions of the individuals from which seed is collected to permit return to the same area for future collections; individual seed trees are sometimes permanently labelled and they are usually photographed;

(v) collect equal quantities of capsules from each sample tree;

(vi) keep seed from each sample tree separate at first;

(vii) obtain a botanical specimen of buds, fruits, and leaves for later confirmation of identity of the sample tree;

(viii) note the size and form of each sample tree.

Following confirmation of species identity, seedlots from individual trees will usually be mixed by provenance to provide the seed for species elimination trials and for detailed provenance studies. For the special purpose of combined provenance and progeny trials, they are kept separate by parent. In any case, it is wise to store at least some of the seed identified by individual seed tree in case it is required later for progeny tests.

The number of trees sampled to represent a stand (item (ii) above) is of critical importance both in the cost of the whole collection and in the precision of estimates of the characteristics of each population (Cotterill 1990). If some knowledge is available on the amount of variation to be expected between trees within stands in each trait (straightness, diameter, etc), or can be inferred from progeny tests of other species, it is possible to estimate how many trees to sample in a stand to estimate the mean value of each trait at a chosen level of statistical confidence. If the coefficient of variation (standard deviation divided by the mean) can be assumed to be 15 per cent and an estimate to within 10 per cent of the true mean with 95 per cent confidence is sufficiently precise, then only 10 trees need be sampled per stand (Snedecor and Cochrane 1980, p. 53).

In recent years it has been the practice in most eucalypt provenance collections to collect from 10 to 15 well-separated trees per stand as the basis for surveying variation in a species. This is a reasonable compromise between precision in estimating the characteristics of each population sampled and being able to collect from a large number of localities. Larger numbers of trees per locality, 50–100 or more, are sampled after provenance trials have shown which provenances are the best. These large samples become base populations for further selection.

3.4 Land races

When eucalypts have been grown in plantations in a particular region for a long time, the processes of natural and silvicultural selection can lead to the population becoming better adapted to the environment of that region. The new population is called a 'land race', defined by Zobel and Talbert (1984, p.89) as 'a population of individuals that has become adapted to a specific environment in which it has been planted'.

Several qualifications are needed to understand and use the concept of a land race as a population of 'best-adapted individuals with desirable growth and form' (Zobel and Talbert 1984, p. 89). The original provenance must be one of the best available, must have a broad genetic base including some of the best individuals, natural and artificial selection in the past must have been intense enough to change the population genetically (towards better adaptation), and the plantations must be large enough to allow intense selection in future. A land race derived from a small number of inferior trees of an inferior provenance is of little value, even though it might be better adapted now than it originally was when first introduced.

Much is said, though there is not much evidence in writing, about the improvement in eucalypts and other trees one generation after introduction from natural stands. For example, the Centre Technique Forestier Tropical (CTFT) in the Congo have compared the growth of two seedlots of *E. urophylla*, one from a natural stand in Timor and the other from trees grown from the same natural seedlot in a Congo trial. The better growth of the Congo seedlot (M. Corbasson personal communication), even without intense selection of seed trees, suggests there has been more recombination by outcrossing in the plantation than in the natural forest; the neighbourhood inbreeding effects of the natural stands have been broken down in the plantation (Section 19.6.2).

Neighbours which provide pollen for a seed tree in a natural stand are likely to be very close relatives and some of their progeny will grow slowly due to inbreeding depression. Neighbours in the plantation (Fig. 19.7, p. 196) are less likely to be closely related because of the mixing of the seeds from the several widely separated unrelated seed trees of the natural seedlot (provenance collections are usually from ten or more unrelated trees). Seeds from plantations are attractive in this circumstance (Griffin 1985). However, this does not mean that land races are always better than newly imported seed.

In South Africa *E. grandis* has been grown for several generations from an uncertain Australian origin since some time before 1900 (Poynton 1979, p. 357). One might have expected that local South African seed, especially from a seed orchard, would have been much superior to the original imports and also to new imports of wild Australian seed. However, trials in South Africa and Australia (Darrow 1983c; Ades and Burgess 1982) have shown that open-pollinated seed from plus trees selected by I.P. Burgess in plantations established with seed from natural stands near Coffs Harbour, New South Wales, produced better growth than seed from the first South African seed orchard.

One reason for the unexpectedly inferior growth of the South African seed orchard seedlot may be that the plus trees for the orchard were selected mainly for lack of log-end splitting and not for fast growth. Another reason may be that the South African land race in which the plus trees were selected may have been based on original imports of seed which were not from the best provenances. Also, the original introduction from which most of the 300 000 ha of fine plantation has descended may have been from a very small number of closely related unselected trees. Not enough of the best genes found their way to South Africa last century.

By contrast, the case of *E. grandis* in Florida, U.S.A. is quite different. New introductions of Australian seed from Burgess' New South Wales plus trees have proved inferior in growth and survival to the second and third generation selected progeny of Florida trees (Meskimen 1983). The original seed trees for Florida were likely to be of about the same quality, or inferiority, as the original South African introductions. The reason for the superiority of the selected Florida land race of *E. grandis* seems likely to be the intense selection for early growth rate and resistance to frost in Florida. There are some frost-resistant individuals among the new introductions from New South Wales, but a far higher proportion among the third and later generations which have survived and grown in that demanding environment. However, it should be noted that new introductions from the Atherton region in north Queensland performed very well in the first generation (Franklin and Meskimen 1984). The south Florida environment differs greatly from the environments in which *E. grandis* grows in Australia and South Africa in having occasional severe and sustained frosts following relatively warm nights and days for several previous weeks.

Another set of land races worth discussing is *E. globulus* in Portugal, Spain, Chile, and other countries. After several generations as exotics, these populations appear to be well-adapted to the plantation environments, and they probably are better than the original introductions. But does that mean that one should be satisfied with the land race as a current source of seed and as the genetic base for future selection? Until there has been adequate study of provenance variation in *E. globulus* there is no justification for assuming that the land race is better than newly introduced seed of the best natural provenances. The original introductions before 1860 were probably at best haphazard and at worst derived from a single, convenient tree as appears to have been the case in South Africa (Poynton 1979, p. 328; and see Section 1.3).

Generalizations about land races of eucalypts should be made with caution. It is wrong to assume that genetic changes in a plantation population must have been large and beneficial just because the plantations have been there for several generations. Sometimes it appears that a strong commitment to the value of a land race may be based more on nationalistic than on scientific grounds. Information is needed from well-designed field trials comparing seed from the land race with natural provenances and, if possible, from records of the original introductions. The important point to find out is whether and by how much the land race is a better seed source compared with alternatives.

Under the best conditions of a good provenance, a wide genetic base, and effective selection, a land race is of great value in providing good seed immediately and also plus trees for a breeding program. Under the worst conditions of an inferior provenance and very few original trees a land race is deceptive, although the extensive plantations producing a lot of wood seem to be suitable as a seed source and as a basis for selection of superior trees. In this case our best may be none too good; a fresh start is needed with comprehensive provenance trials leading to a more suitable genetic base for future selection.

3.5 Conservation of genetic resources

The continued, long-term availability of broad-based genetic resources is essential for efficiency and flexibility in future breeding. This is because, firstly, there is a need to strengthen and expand many of the base populations already assembled for eucalypt breeding. Secondly, there will always be a need for a source of 'new' genes to supplement those being lost due to the inevitable narrowing of the genetic base and reduced heterozygosity after repeated cycles of selection in a breeding population (Libby 1973). The 'new' genes would be needed for resistance to new pests and diseases and to meet the demands of changes in silvicultural practices, of changing and marginal environments, and of new products. Wild relatives in conserved natural forests are the cheapest and most abundant source of such genetic variability.

Not all populations are equally useful, and useful populations are not usually evenly distributed throughout the range of the species they comprise. It is clear that while the risk of extinction of individual eucalypt species may not be great (Section 3.5.1), there is an ever-present danger of the loss of a significant proportion of the total gene pool of many eucalypt species (Turnbull 1978). For example, if only the mountain top Timor *E. urophylla* (Chapter 16) were preserved and the Kangaroo Island *E. obliqua* (Chapter 13) both species would be so depleted as to be almost useless to mankind for wood-producing plantations. Because of the world-wide importance of the genus *Eucalyptus*, Australia has a special and international responsibility for the conservation of the vast array of genetic variability of this genus in line with the *World Conservation Strategy* (IUCN 1980) and general philosophy and strategy of genetic conservation in plants (for example Frankel 1972, 1974, 1978). Australian eucalypt forests have been extensively exploited for timber or cleared for agriculture only for a relatively short time during the period of about 200 years since the beginning of European settlement. In State-owned forests and reserves, the gene pool of most species has been kept nearly intact where management consisted of conservative selective logging and natural regeneration followed by only minor silvicultural alteration of stand composition. In addition to the naturally regenerated forests, National Parks and other publicly-owned lands also have been relatively safe havens for eucalypt gene pools. Outside these protected areas the remnants of eucalypt gene pools (some extensive, some small) continue to be reduced by clearing for urban development, roads, mining, agriculture, and pine plantations.

3.5.1 Threatened eucalypt species and gene resources

There are several eucalypt species with very restricted natural areas which cannot be considered to remain safe into the future (Specht *et al.* 1974; Pryor and Briggs 1981). Forty-eight species are found in only one grid cell (1° × 1.5°, Section 2.2.4) and 47 in only two grid cells (Chippendale and Wolf 1981). The natural populations of 57 species contained so few individuals that Briggs and Leigh (1988) considered these eucalypts were threatened with extinction within 10–50 years if present land use and other causal factors continued to operate. Among these species only *E. benthamii* is of current interest for fast-growing plantations. However, among species which Pryor and Briggs (1981) considered to be of 'relatively limited extent which may be threatened in the near future' there are six species of importance in plantation forestry, *E. dunnii*, *E. gunnii*, *E. nitens*, *E. nova-anglica*, *E. pyrocarpa*, and *E. torelliana*. 'No described eucalypt species, as far as is known, has been lost totally from areas of natural occurrence' (Pryor and Briggs 1981, p. 2).

Pederick (1976b, 1990) has pointed out that forest practices have changed in Australia since about 1960. Direct seeding of cut-over eucalypt forests now carried out in southern Australia using seed collected from other locations adversely affects the mainly intact gene pools in State forests. He suggested that if direct seeding continued at the current rate, the time would come when one would be unsure if any stand in a given forest was of natural genetic origin.

Planting rather than natural regeneration is being increasingly used in regenerating Australian eucalypt forests after harvesting. As interest increases in benefits to be derived from tree improvement, the seed used for planting in future may be of a different species or provenance and from a small number of trees. Thus the natural forest with a broad genetic base will be replaced with a planted forest with a different and probably narrower genetic base resulting in continual erosion of the original gene pool and possibly destroying many of the checks and balances built into the natural breeding system of the genus. It is only the relatively late settlement of the Australian continent and the preoccupation of forest services with pine plantations this century which have prevented the earlier erosion of eucalypt gene resources. Since about 1960 there have been several outbreaks of eucalypt 'dieback' diseases which have killed susceptible eucalypts over a wide range of habitats in western and eastern Australia. In many cases the cause is *Phytophthora cinnamomi* but other agents are also responsible (Bird *et al.* 1975; Old 1979; Old *et al.* 1981).

Outside of Australia, the most significant forestry species naturally occurring are *E. deglupta* and *E. urophylla*. Pryor (1975) reported that *E. urophylla* had disappeared from Solor Island, Indonesia. However, because *E. urophylla* is fire tolerant and is found in more rugged and remote country, it is less affected by human interference than is *E. deglupta* (Davidson 1973a, 1978a) which, in the Philippines and Indonesia, is threatened by clearing for agriculture and the demand for wood, particularly fuelwood and poles, in the alluvial valleys where this species normally grows.

3.5.2 *In situ* conservation

The concept of 'genetic reserves *in situ*' has been proposed by Yeatman (1972) which would allow in the reserves normal management for wood production, amenity, or protection forest since they would not have the same management criteria as applied to ecological reserves. The 'genetic reserves' would be oriented to applied forestry and tree improvement. The establishment of such genetic reserves in Australia was suggested by Pederick (1976b) for the purpose of conserving samples of gene pools not already protected in other special reserves. The concept of 'genetic reserve' was conservation of genes rather than individual trees. The essential requirement of a 'genetic reserve' was maintenance of the population by:

(i) regeneration naturally or from seed collected within the reserve, and

(ii) new generations derived from a large number of parent trees (Yeatman 1972).

Present views tend towards use of existing protected areas of various types for *in situ* plant gene conservation as advocated by Davidson (1983d) and Prescott-Allen and Prescott-Allen (1983). Thus gene conservation is considered only as an additional objective for protected areas rather than leading to a new category. Consequently the International Union for Conservation of Nature and Natural Resources (IUCN 1983) revised its categories, objectives and criteria for protected areas to exclude the idea of separate purpose-designed genetic management units.

In Australia a variety of reserves has been established and these contain many typical eucalypt species associations. They include National, State, and Forest Parks and Flora, Fauna, Ecological, and Scenic Reserves (Specht *et al.* 1974). State Forest Services also set aside reference reserves for protection of certain vegetation types (Shepherd 1974; New South Wales Forestry Commission 1974). It is unlikely that any large reserves will be made available in Australia for eucalypt gene resource conservation alone.

3.5.3 *Ex situ* conservation

Ex situ genetic resource conservation has the same aim as *in situ* conservation, the maintenance of as many useful genes and gene complexes as possible. The difference is that they conserve genetic material at different levels of biological and ecological organization and therefore involve different strategies for achieving the same aim. *Ex situ* gene banks maintain genetic resources through the preservation in cold storage of seeds, pollen, and micro-propagated clonal material collected from individual plants in a population, while *ex situ* gene resource stands and plantations conserve the plants themselves.

From a genetic view point, conservation *in situ* is superior to other methods because it allows the processes of evolution to continue; it represents the base of the 'iceberg' in Fig. 3.1. It is desirable that the greatest number possible of natural eucalypt stands considered valuable for their gene resources should be preserved in the future. However, it may be unrealistic and impracticable to insist that all eucalypt species should be conserved adequately in this way.

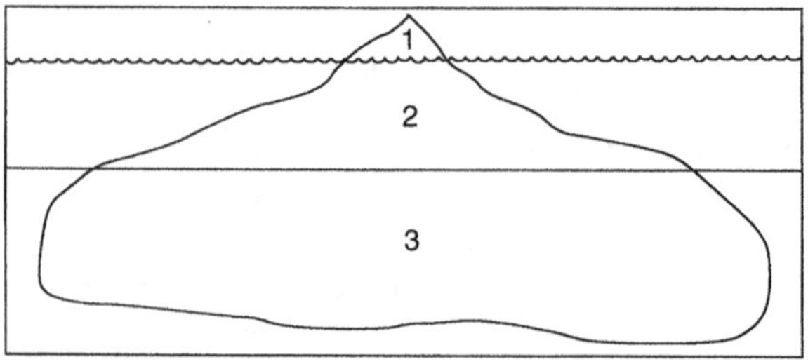

Fig. 3.1. The gene resource conservation iceberg. Off-site protection outside the natural forests (*ex situ* conservation) can preserve only a small proportion of genetic diversity of the eucalypts, the tip of the iceberg (1). Protected areas in National Parks and reserves on site (*in situ* conservation) can preserve much more (2), but this is still small compared with what might be preserved by sound planning allocation and management of natural eucalypt forests and woodlands (3). In practice all three sets of measures are necessary (after IUCN 1980).

For countries or groups of countries having a common interest in a particular species, establishment and maintenance of suitable gene resource collections *ex situ* is most appropriate (i.e. operation in the tip of the iceberg in Fig. 3.1), and would lead to a degree of independence from Australian and other natural habitats in future tree breeding work, though the *ex situ* collection will eventually form a local land race which may diverge with time from the original Australian genetic makeup.

It has been suggested by Guldager (1975) that economic potential, difficulty of reprocurement of seed, and the danger of extinction should be the main criteria used in forming a priority list of species and provenances for *ex situ* conservation. For example, in the case of *E. deglupta*, the species has a wide range from New Britain and Mindanao through three different self-governing countries and often in remote locations making seed procurement difficult. *E. urophylla* has a narrow geographic range but is found in remote, often rugged island locations where logistics are a problem. With *E. camaldulensis* in Australia and *E. tereticornis* in Australia and Papua

New Guinea good provenances are remote from normal forest operations, seed crops are limited, and seed collection activities are expensive.

The establishment *ex situ* of gene resource conservation stands with broad genetic bases of species such as these would enable cheaper and more convenient access to base populations and would act as insurance against any sudden catastrophic loss of the natural stands. Though it may be desirable for some *ex situ* conservation stands to be established in convenient locations in Australia, the main thrust of this work will be in countries with a commercial interest in the species. However, for developing countries some assistance is being made by Australia under its foreign aid programme. Assistance may be available also from international aid agencies such as the United Nations Development Programme (UNDP) or Food and Agriculture Organization of the United Nations (FAO). The FAO panel of experts on Forest Gene Resources made proposals for *ex situ* conservation of selected provenances of *E. camaldulensis* and *E. tereticornis* to be used as prototype stands (FAO 1975). These proposals have been put into effect (Doran and Boland 1978). With the financial support of the United Nations Environment Programme (UNEP), plantations of these species have been established in Thailand, Nigeria, India, and elsewhere, derived from open-pollinated seed of 25 or more trees from natural stands in each of four provenances of *E. camaldulensis* and two of *E. tereticornis*. FAO and UNDP have assisted Bangladesh to establish a 10 ha *ex situ* conservation stand of *E. tereticornis* (75 parent trees of the Helenvale provenance) and two 10 ha stands of *E. camaldulensis* (100 parent trees of the Petford provenance).

Recommended prescriptions for the establishment of these *ex situ* conservation stands are contained in FAO (1977). Unfortunately these recommendations do not prescribe the method of perpetuation of the stands from one generation to the next, do not consider the possibility and desirability of planting a separate single stand of mixed origin covering the entire natural range of a species (Libby 1973), and do not consider that land races might form quickly in eucalypts within a small number of generations (as few as two or three). Thus the ongoing usefulness of these *ex situ* reserves in gene resource conservation and tree improvement may be obscure in the minds of scientists in the host country.

Should conservation *ex situ* be impractical in the field, sufficient is known about seed physiology and biochemistry for preservation of the gene resource in the form of seed. Seeds of most eucalypt species can be stored for more than 10 years, and some up to 30 years, with a little effort to control moisture conditions and temperature (Boland *et al.* 1980a). These periods may be extended by further research on controlled conditions (Roberts 1975) such as the use of the carbon dioxide exchange method now used for eucalypts, acacias and other tree seed by CSIRO, Canberra (Doran *et al.* 1983).

Storage of pollen under controlled conditions could also be effective (Roberts 1975). Eucalypt pollen has been stored successfully for one or two years (Griffin *et al.* 1982b) and it may be possible to extend the storage period to a decade or more.

Both seed and pollen storage have the advantage that space requirements are minimal. Eventually the seed has to be sown and pollen used before viability is lost but these methods do provide insurance against the sudden accidental loss of valuable *in situ* gene resources (Turnbull 1978).

Tissue culture of meristems and cellular material (Hartney 1980; Hartney and Barker 1983) has opened up new avenues for conservation of gene resources of eucalypts. Tissue culture techniques provide an opportunity to maintain very high multiplication rates and the possibility of *in vitro* selection may solve several of the problems of maintenance of genetic bases in *ex situ* material, since unlike seed and pollen, cultures can be maintained indefinitely by periodic reculturing and cool storage under appropriate controlled conditions.

3.5.4 Sampling strategy for gene resource conservation

The population size needed to reduce the probability of extinction to very low levels while retaining a substantial part of the variability in the natural population is the subject of considerable debate and controversy. It is generally accepted that while much of the genetic information represented in a population and having commercial importance at present may be contained in as few as 50 trees (Namkoong 1979), quite large populations are required to protect the total gene pool, including perhaps attributes which are not presently recognized to be of use or value.

Also, in nature, fluctuations in numbers (Slatyer 1975) and health will inevitably occur which could seriously erode a small reserve. The arrangement and shape of reserves is also subject to considerable debate (IUCN 1980).

Sampling strategy for both *in situ* and *ex situ* conservation must ensure the greatest amount of within-population variation is conserved in a limited number of individuals. By its very nature in sampling only parts of populations, *ex situ* conservation is more prone than *in situ* conservation to go wrong. *In situ* gene banks (usually being protected areas) maintain gene pools by conservation of entire populations together with their local environment in ecosystems large enough to be self-perpetuating.

It is difficult to decide the minimum number of individuals to be kept in a population for the purpose of long term genetic conservation of trees *in situ* or *ex situ*. A conserved population needs to be numerous enough to retain its genetic variability through the processes of organic evolution over future generations (Section 2.2.2). Roche (1975b) has suggested areas of 10–30 ha per provenance or population are appropriate for *ex situ* conservation of forest trees with each population planted on at least two sites. It is doubtful if precise recommendations for conservation of genetic variability based on quantitative studies could be made reliably for eucalypts or any single tree species at present, but considerable progress is being made through use of isoenzyme investigations to determine the distribution and levels of genetic diversity within and between populations (Moran and Hopper 1987).

4

Matching species and provenances to site

4.1 Introduction

Scientific methods of making a good match between species and site have now become more precise, and are making an increasing contribution to species–site matching, but are still far from perfect. The most successful eucalypt introductions to date have been made by people with thorough knowledge and experience of eucalypt growing and of eucalypts in their natural habitat, and able to travel to make on-the-spot value judgements in order to match species to sites.

In this chapter we consider the methods available for matching species and provenances to plantation sites. We assume that well-conducted land-use planning has shown that plantations of eucalypts are appropriate. In other cases eucalypts would not be appropriate and plantations of other genera, particularly nitrogen-fixing trees, might be considered. In some instances, particularly in dry areas, it may be that no large plantations should be made because fast-growing, close-spaced trees of any species would be likely to lower the water table to unacceptable levels.

Site assessment should determine the areas within a region which are suitable for eucalypt plantings. It will also provide information on local climate and soils that, when analysed together with the available knowledge on site requirements of different eucalypt species, will reduce to a manageable level the number of candidate species to be included in initial species trials. The end uses required of the eucalypts also have a bearing on species selection and must be considered in species–site matching.

4.2 Assessment of the exotic environment

A government forestry agency, private company, or non-government organization promoting tree planting will often have an interest in a large area of land. The site may encompass hundreds or thousands of square kilometres, and there will usually be substantial variation in environmental and social conditions within the area. Usually, only part of the site will be suitable for eucalypt planting and there will be other competing land uses within the suitable area. Furthermore, in that part of the site where eucalypt plantings are judged the best land use, it may be necessary to zone the planting area for two or more eucalypt species. For example, it appears that at least six eucalypt species will be required to achieve good species–site matching for the range of eucalypt planting programs within Ethiopia.

4.2.1 Land classification

Land classification describes objectively the biophysical character of land. The general procedures used are decribed in the papers edited by Carpenter (1981). Rapid advances in remote sensing from aeroplanes and space-orbiting craft using infra-red and colour photography and radar images are having a major impact on the classification of land but must still be combined with 'ground truthing' (field investigation of soils and vegetation) and climatic analysis. Land classification is conducted at different levels of detail. For example, Davidson (1983c) described a detailed land classification carried out prior to the establishment of eucalypt plantations in the Gogol Valley, Papua New Guinea. His three-part hierarchical terminology consisted of land zones and systems (used at an overview scale of 1:250 000 to 1:2 000 000), land units (intermediate scale of 1:50 000 to 1:250 000), and land components (detailed scales of 1:50 000 or less).

Knowledge of the distribution of land classification units or components is valuable for deciding where species and provenance trials should be located, so that results can be readily and reliably extrapolated to other similar areas. There would be little value in siting a species trial in a land unit that was not well-represented in the region potentially available for eucalypt planting.

4.2.2 Land use planning

Land use planning attempts to allocate land units and components identified by land classification to uses that will be appropriate and sustainable with available technology and management.

Because there will usually be more than one suitable land use option for particular land units defined in the classification, land use planning must make judgements about the most appropriate use for land units. A rating team can be assembled, with individuals having skills relevant to the specific disciplines and proposed land uses. Ecological guidelines related to each activity should be consulted (for example Davidson 1987a). The team can then provide advisory forecasts, based on the best scientific evidence, of the consequences of the competing land uses on each land unit. The suitability of each unit for each plausible land use can be shown by a rating, and the ratings displayed in tabular form on the maps of landscape units (see Map 2 of Davidson 1987b). Two or more different land use scenarios may be prepared for a region, each giving a different feasible array of land uses, with, for example, different proportions of the land being targeted for eucalypt plantations (Davidson 1981, 1983c).

It should be emphasized that involvement of the local people in the land use planning process, and their agreement with the preferred land use zoning, are fundamental to the acceptance of eucalypt plantations in an area, whether for large industrial plantations or for small woodlots in agroforestry. For example, particular land units might be well suited to eucalypt plantations, but unavailable because they are used to meet food requirements for the region.

Having made the well-reasoned decision to plant eucalypts on a particular land type, the next step is to compile a detailed site description for that land type. This is a prerequisite for species selection to match the type.

4.2.3 Site assessment for species selection

This section covers the data needed to assist the matching of eucalypt species to available plantation areas determined by initial land use classification and planning. If a generalized land classification has been carried out, much of this information will be available for each of the land units classified. It may also be possible to make use of existing maps or computerized databases describing agroclimatic or agroecological zones. In a few countries such as South Africa, Zimbabwe, and Brazil major silvicultural zones have been mapped using climate and vegetation data and past experience with exotics. Once a site assessment is completed for a particular land unit or land component, it can be applied wherever that unit or component occurs in the study area.

Location of the planting site, and climate

Country, state, latitude, longitude, and altitude are necessary data. If no local climatic records are available, modern methods of computer-based interpolation can provide a reasonably accurate estimate of the local climate (Hutchinson 1984; Booth *et al.* 1988). Ideally, at least the following climatic data should be obtained either directly from local records or by interpolation, for each month of the year:

- mean rainfall,
- mean daily maximum temperature,
- mean daily minimum temperature.

For months in which frosts occur, mean monthly extreme minimum temperature, and absolute minimum temperature (i.e. the lowest temperature recorded for that month in any year on record) should also be collected, because frost can be a critical factor in determining species choice and the success of eucalypt plantations.

This information provides the climatic parameters used in modern homoclime analysis, described in Section 4.3, and also can be presented in the form of climatic diagrams (Walter and Lieth 1960). Additional useful information includes the mean monthly wind speed 2 m above ground level, and the incidence of severe wind storms, hail, and snow. The climatic information will have been used in any general land classifications done for the area.

Existing vegetation

A description of any remaining patches of natural vegetation usually provides a useful guide to the capability of the site for growing eucalypts. If no natural tree or shrub vegetation remains, interviewing local people and consulting historical records concerning the former vegetation may achieve the same result.

It is rare for any country in the warmer parts of the world not to have some planted exotic eucalypts (Zacharin 1978) or other Australian genera such as acacias and casuarinas, in botanic gardens, agricultural experiment stations, along streets, in parks and gardens, and around homesteads. The performance of these trees may provide information on possible species choice. Even information on non-Australian species may be of value.

Topography and soil

Slope (flat, gentle, intermediate, steep) aspect (north or south), and local relief (flat, hollow, undulating, ridge crest) are important in relation to exposure to the sun, wind, precipitation, and frost, to the degree of water supply and waterlogging of the soil, and to the depth of the soil profile available for tree roots. A recurring pattern of ridges and hollows led to the zoning of land for planting of three eucalypt species in Tokoroa, New Zealand (Lembke 1977).

Information should be collected locally on parent material, soil texture, depth, colour, stoniness and water relations (prone to waterlogging, free draining or intermediate). A very important parameter is pH (less than 4.5 very acid, 4.5–6.0 acid, 6.0–7.5 neutral, and more than 7.5 alkaline). In general, eucalypts do not perform well on alkaline soils. Soil salinity (if applicable) is also important in matching species to site. If detailed local information is not available some help could be obtained from the appropriate FAO-UNESCO soil map of the world (FAO 1971–1982) (Section 4.4), but these are on such a broad scale (1:5 000 000) they may be of limited use in a hilly region with highly variable soils.

Management regime proposed

A summary of the principal factors affecting tree growth and a synopsis of the prescriptions to be adopted in site preparation, weed control, fertilizing, irrigation, protection, initial spacing, and thinning (if any) should already be available since these considerations should have been taken into account in the earlier land use planning study where the decision to plant eucalypts was made in the first place.

Biotic problems may affect species selection: the prevalence of termites, root rot fungi, cankers, and insect attack observed on previous eucalypt introductions should be noted, and species selection and management adapted to minimize damage. For example, in parts of Ethiopia where termites are a problem, *E. tereticornis* has proved more resistant than other species, while over large areas in southern Africa *E. viminalis* has proved more susceptible than other species to the eucalyptus snout beetle (*Gonipterus scutellatus*). Incidence of cyclonic winds, flooding, and fire are other factors affecting the establishment and management of eucalypts.

4.2.4 End-use requirements

The desired end-use is often neglected when choosing species for a site. In some cases plantations have been of little value because markets for the wood have not been found (for example Saxena 1991). The quality of wood, oils, honey, and other products varies from species to species (Hillis 1984) so the end use or uses proposed will affect species choice (see Sections 1.2.2 and 20.1). For example, in tropical subhumid climates, *E. camaldulensis* is less suited for pulp production than *E. grandis*, because of the former's higher density and greater extractives content (Higgins 1984). When choosing species to meet a capital-intensive and large-scale end-use such as pulping, it would be sensible to carry out an economic evaluation that took into account both pulping quality and anticipated wood yields for the candidate species.

4.3 Climatic analysis for species selection

The purpose of climatic analysis is to reduce the number of candidate species to be tested at a particular location. This is done by eliminating species for which the climate at the proposed planting location is clearly different from the range of climates in which these species perform satisfactorily, either within their natural range or in plantations outside their natural range. The reduction in the number of species to be tested will reduce the size of the species trial needed, and eliminate the expense and effort involved in testing species very unlikely to succeed at the location in question.

Climate has been used for over 50 years as the principal basis for the more scientific methods of selecting new species of eucalypts for trial. Descriptions of the climatic requirements of several species were included in reviews such as those by Troup (1932), Streets (1962), and Hall *et al.* (1963). Attempts have been made to match climates in the proposed planting region with that in the area of natural occurrence in order to choose short lists of species for introduction and testing, notably in Brazil (Golfari *et al.* 1978). Summaries have been compiled of climatic requirements based on knowledge of the natural distribution alone or together with results and experience of trials outside the natural range (Metro 1955; Jacobs 1981; Turnbull and Pryor 1984). Maps showing areas with similar climatic conditions (Golfari *et al.* 1978; Poynton 1971) usually have been developed from analyses based on water balance calculations (Thornthwaite 1948; Thornthwaite and Mather 1956).

Climate was the first element introduced in computer-accessible species digests for forest trees, for example by Webb *et al.* (1984). Although soil factors may be just as important, they have proved difficult to incorporate into computer systems other than on a very broad scale because it is so much more difficult to interpolate between soil sites than between climatic sites. A new level of efficiency has been achieved in computer-assisted climatic analysis to select tree species that are appropriate for trials at sites outside their natural range (Booth 1985, 1991). This type of analysis will become increasingly available to foresters as software programs and climatic databases usable by micro-computers are distributed.

4.3.1 Overview of the Australian climate

In an introduction to the climate to which Australian trees are adapted Brown and Turnbull (1986) noted that over much of the continent the climate is generally one of low and erratic rainfall, frequent droughts, high temperatures, high solar radiation, and high evaporation. The interior and much of the west coast are particularly dry and the wetter areas are mainly limited to the coastal strip in the east and south-east. In general, the south has winter rainfall and the north has summer rainfall. Tropical cyclones from the surrounding oceans are frequent in northern Australia during summer and autumn and sometimes bring heavy rain well to the south on the east coast and less frequently on the west coast.

Some of the highlights of the account by Brown and Turnbull (1986, pp. 14–17) are:

About half the country has a median (50 percentile) annual rainfall of less than 300 mm and almost one-third has less than 200 mm.

More than 80 per cent of the continent has at least three months that are without effective precipitation.

The annual evaporation from an open tank exceeds annual rainfall everywhere on the mainland except in the eastern highlands, and exceeds 2000 mm over 75 per cent of the total area.

4.3.2 Climatic regions of Australia and their counterparts overseas

Better comparisons can be made with the rest of the world if the Australian continent is divided into climatic regions (Fig. 4.1; ADTR 1982a-e). Similar climates to the small areas of Australian humid tropics are found in parts of south-east Asia, the Philippines, most of Indonesia, most of Papua New Guinea, the Congo Basin in Africa, the Amazon Basin and north-east coast of Brazil, and in Central America and the Caribbean (Fig. 4.2).

The large areas of wet-dry (or monsoonal) tropics of Australia are comparable with southern Papua New Guinea as well as south-east Irian Jaya and the Banda Arc (East Timor and the Nusa Tenggaras) of eastern Indonesia. Other similar climates occur in Thailand, parts of Indochina, Burma, southern and eastern India, much of central Africa, western Africa south of the Sahel, north-eastern Brazil, Venezuela, and parts of the Caribbean (Fig. 4.2).

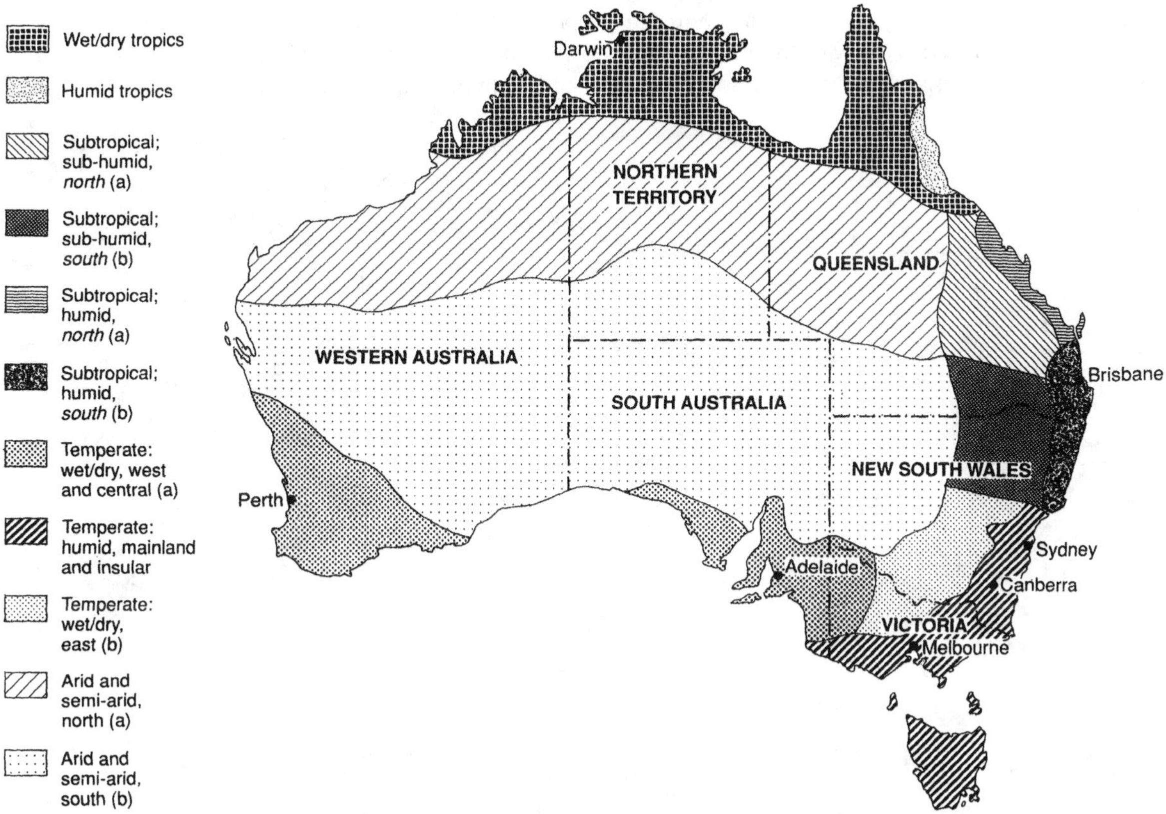

Fig. 4.1. Climates of Australia (after ADTR 1982a).

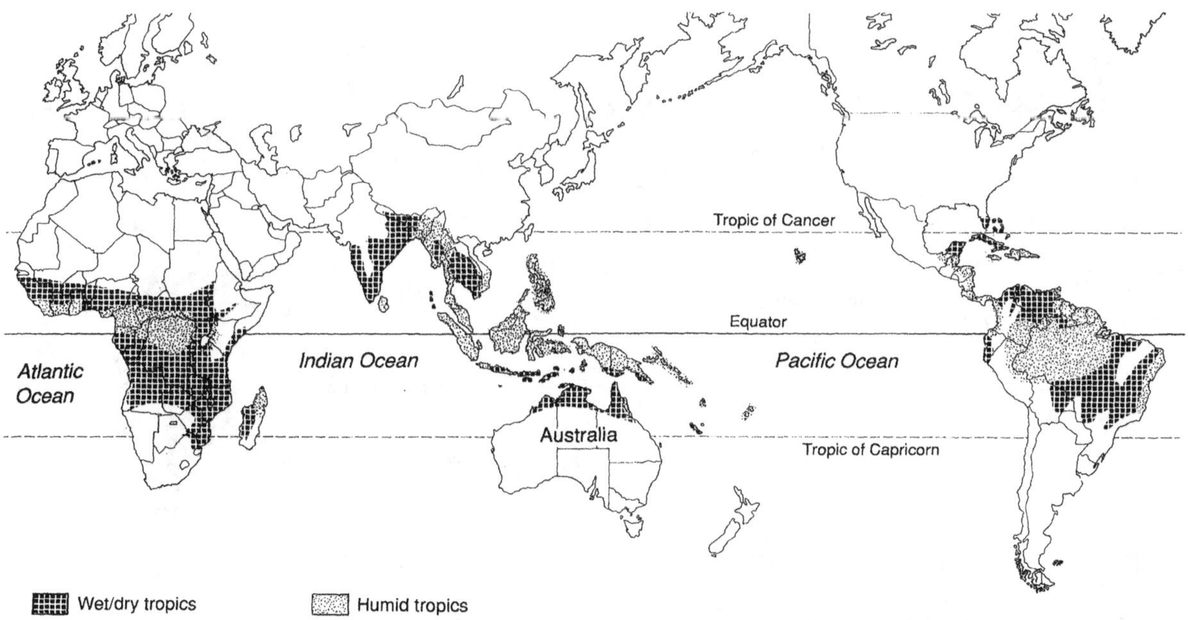

Fig. 4.2. Climates of the world: the tropical region (after ADTR 1982b).

Subtropical humid climates (Fig. 4.3) occur along the eastern seaboard of Australia in southern Queensland and northern New South Wales, with analogous climates along the eastern sides of South Africa and Madagascar, Uruguay and the adjacent parts of Brazil, Paraguay and Argentina, North Vietnam, coastal south China and Taiwan, Florida and a small part of eastern coastal Mexico. Sub-

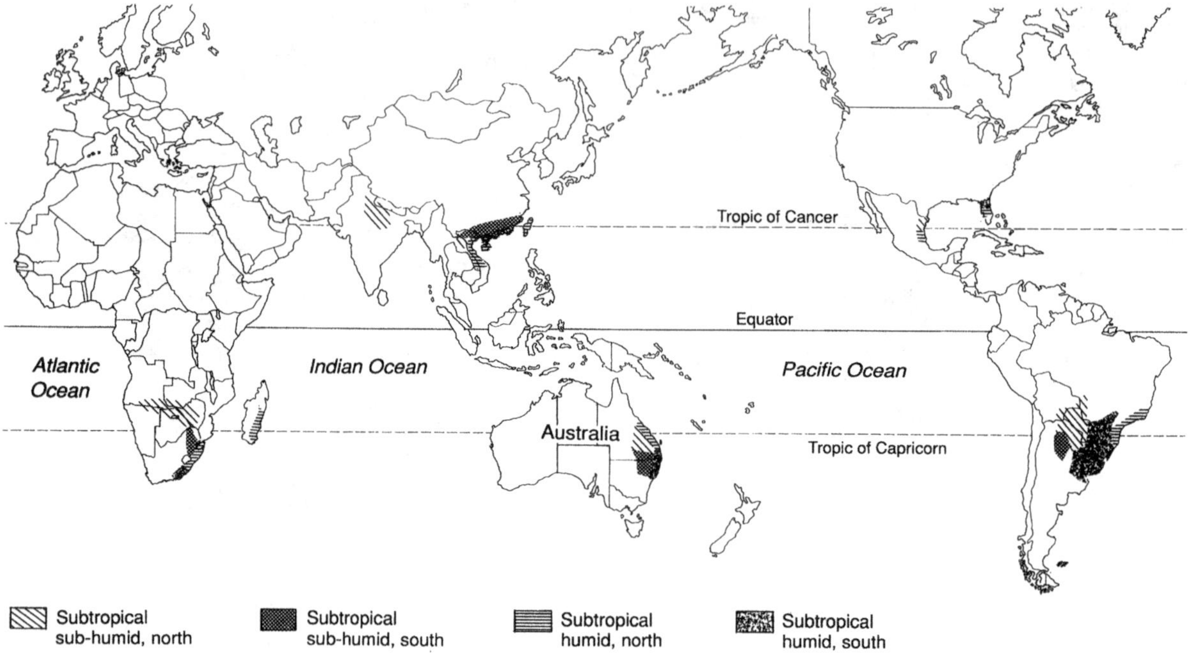

Fig. 4.3. Climates of the world: the subtropical region (after ADTR 1982c).

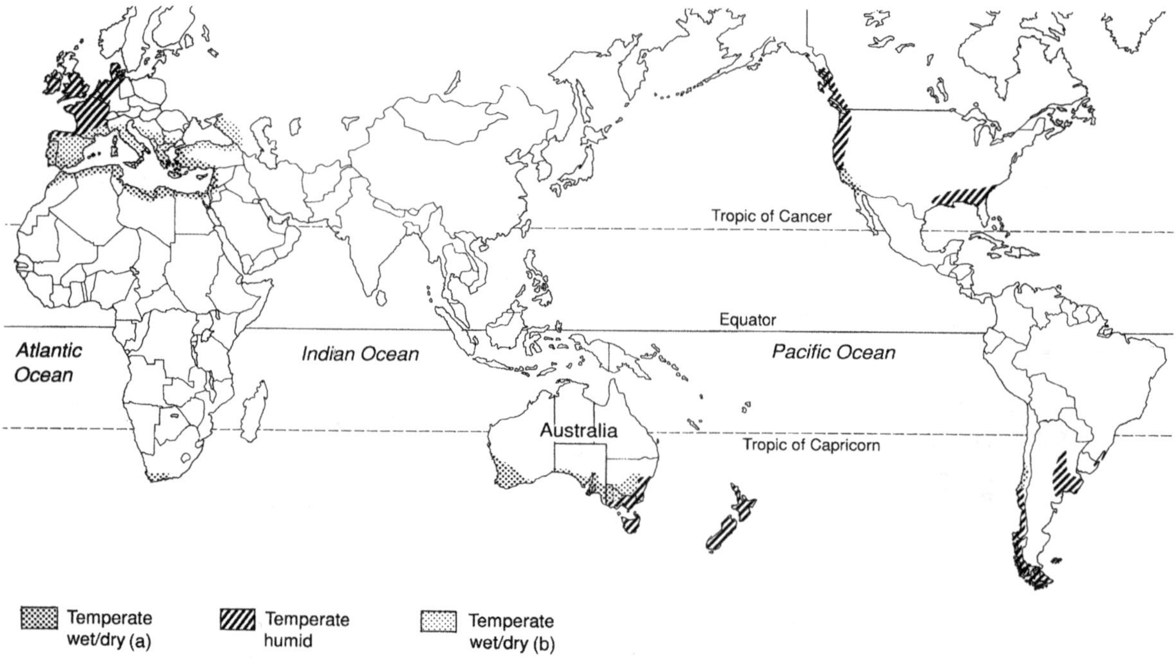

Fig. 4.4. Climates of the world: the temperate region (after ADTR 1982d).

tropical sub-humid climates occur in a wider belt inland from the subtropical humid climates in eastern Australia, and similarly inland from the subtropical eastern coasts in the other continents.

South-west Western Australia, southern South Australia, northern Victoria and south-central New South Wales have temperate wet–dry climates, which find their analogue in coastal regions around the Mediterranean Sea, Portugal, California, and coastal Chile between latitudes 30° and 40° (Fig. 4.4).

Southern and eastern Victoria, south-eastern New South Wales and Tasmania have temperate humid climates, which are also found in New Zealand, north-western Europe, coastal Oregon, Washington and British Columbia, coastal regions of the south-eastern United States north of Florida, eastern Argentina, and southern Chile (Fig. 4.4). However, the northern hemisphere temperate humid climates generally experience lower extreme minimum temperatures than their Australian analogues, and these extreme minima are frequently lethal to eucalypts, making their growth in such regions a risky proposition (Turnbull and Eldridge 1984). There are large parts of Europe, Asia and North America with climates for which there is no Australian analogue, because they experience much more severe winters than anywhere in Australia.

Arid and semi-arid climates (Fig. 4.5) cover the greater part of central and western Australia, and very large areas in southern and northern Africa, the Middle East, Pakistan and north west India, Mexico and the southern United States, western and southern Argentina and a part of north-eastern Brazil.

These climatic matchings are still very broad and can provide only a general indication for species selection for introduction by reducing the number of eucalypt species for consideration to perhaps between 10 and 50, depending on the region, out of about 100 species most used for wood production.

Climatic matching is improved if some account is taken of the seasonality of rainfall, the severity of the dry season, and mean temperature variation throughout the year. The publication over the past few years of summarized meteorological data for most of the world by FAO (1984, 1985, 1987) and the availability of similar detailed records for nearly 1000 stations across Australia (Hall *et al.* 1981; Bureau of Meteorology 1988) provides data for more detailed comparisons between stations in Australia and those located elsewhere. Climate diagrams can assist these comparisons (Walter and Lieth 1960), but monthly mean maximum and minimum temperatures should be considered rather than mean temperature of the day as diurnal patterns may be quite different for sites with the same mean temperature.

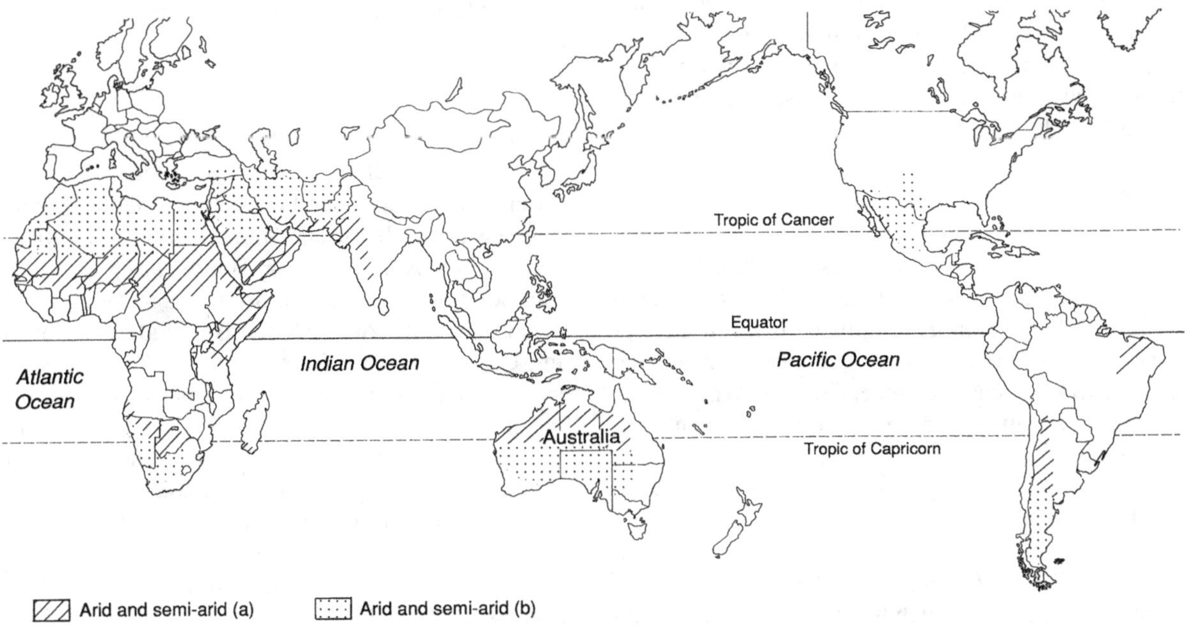

Fig. 4.5. Climates of the world: the arid and semiarid region (after ADTR 1982e).

4.3.3 Computer-based climatic analysis

Climatic analyses can be greatly speeded using computer-based methods of analysis. Methods have recently been devised to assist both comparisons with Australia and analysis of species' climatic requirements.

Climatic comparison with Australia

Booth *et al.* (1987) have described a program which can compare climatic conditions at any potential planting site with conditions at 2795 locations in a regular half-degree grid across Australia. Data for each month for mean daily maximum temperature, minimum temperature and rainfall are entered for the target location. The output of the program consists of details of conditions at the most similar location in Australia and indicates whether the match was good, fair, or poor. It also produces a map indicating the most similar regions in Australia. Considering species which are naturally found in these regions can assist the selection of species for trial at the target site. However, it is well known that many species can thrive under conditions somewhat different from those in their natural distribution. As Willan (1979, p. 158) has noted, 'Precision in climatic matching should not be carried too far. Most species and provenances have some degree of plasticity, a capacity to adapt to environments which differ somewhat from the natural one. Assessment of plasticity is one of the purposes of species and provenance trials'. It is therefore desirable to know a species' particular climatic requirements both in its natural range and in successful plantations throughout the world (Booth and Pryor 1991).

Analysis of species climatic requirements

Booth *et al.* (1988) determined the 'bioclimatic niche' for 13 important eucalypt species. The first step was to examine the species' natural distributions, using locational data (latitude, longitude, and altitude) for hundreds of natural sites obtained from the EUCALIST database (Chippendale and Wolf 1984). Where only latitude and longitude information were available, altitude was estimated from topographic maps. The BIOCLIM computer program (Nix 1986) was then used to estimate twelve important attributes of the climate at each point of occurrence for each species. BIOCLIM determines the climate of any point in Australia for which the latitude, longitude and altitude are known by reference to 'surfaces' derived from analysis of data from several thousand meterological stations across Australia (Booth 1985; Hutchinson 1984). In this way, the range of climates occuring across the natural distribution of each species was determined. The analysis of the natural distribution provides a first indication of a species' climatic requirements. For some lesser-known species it may provide the only information available, but most eucalypts have been evaluated outside their natural distribution.

The next stage of the investigation was to collect information on performance in trials outside Australia. Booth *et al.* (1988) gathered data for 13 species from 132 trial sites in Africa. For each of the trial sites, the twelve climatic attributes were either derived from on-site records of climate, or estimated using interpolation 'surfaces' which were specially developed for Africa. For those sites where a species had grown successfully, the climatic attributes were examined and the climatic niche of the species modified if climatic attributes of the successful trial sites fell outside the realized niche derived from the natural range. If sufficient data are available, description of climatic requirements for a species can be based solely on conditions at successful plantation sites.

The broad climatic requirements of 21 eucalypt species, including the species considered in Chapters 6–17, are summarized in Table 4.1, from Booth and Pryor (1991). Areas of optimal growth occupy a somewhat narrower range.

Mapping climatically suitable areas

The climatic niche for a species, derived from its natural distribution and/or successful performance overseas, can be used to map areas climatically suitable for its introduction. Booth *et al.* (1989) used interpolation techniques to produce a half-degree grid of climatic data for Africa. This included the following climatic factors for each of 10 187 grid points: mean annual rainfall; rainfall regime (uniform or biomodal, summer, or winter); dry-season length (number of consecutive months with less than 40 mm rainfall); mean maximum temperature of the hottest month; mean minimum temperature of the coldest month; and mean annual temperature. This data set was used to develop a microcomputer program for species–site matching.

Table 4.1
Climatic requirements of 21 eucalypt species estimated by Booth and Pryor (1991).

	Mean annual rainfall (mm)	Rainfall regime	Dry season (months)	Mean max. temp. hottest month (°C)	Mean min. temp. coldest month (°C)	Mean annual temp. (°C)	Absolute min. temp. (°C)
E. camaldulensis (northern provenances)	250–2500	s	2–8	28–40	6–22	18–28	n/a
E. camaldulensis (southern provenances)	400–1200	w,u	0–7	21–40	3–14	13–22	>–8
E. citriodora	650–2500	s,u,w	0–7	28–39	8–22	17–28	>–3
E. cloeziana	700–2000	s	0–5	27–36	6–17	16–27	>–3
E. dalrympleana	600–1500	w,u,s	0–5	18–29	0–7	8–17	>–12
E. deglupta	1200–5000	s,u	0–4	24–33	16–26	20–32	n/a
E. delegatensis	700–3500	w,u	0–6	18–25	–1–6	7–17	>–12
E. diversicolor	500–1500	w,u	0–6	23–29	5–9	12–18	>–4
E. fastigata	750–1900	w,u,s	0–5	22–29	–1–6	12–18	>–10
E. globulus	550–1500	w,u,s	0–7	19–30	2–12	9–18	>–8
E. grandis	700–2500	s,u	0–7	25–34	3–18	14–25	>–8
E. gunnii	800–2400	w,u	0–2	15–26	–3–3	5–13	>–14
E. maculata	620–2000	u,s	0–5	24–32	0–13	12–23	>–8
E. nitens	750–1500	w,u,s	0–6	20–28	–1–7	9–18	>–12
E. obliqua	700–2000	w,u,s	0–5	22–32	–1–7	9–17	>–10
E. pauciflora	600–1600	w,u,s	0–4	16–30	–5–6	5–16	>–14
E. pilularis	750–1850	u,s	0–6	22–31	5–12	15–22	>–6
E. regnans	700–2000	w,u	0–5	18–29	0–10	10–20	>–7
E. saligna	700–1800	s,u,w	0–6	22–32	1–14	14–23	>–8
E. tereticornis (tropical provenances)	500–1500	s	4–8	22–42	6–19	17–27	>–6
E. urophylla	1100–2700	s,u,w	0–6	20–34	8–22	18–28	>–4
E. viminalis	500–2500	w,u,s	0–5	16–32	0–8	10–18	>–12

u, uniform or biomodal; w, winter; s, summer
n/a, not applicable because frost should not be a problem in these areas, as indicated by the minimum temperature of the coldest month

Entering data for the climatic limits for a species produces a map on the computer screen showing the grid points in Africa climatically acceptable to that species. It was shown that the climatic limits given for *E. grandis* by Webb *et al.* (1984) resulted in a map that was a poor match with the areas of Africa where the species actually succeeds. The new climatic limits derived for the species by Booth *et al.* (1988) from analysis of 107 locations within its natural distribution and 52 trial sites in Africa resulted in a map with much better agreement (Fig. 4.6). Booth *et al.* (1988) noted that in southern Africa the lower limit of daily minimum temperature for the coldest month for successful growth of *E. grandis* plantations was 1.3° and 2.9°C at two trial sites. Schönau and Schulze (1984) suggested 4°C as the lower limit for planting the species in South Africa. Accordingly, Booth *et al.* (1989) suggested raising slightly the minimum limit of −1°C which they had calculated for *E. grandis* in the natural range.

The same methods can be applied at a more detailed level to enable finer-scale mapping of areas suitable for species within individual countries.

Fig. 4.6. Areas in Africa suitable for growing *E. grandis* as indicated by climatic requirements described by Booth *et al.* (1988) (after Booth *et al.* 1989).

Booth *et al.* (1990) developed a climatic database for Zimbabwe with climatic estimates for 4999 locations on a 5 minute grid (about 10 × 10 km). It is anticipated that similar climatic databases will be developed for other parts of the world within a very few years, and will be widely available for interrogation using personal computers (Booth 1991).

4.3.4 Limitations in the use of homoclime analysis

Although modern climatic analysis is a powerful tool, certain limitations should be kept in mind.

1. If a species' climatic niche is derived solely from analysing its natural distribution, without reference to overseas trial plantings, the range of climates that the species can succeed in will usually be underestimated. Fire, competition and edaphic factors, rather than climate, often set the limits to natural distributions (Pryor 1976).
2. Another factor which may lead to errors in niche determination based on natural distributions is that the remnants of ancient drainage lines are common in inland Australia. The water gathering and storage capacity of these old systems, and the beds of current water courses subject to only infrequent flooding, provide important ecological niches for some inland eucalypt species such as *E. camaldulensis*. Much more moisture is available to trees growing in these habitats than the rainfall figures in the local region would suggest.
3. The systems developed to date do not deal very well with frost as a limiting factor in eucalypt establishment. While frosts are most damaging to small seedlings and the frost climate of an average year may form the barrier to plantation establishment, it can be the extreme frost event that determines the success or failure of eucalypt plantings. For example, thousands of hectares of eucalypt plantations were killed by temperatures of −11°C at sites around the Black Sea in the USSR (Linnard 1969), despite having grown well in the preceding few years when mild winters were experienced.

Simple 'cut offs' based on the extreme minimum temperature experienced at a site are of some help in determining the areas where species should be planted. Gentilli (1961) suggested that, as a practical rule, adult eucalypt trees could survive extreme minimum temperatures 11°C below the July mean daily minimum temperature at their point of origin in Australia. Booth (1990) has suggested as an initial measure an absolute minimum of −8°C as the limit below which *E. grandis* cannot succeed.

Eucalypts do not have seasonal winter dormancy and hardening determined by photoperiod, their level of frost hardiness being determined primarily by the minimum temperature regime in the few days and weeks prior to a potentially damaging frost (Eldridge 1968; Harwood 1981; Tibbits and Reid 1987). They are very susceptible to frost damage in climates where prolonged, severe advective freezes (arising from an influx of freezing air from a colder region) follow mild weather with night minima higher than about 5°C, for example in south-eastern USA and southern France. Such climates may be contrasted with those in Australia where a gradual onset of radiation frosts of increasing severity in autumn and early winter enables progressive hardening, and sub-zero temperatures only prevail at night, except at the highest altitudes (Turnbull and Eldridge 1984).

Records of mean minimum and maximum temperatures are good predictors of average growing conditions. However, records of extreme minima (lowest on record) are not good predictors of damaging frosts; damage may result from a moderate frost after relatively mild nights. Also frost-hollow effects produce substantial local variation in minimum temperature, with differences of several degrees Celsius over distances of a few hundred metres in rolling terrain (Moore and Williams 1976). For these reasons, homoclime analysis of frost as a limiting factor in eucalypt plantings will always have to be carefully supplemented with local judgement and experience.

4. Homoclime analysis (and indeed all species–site matching methods that work at the species level) need to be interpreted with provenance variation kept in mind. The provenances from the extremes of a wide-ranging species (for example the Petford and Lake Albacutya provenances of *E. camaldulensis*) may differ substantially in their climatic requirements. In such cases where provenance variation is already known it is best to determine a bioclimatic niche for a particular provenance or a set of provenances of the species, for example tropical *E. tereticornis* provenances (Booth *et al.* 1988). In general, it cannot be assumed that any one provenance of a species will perform satisfactorily in all the climates defined as suitable for the species by homoclime analysis; provenance trials will usually be needed to determine the most appropriate provenances for particular sites.

4.4 Soil attributes and their influence on species selection

Soils, as well as climate, have had a major influence on the distribution and adaptation of the eucalypts. Now computer-based systems are being developed to match species to sites by taking into account soils, geology, landforms, and climate.

The most comprehensive world-wide standardized soil classification and maps available are by FAO (1971–1982). These include ten volumes of reports and 18 accompanying maps at 1:5 000 000 scale. The United States Department of Agriculture Soil Survey has produced a Comprehensive Soil Classification System which is widely used in developing countries as well as in USA although its coverage is not global (Money 1988, p. 89).

The widely used soil classification in Australia (Stace *et al.* 1972) recognizes 43 'Great Soil Groups' based on the degree of profile development and amount of leaching. This wide range of soils has developed as a result of the influence of the ancient parent materials rather than recent or present climates. Parent materials include mixtures of the deeper weathered horizons of ancient profiles, alluvial and colluvial materials, and parent rock. Most of these Australian groups have equivalents in other parts of the world as can be shown by comparison with FAO's standard terminology and maps (FAO 1971–1982). However, there are many differences between the soils of other continents and those of Australia (Sanchez and Isbell 1979). Parts of Africa come closest to a good match. More than 25 per cent of Australia's land area has strongly differentiated soils with sodic clay B horizons (mainly podzols and podzolic soils). Such two-layered (duplex) soils are less common in Africa and absent in South America.

Over much of Australia, the soils (or at least their A horizons of sandy loam) are more acid than pH 6.0. Because of this, only a very small number of the eucalypts are adapted to alkaline soil conditions. One such species is *E. gomphocephala*, which occurs naturally on calcareous sands and heavy textured alkaline soils on the coastal strip of south-western Western Australia. This species has proved outstanding in plantations around the Mediterranean, especially in Algeria, Cyprus, Israel, Jordan, Libya, Morocco, Syria, and Tunisia. *E. microtheca* also grows well on heavy, alkaline or calcareous soils, for example in the Sudan. Nevertheless, the range of such candidate species for fast-growing eucalypt plantations under alkaline conditions is severely limited.

High concentrations of salt (sodium chloride and sodium sulphate) are found frequently in Australian coastal and inland soils. *E. robusta* occupies swampy, somewhat saline estuarine sites in New South Wales and southern Queensland. Inland, *E. camaldulensis* occupies shallow valleys of old river systems and wide, shallow depressions where low run off and high evaporation have apparently raised the concentration of salt in surface soils. Thomson *et al.* (1987) have demonstrated considerable intraspecific variation in salt tolerance of the seedlings of both *E. camaldulensis* and *E. tereticornis.* Although *E. tereticornis* was generally found to be less salt tolerant, some *E. tereticornis* provenances had high salt tolerance. Thomson *et al.* (1987) also provided a provisional list of 21 eucalypt species known to occur on saline soils over all or part of their natural distributions. All but five of these species are untried overseas and unlikely to be important for commercial forestry, being of mallee or woodland tree form, but may become important for reclamation of saline sites and provision of fuelwood.

Because data are lacking for matching eucalypt species to soils at a fine scale in the natural habitat, it is inappropriate to present here a detailed description of the Australian soils and physical environment for species–site matching. The more sophisticated methods discussed later in this chapter are based on the rapidly increasing bank of knowledge of the performance of eucalypt species overseas in trials and plantations.

4.5 Species–site matching systems which integrate different types of data

Geographic information systems (GISs) and other synoptic inventories of physical land resources are rapidly becoming available for more and more countries, because of their value in land use planning. Although particular emphasis is placed on those attributes of land which influence agricultural production, the results can also be used in connection with suitability for tree planting.

Matching a eucalypt species to a site is much the same task as choosing suitable food crops and predicting their yield for particular sites. In Papua New Guinea 4566 resource map units were matched with 17 food crops using semi-quantitative descriptions of the plant–environment relationships (Hackett 1988). Hackett's approach describes objectively the ecophysiological character of the crop species to be planted, and matches it to the site, which has already been subject to land classification and site assessment (Sections 4.2.1, 4.2.2).

Twelve land factors (eleven soil factors and one related to length of growing season) were used to match four eucalypt species to site in Bangladesh and compare their suitability against several other species on the same land units (Davidson and Khan 1986). The printout indicates the area of land within a local region suitable for each species, according to soil association and four levels of suitability (S1 : very suitable; S2 : moderately suitable; S3 : marginally suitable; and NS : not suitable).

More complex methods of species–site matching may use computer databases of primary or summary information. For example, the TREDAT database (Brown *et al.* 1989) stores detailed information on site, management, and performance for many field trials of many species and provenances. In contrast, Webb *et al.* (1984) have provided summary information for 173 species, including 40 eucalypts. This system used 21 criteria relating to altitude, climate, soils, production, utilization, and silvicultural characteristics, with the user required to choose among several categories in each criterion to describe the site under consideration (for example soil texture would be classed as either light, medium, or heavy). When the site characteristics and user requirements are specified, the computer program checks them against the species database and prints out a list of candidate species: those fully meeting the specified criteria and those marginally suitable.

Such approaches to matching species and site, employing methods which are effective, quick, and cheap, will become increasingly useful for planning successful eucalypt plantations as the amount of accurate information on species and provenance performance accumulates.

5

Testing species and provenances

5.1 Need for species and provenance testing

Although eucalypt databases (Chapter 4) are rapidly becoming more comprehensive, this information will not supplant the need for species elimination and provenance trials. There are several reasons for this.

1. Databases derive their effectiveness from an increasing body of experience with eucalypts already growing or being tested as exotics across a wide range of ecological conditions (Table 5.1 and Chapters 6–17). Little-known and rarely tested species are not represented adequately enough for the database method to be used for species selection and site matching. More likely one of the homoclime techniques described in Section 4.3.3 would be used to select a limited number of species for elimination trials.
2. The climatic databases cannot account for the fact that some species are more adaptable when fitting into new environments than previous experience would indicate (Willan 1979; Booth and Pryor 1991).
3. One would expect quite a precise correspondence between the performance of provenances of a particular species at a well-documented site and the predicted performance at a comparable new site. This is not always the case. Genotype–environment interactions are very prominent in some species of eucalypts and much less in others (Table 5.2).

There always will be some need for well-conducted field trials for precise matching of species and provenances to particular sites of a planting project, even if only to validate predictions and refine prediction techniques. The numbers of species or provenances required to be tested is likely to decrease in future years as databases improve and other means of predicting species and provenance performance become available.

Screening tree species and provenances using small plants in a nursery or controlled environment facility, as seedlings or as physiologically older cuttings, has been attempted many times but as yet with limited success in predicting growth rate in the field. Probably the first attempt was by the pioneer Californian tree breeder, Lloyd Austin, a protege of Luther Burbank, who screened 729 seedlots of *Pinus ponderosa* in a well-designed nursery experiment (Day and Austin 1939). More recently, success has been achieved in predicting the frost resistance of eucalypts in the field from studies on seedlings in the laboratory (for example, Harwood 1981; Rook *et al.* 1980; Raymond *et al.* 1986) and salt tolerance (Thomson *et al.* 1987; Marcar 1989). Further research on physiological processes and experimental techniques may well allow rapid and accurate screening in the nursery and laboratory for species, provenances, and individuals tolerant of frost, drought, high concentrations of salt or lime, and low concentrations of phosphorus.

5.2 Sequence of trials

When deciding what species to plant for a new plantation project there are usually several species and many provenances of each species to be considered, even after employing climatic matching and other means to confine the list to the most promising seedlots (Chapter 4).

A sequence of field trials is required. The first, a species elimination trial, will test the survival and early growth of a large number of seedlots—perhaps 50 or 100 in small plots—and will last for a short time, one to three years in the tropics. The next trial will have provenances of only the better species and last for about half the expected rotation. The process of elimination on the basis of survival and growth continues until one feels assured the most suitable provenances of the best species have been chosen. For example, Burley and Wood (1976) identified an arboretum phase followed by a sequence of four phases of species and provenance testing. Then the growth and yield of chosen species and provenances would be assessed in growth trials and pilot plantations. Finally, plantation trials can be established to choose the best management option, including deciding on initial spacing, thinning practice, and rotation length (Davidson and Das 1985).

In each trial in the sequence the purpose is to eliminate the seedlots with poorer performance and to identify the best seedlots, those showing enough promise for further testing. Each successive phase tests fewer seedlots and employs larger plots than the last. Whatever sequence of trials is followed, some may be conducted simultaneously, in combination, or in parallel.

In this chapter the organization of species elimination and provenance trials will be described and no further reference made to growth and plantation trials, since the latter involve mostly silvicultural and management aspects rather than being directly connected with selection and breeding of trees.

While arboreta certainly are valuable for demonstrating botanical differences between species and for public enjoyment of a wide variety of trees, we consider that for testing eucalypts the arboretum phase is not necessary or appropriate for two main reasons.

1. The arboretum phase is time-consuming, more or less wasting one generation of testing and development time. This phase was of some use in the past when knowledge about differences between species of eucalypts in exotic situations was minimal. With the development of techniques for more precisely matching species to site, this step is no longer necessary and species elimination trials should form the first phase in any new program.
2. Eucalypt arboreta have been used as seed sources nearly everywhere they have been established. This has led to later problems with initial material for planting because the seed collections were made without recognition of the effects of hybridization and inbreeding. The famous arboretum created by Navarro de Andrade at Rio Claro in the state of São Paulo, Brazil (Gurgel and Cavalcanti 1978) has been one example mentioned several times in this book (Sections 1.3, 16.2). Another is the large arboretum at Butare in Rwanda where 67 species of eucalypt have been planted since 1934, in 50×50 m plots (Anon. 1987). The arboretum at Asella in Ethiopia was established by the Chilalo Agricultural Development Unit in 1965 assisted by the Swedish International Development Authority. Similar examples exist in a number of other countries.

A major constraint with designing species elimination and provenance trials is the great difficulty experienced in procuring eucalypt seed which is authenticated as regards botanical identity of species and geographic origin (the precise provenance locality in the natural range). It is wise to pay a higher price to obtain seed of known genetic history at this initial stage of testing if doubt is not to be cast on the results later. The principal sources of eucalypt seed for research, with authenticated botanical identity and geographic origin, are the CSIRO Australian Tree Seed Centre in Canberra and Centre Technique Forestier Tropical (CTFT) in Paris.

5.3 Selecting trial sites

Sites for species elimination trials can be selected in advance of species introduction by reference to land classification maps (see Section 4.2.1) which partition land into homogeneous ecological units. Normally, trials will be repeated in time and space in each land type. Positions may be determined randomly in each land type or chosen systematically because of other factors or constraints such as ease of access, vacant land available, ease of protection from fire and browsing, degree of fragmentation of the land type, etc.

Sites for provenance trials should be representative of the bulk of the project area to be planted and of a site type already proven to be reasonably suitable for the species.

For an effective trial it is most important to be able to characterize the environmental variation statistically over the trial site. Thus it is important to avoid sites with extreme variability. The site should have sufficient uniformity in soil conditions to allow all the plots within each block of the statistical layout to have approximately the same environment.

5.4 Raising plants for trials

A great deal of careful, well-organized work is required to carry out species and provenance trials successfully. Each step must be done correctly or the trials will fail to produce useable results, though almost every step is relatively simple, except for some aspects of statistical design and analysis. There are two users' manuals giving the detailed instructions needed for seed handling, nursery practice, site preparation, planting, and the design, layout, and assessment of trials (Burley and Wood 1976; Carter 1987). Supplementary information is given in other publications. Seed characteristics including conditions for germination of eucalypts are described by Boland *et al.* (1980a). Nursery practice for eucalypt seedlings is covered in detail by Jacobs (1981) and Doran (1986).

Growth conditions of plants should be as uniform as possible within a nursery (Burley and Wood 1976). However, because there are usually systematic differences between beds in nurseries, it is important to use an experimental design in the nursery which is compatible with the proposed field design. In the case of a randomized complete block design (RCB) the plants for a particular field block (replicate) should all come from the same nursery block. Since eucalypt seedlings are usually grown in containers, often held in plastic baskets, it may be possible to increase uniformity within blocks by periodically rotating plants from the middle of the bed to the outside edges. During such manipulations there is a danger that plant identities will be lost and the great importance of adopting an effective labelling system must be emphasized.

Labels which are easily detached from containers or pulled from the germination trays or which fade in sunlight should be avoided. Commonly, nursery staff are illiterate or semi-literate in many of the countries where eucalypts are widely used or have potential. They do not realize whether labels have been switched around or not. To avoid this problem colour-coded plastic flagging can be used to identify batches of plants and the corner pegs of field plots. This flagging is commercially available in many bright colours and patterns with more than enough variety to tag a large number of treatments without confusion.

5.5 Experimental design

The purpose of experimental design in variety trials is to be able to decide with confidence whether the differences between seedlots in growth and other characteristics were due to real genetic differences in performance and not simply due to chance. In statistical terms the purpose is to minimize the error variance in analysis of trial results or, put another way, to minimize the standard error of the mean estimated for each seedlot. The statistician's contribution is to try to 'cater for the perceived areas of variation across the site. The philosophy is to design for known and suspected trends and to use randomization to try to overcome biasses as a result of unknown areas of variation' (E.R. Williams, personal communication). This can be done by:

(a) appropriate experimental design, for example a randomized complete block (RCB) design for comparing a small number of seedlots and an incomplete block design for a large number;

(b) use of site information, such as allowing for increased fertility down a slope (Fig. 5.1);

(c) randomization of the position of the seedlots over the trial site;

(d) sufficient repetitions of each seedlot, with each seedlot appearing only once in each replicate;

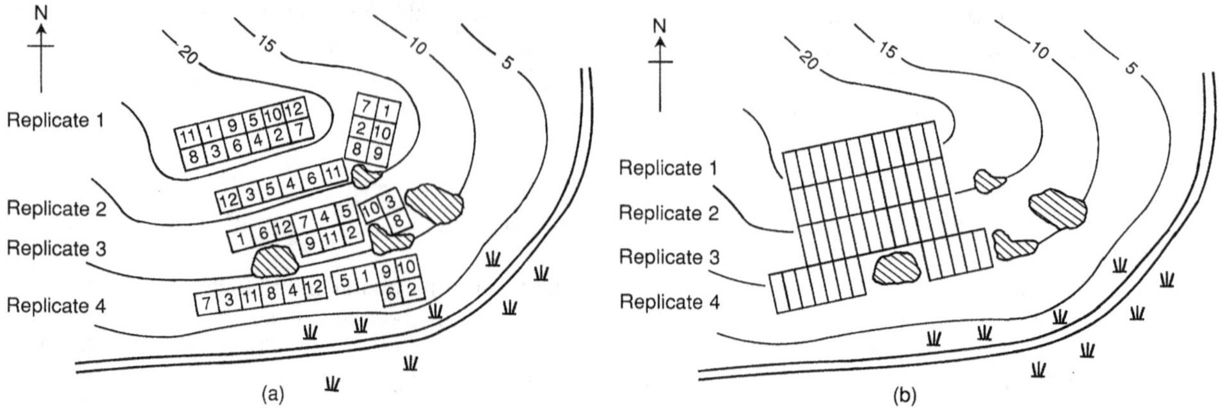

Fig. 5.1. Hypothetical layout of a field trial (a) with square plots, (b) with row plots. Layout of plots in a field trial must take into account gradients of the environment, for example slope, fertility, and soil moisture in this case. The plots within replicates are arranged across the slope, and row plots are oriented down the slope. The contour lines have heights in metres, the shaded areas are outcrops of rock, and there is a river running through a swamp (after Carter 1987).

(e) appropriate plot size (number of trees and area) in relation to total number of seedlots, spacing between trees, variation expected within seedlots, variation over the trial site, and number of years the trial will run;

(f) minimum area and maximum uniformity for each block, so that all the seedlots within a block have much the same environment.

The experimenter can achieve a balance of these somewhat conflicting requirements by applying common sense and some knowledge of basic principles to produce trials in which differences in performance of seedlots are due to inherent genetic differences and not to chance environmental effects. Experimental design is the way the experimenter allocates seedlots to plots, employing the essential procedures of randomization, replication, and blocking (Roger and Rao 1990).

Sections 5.6 and 5.7 give an account of experimental design, and procedures to implement a design, which are widely and successfully used without giving much detail of statistically more efficient alternatives which require more statistical knowledge and experience.

5.6 Species elimination trials

Testing of several species each represented commonly by three or more provenances can be accomplished rapidly in species elimination trials. They are of short duration and are followed by trials of a large number of provenances of the best species.

5.6.1 Spacing and plot size

There has to be a compromise between overall size of the trial and finding sufficient reasonably uniform land in a compact representative site. One of the most important points is that the total area of the block must be small enough to have very little environmental variation within it. One contribution to achieving this is to use incomplete block designs (Section 5.6.2) and another is to use closer than normal spacing between the trees giving more trees without occupying an excessive area. For example, 25 trees at 1.5×1.5 m occupy 56 m^2 and at 3×3 m occupy 225 m^2. Close spacing is justified because the trial is intended only to rank the seedlots not to accurately represent the expected yield. A species elimination trial is not meant to proceed for a full rotation. Spacing should be constant among similar trials repeated in time and space.

In many cases the investigator will not have been able to reduce the number of promising species to be tested to less than about 25 and with the number of provenances, say, an average of four, there could still be 100 or more seedlots for a field trial. With wide spacings (3 × 3 m) and multiple-tree plots (25 trees), each replicate could occupy more than 2 ha, which is far too large for the randomized complete block design. However, if the investigator is prepared to plant at very close spacing (1.5 × 1.5 m) and with fewer trees per plot, the space required per replicate can be reduced to a small and relatively uniform area. An assessment after one or two growing seasons could reduce the number of seedlots to a manageable number, perhaps the best 20, to start new, well-replicated trials to be run for half the rotation with larger plots at normal spacing.

The kind of information expected from a species elimination trial determines the plot size. If the trial is purely for ranking a large number of seedlots (say 100 or more) and will be abandoned after two or three years when a new trial is started with the better seedlots, then small plots may be justified, perhaps 6 trees in single-row plots replicated 10 times, or 12 trees in rectangular plots replicated 5 times. A relatively simple experimental strategy of several randomized complete block experiments, with a few seedlots common to all experiments, could be used if spacing is close and the site reasonably uniform within replicates. An undesirable consequence of small plots is that some slow starters which will be suppressed in the first year or two may have performed much better in later years ('stayers' in horse racing terminology) and some which were the tallest in the first years may grow much slower in later years ('sprinters'). However, a slow starter is of no interest for very short rotation eucalypt crops for fuelwood or pulpwood.

Sometimes the investigator is required to provide some indication of the expected yield of the better seedlots, as well as ranking all seedlots, and to run the elimination trial to at least half the expected rotation. This case requires square plots with a large number of trees (at least 25 and sometimes 49) and experimental design becomes much more difficult because of the large, and therefore expensive, area of fairly uniform site set aside for a trial much of which will not produce a useable crop of wood. Also there is the need for a higher level of statistical expertise for the complex design and analysis. With such large plots the number of seedlots is likely to be limited, thus foregoing the opportunity to test many promising species and provenances. An advantage of large plots for long rotation crops for sawn timber is the opportunity to differentiate between 'sprinters' and 'stayers' since internal buffers are available to delay inter-plot competition and suppression.

5.6.2 Statistical aspects

Likelihood of plots being lost during the course of the trial should be considered when selecting an experimental design. Lost plots cause least problems in simple designs. For this reason complete randomization (without blocks) and randomized complete blocks (RCB) are preferable to the more complex designs when missing whole treatments (seedlots) are expected due to death or very slow growth of some seedlots, as in species elimination trials. If several missing plots are likely to be caused by non-treatment effects such as damage by animals, fungi, or people, the number of replicates should be increased accordingly. Four replicates is the absolute minimum for well-conducted trials with good survival on 'uniform' sites.

The RCB design has been the most frequently used in variety trials in forestry. It is easy to analyse using instructions in such books as Dawkins (1975), Little and Hills (1978), Mead and Curnow (1983), Freese (1980), and Andrew (1986). Completion of the analysis requires an estimate of the values of any missing plots. Too many missing plots reduces the sensitivity of the statistical test of differences between seedlot means.

In the RCB design the analysis allows for variation between blocks: the blocks (replicates) are recognized as an identifiable source of site variation that is isolated in the analysis of variance. If successfully applied, blocking reduces error variance and hence gives a more sensitive test of differences between seedlots than could be obtained with a completely randomized design.

Although the RCB analysis allows for variation between blocks, this variation should not be too large. In some species of eucalypts, genotype–environment interactions may be sizeable, giving rise to different rankings of species in each block if the site quality of blocks differs too much. Important precautions to be taken in the spatial layout of blocks in the field (see for example Fig. 5.1) have been illustrated by Burley and Wood (1976), Zobel and Talbert 1984, and Carter

(1987). Row and column designs (John 1987) can be used whenever there is two-way heterogeneity in site conditions that cannot be controlled by blocking.

When the number of seedlots to be compared is large, more than about 10 with large plots and about 20 with small plots, the RCB design is not efficient. This is because the area of each block becomes too large and variable. The use of incomplete block designs such as alpha-designs (John 1987) greatly increases statistical efficiency and the chance of detecting, with confidence, the differences between seedlots. It is wise to use only 'resolvable' designs, that is, incomplete block designs which can be analysed as RCB, in case something goes wrong!

For an experimenter with sufficient statistical and computing skills, incomplete block designs are always recommended, even for smaller numbers of seedlots. Provided every aspect of design, layout, analysis, and interpretation is done correctly it is always preferable to use an appropriate incomplete block design. However, such skills often are not available in the field and the help of a well-qualified statistical consultant is required. Without such help a simpler design is preferable because, as Roger and Rao (1990) warn, 'it is better to study fewer treatments (seedlots) in a smaller experiment than a larger one that answers nothing'.

5.6.3 Site preparation and field layout

Effects of ground preparation on eucalypt seedling establishment and growth are very marked and last throughout the life of the plantation. In order to reduce as far as possible the effect of unequal establishment procedures in masking genetic differences between species, uniform site treatments that promote high survival and rapid early growth are essential in these trials.

Precautions and procedures to be taken in transporting plants to the field and storing them at site, control of competing vegetation on site, removal of physical obstructions to tree growth, cultivation of the soil, modification of fertility and drainage, weed control for at least the first growing season, and control of pests such as termites are discussed by Burley and Wood (1976), Jacobs (1981), Evans (1982), Cremer *et al.* (1984), Doran (1986), Zobel *et al.* (1987), and Carter (1987).

Every effort should be made to raise enough good plants of every seedlot to be tested in a trial. However, inexperience in raising eucalypts for the first time, attacks by diseases and pests, and accidents during the germination, pricking out, nursery, and transport to the field more often than not lead to some seedlots being short of plants because of mortality. This is unfortunate as the sensitivity of the test for differences between seedlots will be reduced. However, it is still possible to proceed with a statistically valid trial but it will be incomplete or unbalanced, and will require skilled help for analysis, using the program REML (Robinson 1987) or other statistical methods (White and Hodge 1989).

It is important that the field layout is prepared on the spot when the number of plants available for each treatment is known exactly. Complete designs should not be laid out in the field weeks in advance, then plots left blank when stock becomes unavailable later. This means individual blocks and the whole trial will not be as compact and evenly balanced as they could be.

Immediately after planting, a layout plan should be prepared accompanied by a sketch map of the location and natural features nearby. North point, block and plot numbers should be indicated (Fig. 5.1). The trial should be clearly signposted and plots indicated by stakes with labels. If stakes and labels are likely to be impermanent or stolen, shallow trenches one spade width wide can be constructed to mark borders of plots and blocks. The layout plan must be of sufficient detail for an assessment crew to determine without doubt the layout and measurement sequence at each visit. It is a bad practice to leave single rows blank between plots and double rows absent between blocks since this increases the area occupied by the trial and increases within block variation. For example, in a square block of 25 plots or treatments, each of 5×5 trees at 1 m spacing, leaving one blank row between plots increases the size of the block by 35 per cent. Similarly, unnecessarily large plots or internal buffer rows should be discouraged for the same reason. External buffer rows, however, are essential around the outside perimeter of every trial to ensure that plots on the outside edge are not disadvantaged by exposure.

5.6.4 Aftercare

Seedling mortality at planting or shortly thereafter (say up to the end of four weeks) may be due to poor physical condition of the seedling caused by rough handling, damage during transport, or poor planting technique. None of these are connected with the suitability of the species to the site, so replacement planting is permissible. Later mortality, however, is generally due to poor site adaptation and percentage survival is one of the most important parameters assessed in species elimination trials. Replacement, therefore, should only be done once during the first four weeks and replacements should be indicated with an 'R' or other symbol on the trial layout plan.

Following the replacement phase, surplus stock, including that of species for which not enough trees were raised for one replication, should be planted around the margins of the trial site for observation and as an external buffer. If species not planted in the trial proper do well, they can be included in a future trial on a similar site.

Trials should be fenced or guards employed if there is possibility of damage by animals or humans. Firebreaks should be constructed around the trial in areas prone to fire. If the experimenter is not confident that cultural practices, protection, and aftercare will be adequate, it is better to postpone the trial so that the expensive seed supplied will not be wasted (Burley and Wood 1976).

Fertilizers and/or insecticides may be applied to species trials according to routine practice, particularly where many species cannot be planted without them, as in termite-infested areas of Africa. However, it must be remembered that natural resistance to termites, other pests and diseases, and good performance on nutrient-poor soils may be important attributes required of the species being tested. In many cases eucalypt species will be used by rural people who do not have access to agro-chemicals. Furthermore, most pesticides are extremely dangerous, particularly if used carelessly or in ignorance.

Eucalypts are very sensitive to weed competition in the first year. It is important to clean weed and cultivate at least a one metre diameter circle centred on the seedling. Cutting grass above ground is not enough since competition from grass is mainly from the roots, particularly the rhizomes of such species as *Imperata cylindrica*.

5.6.5 Assessment

Species elimination trials for eucalypts normally will have a short lifespan. They are but the first stage in a genetic improvement program. In the tropics some of the unsuitable species can be identified after one season of growth and the results confirmed by year two. In temperate and dry climates results available at 3–4 years can be used to eliminate the least successful species from further testing. No thinnings are carried out and later onset of competition or edge effects usually are not important as they exaggerate differences but do not change the ranking. Whole trials are often felled after 3–5 years to provide early data on relative coppicing ability and biomass production of different species and to develop biomass and stemwood volume tables for small-sized material of the best species (for example Davidson *et al.* 1985). These tables can then be used later for more accurate assessment of provenance trials of the same best species.

Normally measurements of tree height will be made after planting, at six months, at one year, and yearly thereafter. Survival would be assessed at the same time. Diameter measurements are usually not taken until diameter at breast height reaches 5–10 cm coinciding with one of the height measures. All that is required from the species elimination trial is to be able to pick with some certainty the small number of winning species, often only two or three. Because of the likelihood of unwanted hybridization within the trials, seed should not be collected from this source for industrial plantations or for further breeding work.

5.7 Provenance trial design

Provenance testing sometimes is carried out concurrently with species testing in a combined trial. However, with many eucalypts each of which may be represented by a large number of provenances, trials of more than a few species readily become unwieldy in size and the special requirements of provenance testing may cause the two objectives of choice of species and choice of provenance within species to be incompatible. Provenance testing and progeny testing are compatible when sufffficient resources are available for large trials, such as the trial of 216 open-pollinated families of 22 provenances of *E. obliqua* set out in a cubic lattice design ($6 \times 6 \times 6 = 216$ seedlots) on three sites in Australia (Brown *et al.*

1976). Plot size was 6 trees and there were 6 replicates.

5.7.1 Spacing and plot size

Spacing between trees needs to be larger than that used in species elimination trials because the provenance trial should run for at least half the length of the expected rotation. Spacing therefore will be governed by normal practice. As a guide 1×1, 1×1.5, 1.5×1.5, and 2×2 m spacings are used for fuelwood and small round posts and poles grown under short rotations on good sites without thinning. Wider spacings of 3×3 m or more are used for pulpwood and longer rotations with thinnings. With very close spacings larger numbers of trees are required in each plot, a minimum of 7×7 (49).

Because provenances should be grown without excessive competition from neighbouring provenance plots, a buffer row may be required around the trees to be measured in each plot. For provenance trials at wider spacings running over half the rotation age, and where large differences between provenances are expected, a minimum plot size of 6×6 (36) trees is recommended with the inner 4×4 (16) assessed. However, much smaller plot size can be used when the objective is a quick screening to eliminate unsuitable provenances.

5.7.2 Statistical aspects and layout

Similar considerations to those for species elimination trials apply. A minimum of four and preferably five or more replicates are recommended. Trials of the same provenances and design planted on different sites will give information on whether important genotype–environment interactions are present. For site preparation, field layout, and aftercare the reader is referred to Sections 5.6.3 and 5.6.4.

5.7.3 Assessment and management

Assessment would start off as described for species elimination trials but as the trial approached half rotation age the principal assessment of yield and quality parameters in addition to survival, height, and diameter growth is made. Straightness and cylindricity of the stem, volume of the branch-free bole, and wood quality determined from thinned stems or by non-destructive sampling methods are examples of the kinds of data collected from provenance trials. The trial may be permitted to grow on to full rotation age for further observations if plot size and resources permit, even if the decision on which provenances were the best was taken at half rotation age.

One often observes in the literature comparisons between 'wild' provenances and land races of the same species in the same trial. It was pointed out in Section 3.4 that differences in growth may be due to breaking down neighbourhood inbreeding effects. Comparison of the performance of newly introduced provenances with land races and the progeny of seed production areas and seed orchards must all be interpreted with care because of the widely different genetic histories or pedigrees of the material.

Provenance trials can be converted for seed production by removing all but the best provenances, so that pollen from genotypes in the poorer provenances will not degrade the quality of seed collections from the best. Also there must be sufficient isolation from other possible sources of hybridization with the species in the provenance trial. Even then the genetic base may be narrow since each provenance is usually represented by no more than 10 open-pollinated families. A better procedure after choice of several of the best provenances is to return to the original seed source areas of the best provenances and collect from 50–100 or more parents from each provenance region to establish seed production areas.

5.8 Introduction to 12 chapters of provenance information

Provenance studies are one of the most cost-effective forms of investment in forestry research and development because, as soon as superior provenances are identified, they can be put to use immediately in establishing plantations simply by obtaining large quantities of seed from the chosen natural forests, or better still from a plantation grown from a large number of mother trees of each provenance. There is so much to be gained, and relatively cheaply, from choosing the best provenances that we have devoted a considerable part of the book to this subject.

There are several cases where choosing the best provenances for particular planting sites has made a eucalypt planting project feasible and profitable instead of a failure. A good example comes from a provenance trial in northern Nigeria (Jackson and Ojo 1973; Eldridge 1978; Otegbeye 1985). Before 1973 when results became available from a trial of 16 *E. camaldulensis* provenances from throughout Australia it was thought that *E. camaldulensis*, and eucalypts in general, were unsuitable for the semi-arid savanna region of Nigeria because of their slow growth. Comparison of southern and northern Australian provenances in the trial soon showed that the previously planted *E. camaldulensis* was from an unsuitable southern Australian provenance and that the newly introduced northern Australian provenances, especially Petford, were well adapted and produced wood many times faster (Fig. 5.2).

In the following 12 chapters we present detailed information on provenance variation of the major plantation eucalypts from a review of world literature and from personal experience of the authors.

There is general agreement that four species are more widely planted than all others: *E. camaldulensis, E. globulus, E. grandis, and E. tereticornis*. For all four there has been extensive provenance testing, though by no means complete as some promising natural populations of each species remain to be either adequately sampled or adequately tested.

The other species were included among the digests of provenance information because they were either increasingly important for plantations or had been thoroughly studied in provenance trials or both (Table 5.1). *E. viminalis* is an example of an important plantation species not adequately sampled throughout its range for comprehensive provenance trials.

Among the 12 species only three, *E. delegatensis*, *E. nitens*, and *E. regnans*, have been sampled and tested in a comprehensive, once-and-for-all manner. Resources were available for collection of seed from several hundred trees at large numbers of localities adequately representing the full natural range of these three species. All are economically important wood-producing species of south-eastern Australia but not major plantation species in other countries. Several well-planned field trials were established for each species in the 1970s.

Fig. 5.2. A 21-year-old provenance trial of *E. camaldulensis* at Afaka, near Kaduna, Nigeria, showing the much greater growth of the Petford, Queensland, provenance (left) compared with the adjacent plot of the Lake Albacutya, Victoria, provenance (right). Similar results have been obtained in several trials in tropical, summer rainfall areas. In Mediterranean, winter rainfall areas the situation is reversed — the Lake Albacutya provenance is greatly superior to Petford (photo L. A. J. Thomson).

Other species with almost as comprehensive study include *E. camaldulensis, E. cloeziana, E. globulus,*

Table 5.1
Eucalyptus species for which provenance information is reviewed in Chapters 6 to 17. Informal subgenus names according to Pryor and Johnson (1971).

Chapter	Species	Subgenus
6	*E. camaldulensis*	*Symphyomyrtus*
7	*E. deglupta*	*Telocalyptus*
8	*E. delegatensis*	*Monocalyptus*
9	*E. fastigata*	*Monocalyptus*
10	*E. globulus*	*Symphyomyrtus*
11	*E. grandis*	*Symphyomyrtus*
12	*E. nitens*	*Symphyomyrtus*
13	*E. obliqua*	*Monocalyptus*
14	*E. regnans*	*Monocalyptus*
11	*E. saligna*	*Symphyomyrtus*
15	*E. tereticornis*	*Symphyomyrtus*
16	*E. urophylla*	*Symphyomyrtus*
17	*E. viminalis*	*Symphyomyrtus*

and *E. obliqua*. Several trials planted with seed from the range-wide collections of *E. globulus* made in 1987 and 1988 will soon remedy the incompleteness of information on provenance variation in that important species.

It has been difficult to obtain sufficient resources for comprehensive, once-and-for-all treatment of species of very wide natural distribution, *E. camaldulensis, E. deglupta, E. microtheca, E. tereticornis*, and *E. urophylla*, however desirable this might appear to be. Some of the natural stands of these species are in such remote places that it is difficult for seed collectors to repeat a visit which failed due to irregular seed crops. Besides, a single large field trial of provenances adequately representing the full range of one of these species is likely to include many predictably unsuccessful plots of ill-adapted provenances. To give an extreme example, there is no point in testing Victorian provenances of *E. camaldulensis* at tropical lowland sites because it is well established that their growth will be inferior.

While detailed presentation of up-to-date information on provenance variation of the 12 species was obviously an important objective in preparing the digests, the opportunity was taken to use these examples to emphasize the fundamental principles of testing provenances and interpreting results, and of setting up base populations for selection. One can expect some of the information to be superseded as new results from recent provenance trials and other research becomes available, but the principles developed through discussion of numerous examples should endure and influence the quality both of future provenance investigations and the application of their results by plantation managers and eucalypt breeders.

We are reluctant to generalize and put the 12 species into groups with large and small provenance variation and provenance–environment interaction (Section 18.3.5). However, a start can be made by using the preliminary scheme devised by Raymond (1989) to classify the patterns of variation between broad geographic areas in provenance trials in south-eastern Australia of five intensively studied eucalypt species of the region. Four patterns of variation emerged, as shown in Table 5.2.

Table 5.2
Preliminary classification of patterns of provenance variation in eucalypt species reviewed in Chapters 6 to 17.

Pattern of variation (Raymond 1989)	Species	Variation between broad geographic areas (regions of provenance)	Provenance –environment interaction
1. *E. obliqua*	*E. fastigata, E. obliqua*	very large	small
2. *E. delegatensis*	*E. camaldulensis, E. deglupta, E. delegatensis, E. nitens, E. viminalis, E. urophylla*	very large	large
3. *E. globulus*	*E. globulus*	not detected	small
4. *E. regnans*	*E. grandis, E. regnans, E. saligna, E. tereticornis(?)*	small, irregular	small

There was statistically significant variation between the several provenances within each broad geographic area in almost all the provenance trials of all the species in Chapters 6–17. And in all species there was evidence (of varying quality) of a large amount of genetic variation between trees within provenances.

An important conclusion follows from this consideration. In making a base population of a species for further selection and breeding the selected trees should be from only the best-suited broad geographic areas in some species (patterns 1 and 2, and to a much lesser extent pattern 4, Table 5.2), but for other species (pattern 3, and usually pattern 4) the whole geographic range of the species is a suitable source of selected trees. The only basis for classifying a species into one of these patterns is the results of thorough, well-conducted provenance trials.

Qualifications to the classification in Table 5.2 are soon apparent. Although *E. regnans* has little variation between broad geographic areas (for example Victoria and Tasmania), it does have considerable provenance variation in frost resistance with altitude of seed source. In *E. obliqua* there is strong provenance–environment interaction since provenances from northern New South Wales were far more successful in a small trial in New South Wales (near Canberra) than in Victoria or Tasmania (CSIRO, unpublished data).

It should be emphasized that the large amount of provenance information already available for other species has little value for predicting the pattern of provenance variation of an untested or partly tested species. Nor is its geographic distribution of much help. Who could have predicted the large difference in morphology and growth rate between provenances of *E. nitens* from eastern and central Victoria when the mature stands and the two environments are so similar? Nor could it have been expected that Victorian and Tasmanian *E. delegatensis* would differ so greatly as to be designated as distinct subspecies, while *E. globulus*, *E. obliqua*, and *E. regnans* vary so little across the 200 km of shallow sea separating Tasmania from the mainland of Australia. Also it is hard to imagine any way other than by field trials that the Petford provenance of *E. camaldulensis*, among all the northern Australian populations of the species, could have been selected as an outstanding seed source for planting in tropical monsoon areas.

Clearly, in deciding which provenances best match a particular site there is no substitute for information from well-planned, well-executed provenance trials. In future, provenance trials testing the better known species will be made more efficient by including only those provenances most likely to succeed.

Each of the 12 species digests of provenance information, Chapters 6 to 17, contains information on the distribution and climate of ***natural forests*** and the scale of ***plantations***. Information under the heading ***genecology*** deals with patterns of variation in the natural forests from the botanical point of view and, where necessary, with taxonomy. The main part of each chapter is on ***provenance variation*** with silvicultural information on performance of provenances for wood production in field trials. Often there is some overlap of the sections on genecology and provenance variation. Finally there is a section on the suitability of natural provenances and land races as ***base populations for selection***.

6

Eucalyptus camaldulensis

6.1 Natural forests

The most widely distributed of all the eucalypts, *E. camaldulensis* (river red gum) is also one of the most variable. Its range of latitude is from 12° 48'S on the Mary River in the Northern Territory to 38° 15'S in south-western Victoria (Fig. 6.1, Table 6.1). In the hot, dry inland it is confined to watercourses where the trees rely on ground water stored from infrequent flooding (Fig. 6.2). In some less arid areas *E. camaldulensis* occurs away from its typical riverine habitat on plains and hill slopes. It can be found on the margins of salt lakes and on a variety of soil types. Soils are typically alluvial silts and sands. Except for a few populations in South Australia and Western Australia that occur on shallow soils over limestone, the species is not adapted to calcareous soils. *E. camaldulensis* grows in temperate and tropical climates with rainfall from 200 mm to more than 1100 mm annually, with summer rains in the north and winter rains in the south of the continent. The most extensive natural stands are found along the banks and wide flood plains of the Murray–Darling river system in the inland of south-eastern Australia (Fig. 6.3). In this region large trees, often more than 200 years old, are cut for sawn timber, railway sleepers, and charcoal, but the rate of harvest is declining as the large old trees are cut out (Fig. 6.4).

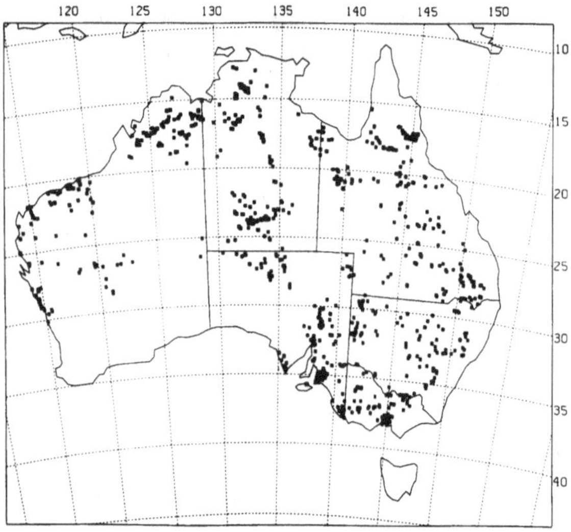

Fig. 6.1. Natural distribution of *E. camaldulensis*.

Fig. 6.2. *E. camaldulensis* at the edge of a seasonally dry watercourse in western New South Wales (photo CSIRO).

Table 6.1
Sites of seed collection of E. camaldulensis for provenance trials mentioned in Chapter 6 (after Midgley et al. 1989).

State and location	Latitude °S	Longitude °E	Altitude (m)
New South Wales			
Broken Hill	31 58	141 27	300
Moree	29 28	149 51	180
Silverton	31 53	141 13	230
Yass	31 51	148 55	500
Northern Territory			
Alice Springs	23 42	133 52	580
Katherine	14 29	132 15	120
Mary River	12 48	131 40	120
Tennant Creek	19 33	134 15	340
Queensland			
Chillagoe (Walsh River)	16 59	144 18	360
Emu Creek near Petford	17 20	144 58	460
Irvinebank	17 24	145 09	680
Quilpie	26 37	144 16	240
South Australia			
Kangaroo Island	35 50	138 03	5
Minlaton	34 47	137 36	60
Port Lincoln	34 35	135 38	60
Mengler Hill	34 29	139 10	630
Victoria			
Douglas	37 02	141 44	150
Lake Albacutya	35 44	142 02	70
Lake Coorong	35 45	142 23	75
Nathalia	36 03	145 12	150
Shepparton	36 23	145 24	100
Whiteheads Creek	37 02	145 14	250
Western Australia			
Agnew	28 30	120 57	480
DeGrey River	20 19	119 15	100
Gibb River	16 08	126 30	400
Murchison River	27 28	115 30	40
Pentecost River	15 48	127 43	60
Wiluna	26 34	120 03	550

6.2 Plantations

E. camaldulensis is perhaps the world's most widely used tree species for plantings in arid and semi-arid lands. In indexes of books on tree planting for dry areas such as Kaul (1970), Hall *et al.* (1972), Laurie (1974), and Goor and Barney (1976) there are more references to *E. camaldulensis* than to any other spe-

Fig. 6.3. An outstanding *E. camaldulensis* 50 m tall in the Barmah Forest near Echuca (photo L.A. Pederick).

cies of tree except *Pinus halepensis*. In addition to extensive but unrecorded plantings for shade and shelter, over half a million hectares of plantations have been established (Jacobs 1981), mainly in the Mediterranean region and particularly in Spain and Morocco using southern Australian provenances. Seed for these plantings comes mainly from local land races established before 1900 using southern Australian provenances. Very high productivity is obtained under favourable conditions, for example, 32 t ha^{-1} yr^{-1} (dry weight of wood) on a fertile, well-watered site in Israel (Zohar 1989).

Planting in the tropics, especially in south-east Asia and Brazil, is increasing with the increased availability of the climatically adapted northern Australian provenances. This increase is reflected by the high world demand in the 1980s for seed from these northern provenances with an estimated 2 tonnes of seed exported annually from north Queensland alone. (The seed is quite small; each tonne will yield about 600 million germinants.) Another measure of the importance of *E. camaldulensis is* the large and increasing demand for small quantities of seed for research; in the 2 years from January 1985 CSIRO's Australian Tree Seed Centre in Canberra sent 71 kg of *E. camaldulensis* to 348 researchers in 79 countries. About 30 per cent of the 1700 packets of seed went to India, Nepal and Zimbabwe (Midgley *et al.* 1989).

The success of this species as an exotic is attributed to its superiority to other trees in production of wood

Fig. 6.4. Cutting railway sleepers from *E. camaldulensis* near Mathoura in southern New South Wales (photo Z. Mazanec).

on infertile and dry sites, its tolerance of drought and high temperature combined with rapid growth when water is available, deep penetration of roots, tolerance of periodic waterlogging and soil salinity, some tolerance of frost, good coppicing ability, and usefulness of the wood. The main uses for the wood of planted *E. camaldulensis* have been for poles, posts, firewood, and charcoal. In Argentina, Israel, Mexico, Spain, and Thailand it is used for hardboard and particleboard, and in Burma, California, India, Morocco, Portugal, and Spain it is used for paper pulp. It is sawn into boards, though sometimes of not very good quality, and is used for fruit cases in Argentina and Turkey (L.D. Pryor, personal communication).

6.3 Genecology

A northern and a southern form of *E. camaldulensis* have been recognized in several studies (Karschon 1967, 1974; Pryor and Byrne 1969; Banks and Hillis 1969; Abd-Alla *et al.* 1980; and others). The northern form typically has rounded or conical opercula on the flower buds (Fig. 6.5), broad glaucous juvenile leaves, fairly straight stems, and white bark. The southern form has rostrate (beaked) opercula, narrow non-glaucous juvenile leaves, few or no lignotubers, rather crooked stems, and mottled grey bark. The southern form is confined to south-eastern Australia as the species does not occur naturally in the southern part of Western Australia (Fig. 6.1).

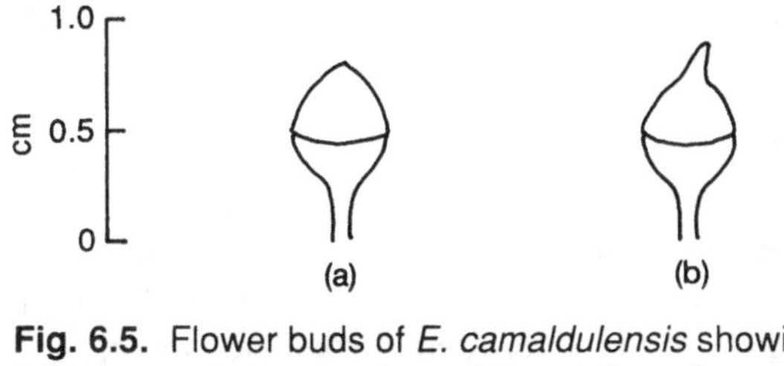

Fig. 6.5. Flower buds of *E. camaldulensis* showing (a) the rounded operculum of the northern form, and (b) the rostrate operculum of the southern form (drawing S.M. House).

A distinct boundary does not appear to exist between the northern and southern forms; most studies reviewed by Eldridge (1975a) and Midgley *et al.* (1989) suggest patterns of clinal variation. Furthermore, clinal variation in several traits has been demonstrated between west and east in the northern populations (Grunwald and Karschon 1982). Nevertheless there are likely to be mosaic patterns of discontinuous variation in some traits of the species from one drainage basin to another as there is less chance of gene flow between basins than within (Turnbull 1973). For example, a few southern populations have highly glaucous leaves, including those near Broken Hill and Silverton in western New South Wales. In a study of eight polyphenols in mature leaves collected at 103 localities, Banks and Hillis (1969) were able to show six phytochemical groups using computer-generated boundaries but these groups have not received taxonomic status. Varietal or subspecific names of *E. camaldulensis* are not used commonly by plantation foresters who seem content to describe their plantation by the name of the place where the seed was originally collected, for example 'Albacutya camaldulensis' and 'Petford camaldulensis'. The common name river red gum is widely used.

Detailed studies of more traits and more samples of the species from the entire natural range are required before drawing definite conclusions about the present patterns of genetic variation and their evolution, and before establishing subspecific taxa.

As noted in a review by Midgley *et al.* (1989), it has been recognized for a long time that where the distributions of *E. camaldulensis* and *E. tereticornis* make contact there are zones of introgression where the shape of buds and fruits is intermediate between the two species. This occurs in eastern Victoria and in southern, central and far-north Queensland. For example, there is speculation (Turnbull 1975) that stands near Petford commonly referred to as *E. camaldulensis* are actually an intermediate form occurring in a zone of gradual change between pure *E. tereticornis*, as occurs at the headwaters of the Walsh River near Mareeba, and pure *E. camaldulensis* west of Chillagoe (about 140 km by road) also on the Walsh River (Fig. 15.3). The intermediate characters include the shape of the operculum of the floral bud (Fig. 15.4) and the presence of black-seeded individuals (typical for *E. tereticornis*) in the Petford population (Doran and Burgess, in press).

Other intermediate populations also occur on Cape York Peninsula north of about 15°55'S latitude (Lakeland Downs) including the important seed sources of Kennedy River, Morehead River and others (Fig. 15.3, Table 15.1). Previously botanists have named these populations *E. tereticornis*, because of operculum shape, but many other characters have stronger affinities with *E. camaldulensis* (e.g. seed colour and juvenile foliage) causing some seed collectors to dispute the botanical wisdom of the time (Doran and Burgess, in press).

In plantations outside Australia *E. camaldulensis* has been shown to hybridize readily with *E. grandis*, *E. tereticornis* and other species of the subgenus *Symphyomyrtus* (Griffin *et al.* 1988). Natural hybrids between *E. camaldulensis* and *E. alba* have been commonly recorded in northern Australia.

6.4 Provenance variation

In the past the features which constrained the wider use of *E. camaldulensis* were poor growth in summer rainfall areas (due to the use of non-adapted provenances from southern Australia), insufficient tolerance of frost, intolerance of very calcareous soils, and crooked stems. Field trials have shown that each of these properties varies with provenance, and choice of the correct provenance is now known to be very important.

Genetic variation in growth rate, lime tolerance, and stem straightness was demonstrated in pioneering research by Franclet (1956), Karschon (1960), and Giordano (1961). These studies led to the decision in 1962 of the FAO Mediterranean Forest Research Committee to begin an investigation of provenance variation in the species. Collections throughout the continent were organized by the late Egon Larsen of the Forest Research Institute, Canberra, now CSIRO Division of Forestry. Joint teams of Australian and Tunisian foresters collected seed in 1964 from 5 to 10 trees at each of 40 localities (Anon. 1964). The only region of natural occurrence not sampled was the Kimberley in the north of Western Australia.

The establishment of 32 well-designed field trials in 14 Mediterranean and tropical countries was organized by J.F. Lacaze (1970, 1978) between 1966 and 1972 through the Centre de Recherches Forestières of INRA, France. Many of the trials were planted with only 16 of the 40 collections due mainly to shortages of seed.

Provenance variation has been recorded for growth rate, wood properties, leaf oil content, lignotubers and coppice, drought tolerance, frost tolerance, and salinity tolerance.

6.4.1 Growth rate

When most of the trials were 8 to 10 years old Lacaze (1978) reported results from 24 of them. Within the Mediterranean winter rainfall zone the growth of the Lake Albacutya, Victoria provenance (CSIRO seedlot number 6845) was outstanding in almost every trial, and in the tropical, summer rainfall zone Petford, Queensland (6953) and Katherine, Northern Territory (6869) were superior for growth and also for stem straightness. This difference in ranking showed that provenance–environment interaction (or genotype–environment interaction, GEI, Section 18.3.5) was extreme between the Mediterranean and tropical environments but that within either of these zones GEI did not appear to be important even over a considerable range of soil types. For example in Italy, on a calcareous site in Calabria and on an acid drier site in Sicily, Sesbou and Nepveu (1978) found similar ranking of 17 provenances. However, further investigations might be expected to reveal cases of significant interaction because, as noted by Turnbull and Pryor (1984), some natural populations of *E. camaldulensis* occur on distinctly alkaline or saline sites. These populations would be expected to be far better adapted to such conditions in plantations than populations from more favourable sites.

It is also interesting to note that the Lake Albacutya provenance was the best of the southern provenances in tropical trials, and that among the tropical provenances Petford and Katherine performed moderately well in Mediterranean trials. These observations indicate the phenotypic plasticity, or adaptability, of these fast-growing populations. Evidently the inherent capacity for fast growth is not completely suppressed in an unfavourable environment.

The preliminary ranking for growth was generally confirmed by later observations (Lacaze 1970, 1978), thus showing strong juvenile–mature correlation. A striking example of juvenile–mature correlation was observed in a provenance trial of 11 provenances of *E. camaldulensis* established on a dry site in Zimbabwe in 1966 (Barrett and Carter 1970). The plots of the two best provenances at $2\frac{1}{2}$ years, Petford and Katherine, were well developed stands at the age of 15 years. The other nine provenances originally in plots of 49 trees each, were represented by a few stunted remnants in a grassy sward (R.D. Barnes, personal communication).

These very successful provenance trials in several countries show that *E. camaldulensis* contains certain populations which are outstanding in some traits of economic importance. Luck was with the seed collectors who chose Lake Albacutya as one of the collection sites in south-eastern Australia. The short boles and spreading crowns of the trees at the edge of this usually dry lake-bed (Fig. 6.6) gave no indication of their remarkable potential as a seed source for fast-growing plantations in Mediterranean regions. Nor

Fig. 6.6. *E. camaldulensis* at Lake Albacutya, Victoria. This tree was one of the first ten of this provenance from which seed was collected by E. Larsen in 1964 (photo CSIRO).

Fig. 6.7. *E. camaldulensis* at Emu Creek near Petford in north Queensland. Five trees from this location were sampled by E. Larsen as part of the first provenance study of this species. It is the most widely planted provenance of the species in the wet/dry tropics. Note that several branches have been cut off by seed collectors (photo CSIRO).

does the form or stature of the trees in the natural stand near Petford in north Queensland (Fig. 6.7) indicate their genetic potential as seed sources for tropical plantations.

In subsequent years additional provenance collections of *E. camaldulensis* have been made in places between those sampled in 1964 and in the north of Western Australia. So far there have been no more surprises like the Lake Albacutya provenance. Collections at Lake Coorong on a parallel drainage line 40 km to the east of Lake Albacutya produced progeny with similar early growth and survival characteristics to the Lake Albacutya provenance in field trials in Israel (Grunwald and Karschon 1977b, 1984) and in Egypt (El-Lakany 1983). However, there are anatomical differences in seedling leaves of Lake Albacutya and Lake Coorong provenances, particularly in cuticle thickness and stomatal frequency (El-Lakany *et al.* 1984). Also, when the trials in Israel reached eight years the best of the eight Lake Albacutya families were distinctly taller than the three Lake Coorong families. Further sampling and field trials are warranted. Seed collections were made in 1985 from 252 trees at 21 localities in the vicinity of Lake Albacutya and Lake Coorong (Thomson and Merwin 1987) (Fig. 6.8). Field trials established with this material in California and elsewhere will show more clearly the amount and pattern of genetic variation in the species in north-western Victoria.

The seedlots collected from several stands within about 30 km of Petford in the Walsh River drainage in north Queensland, including Emu Creek and Irvinebank, have been found to have similar performance when grown in Brazil (Moura 1986), India (Rao 1984), Bangladesh (Davidson and Das 1985), and elsewhere.

Some of the trials discussed by Lacaze (1978) have been reported in detail by other authors whose publications were reviewed by Eldridge (1975a). Since then there have been reports of trials containing additional provenance collections (Darrow 1983b; Davidson and Das 1985; El-Lakany 1983; Mendonza 1974; Moura 1986; Rao 1984; Emery and Ledig 1987) which generally confirm the previous findings and add Gibb River and other provenances of northern Western Australia as tropical seed sources of similar value to those of the Petford and Katherine areas. The search for *E. camaldulensis* provenances to match particular planting sites is certainly not finished yet.

6.4.2 Wood properties

In two of the trials planted in 1967 in Italy, Sesbou and Nepveu (1978) made a thorough study of variation in wood density and shrinkage. The trials were in Sicily and Calabria and each contained 17 provenances in six replications of 25-tree plots. In 1977 two increment cores were extracted from two trees in each plot and shrinkage and basic density were measured. The nine provenances from the Murray–Darling river system had faster growth, lower density and higher shrinkage than the northern provenances.

When averaged over the two trials in Sicily and Calabria, the range of wood density was large, from 588 kg m^{-3} for Tennant Creek, Northern Territory, the slowest growing provenance, to 449 kg m^{-3} for Shepparton, Victoria, one of the fastest growing. In terms of weight of wood produced the denser wood did not compensate for the slower growth. Wood of the Petford provenance was only 8 per cent more

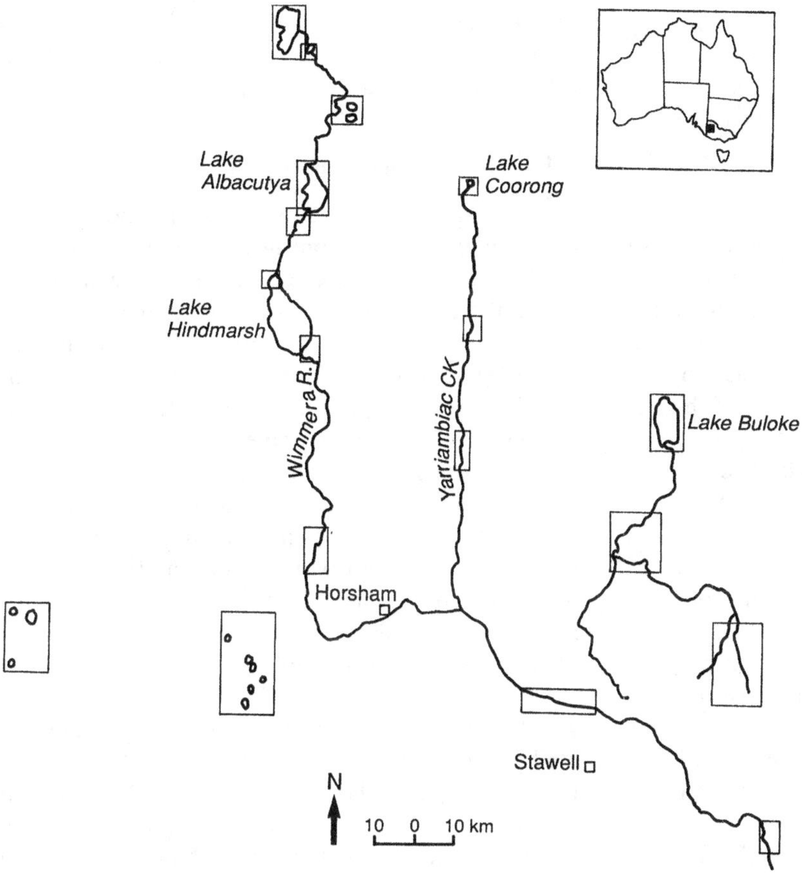

Fig. 6.8. Locations of *E. camaldulensis* seed collections in north-western Victoria in 1986 (after Thomson and Merwin 1987).

dense than that of the Lake Albacutya provenance, and Katherine only 4 per cent more dense. Lake Albacutya produced 77 per cent more dry wood than Petford and 45 per cent more than Katherine in these two trials (Table 6.2).

An indication of variation between trees within provenances is contained in the data presented by Sesbou and Nepveu (1978) as one provenance, Mengler Hill, was represented by seven open-pollinated families. Although variation between families was not examined by the authors their data suggest statistically significant variation for shrinkage and density of wood but not for tree growth.

In an irrigated provenance trial at Peshawar in

Table 6.2

Provenance variation in wood production of E. camaldulensis. *Wood density and tree volume were averaged over two trials at age 10 years in Sicily and Calabria (Sesbou and Nepveu 1978). Volume per tree was calculated from volume tables of Knockaert and Ait Hamou (1979).*

	Lake Albacutya	Petford	Katherine
Basic density (kg m^{-3})	504	542	524
Height (m)	11.1	8.3	10.1
Circumference (cm)	42.2	32.5	35.3
Volume per tree (dm^3)	74	39	49
Dry mass per tree (kg)	37	21	26

Pakistan basic density of wood was measured on samples from four trees from each of six northern provenances at age ten years by Siddiqui *et al.* (1979). The range of density was 546 to 647 kg m^{-3} for the six provenances. Basic density was also examined in seven provenances in an eight-year-old trial at Dehra Dun, India, where three trees were sampled from each of seven provenances (Ghosh *et al.* 1977). Despite the limited sampling both the Pakistan and Indian trials gave similar results to the Italian trials in that higher wood density did not compensate for slower growth, and that tropical provenances tended to produce wood of higher density than southern provenances. In Brazil two trials of 17 northern provenances had basic density at age seven years of between 470 and 570 kg m^{-3} (Moura 1986; Moura *et al.* 1987). Particularly high values for basic density of tropical provenances of *E. camaldulensis*, 610 to 640 kg m^{-3}, were found in Sri Lanka by Vivekanandan (1978).

There are few studies of genetic variation in wood properties of *E. camaldulensis* other than basic density and shrinkage. In a study of fibre length of seedlings Rudman (1970) grew cuttings from seedlings to 1.5 to 2 m in 23 weeks in a controlled environment and found that plants grown in higher temperatures produced longer fibres. He also found a marked difference in fibre length between the seven clones, and a significant negative correlation between height growth and fibre length in one clone. In an irrigated field trial in Egypt El-Lakany *et al.* (1980) examined fibre length of 17 provenances at two years. They found a significant difference between provenances but no indication of a consistent difference between northern and southern forms.

6.4.3 Leaf oil content

Significant differences were recorded between provenances and individual trees of *E. camaldulensis* in the chemistry of leaf extractives, among which leaf oils could become economically important. In recent studies some tropical provenances of *E. camaldulensis*, including Petford, were found to have surprisingly high levels of cineole in individual trees (Doran and Brophy 1991; K.J. White personal communication). Cineole can be simply and cheaply extracted in low-cost stills and the Petford provenance is already widely planted throughout the tropics. Continuing research on *E. camaldulensis* leaf oils foreshadows an important additional product from trees being grown for fuelwood and other purposes.

6.4.4 Lignotubers and coppice

The distinct difference between the northern *E. camaldulensis* populations in which lignotubers form on all seedlings and the southern populations in which there are very few or no lignotubers was first recorded by Karschon (1966, 1967) and Larsen (1967). Experiments were made by Jahnke *et al.* (1983) in an attempt to induce lignotubers in a non-lignotuberous southern provenance by varying the concentrations of soil nutrients. Seedlings of the southern provenance, Lake Albacutya, always failed to produce lignotubers no matter what concentrations of nitrogen and phosphorus were applied, and the northern provenance, Petford, always produced lignotubers. Evidently lignotuber development is under strong genetic control.

Coppicing, by contrast, showed no indication of genetic variation between populations when a replicated trial of 16 provenances was clear felled and the resulting coppice shoots were measured (Grunwald and Karschon 1977a). Every stump sprouted coppice shoots and their number, height, and biomass after 19 months were strongly correlated with the height and diameter of the original tree. Even the trees of the non-lignotuberous provenances coppiced well. In this species none of the fast-growing provenances coppiced poorly, an encouraging finding for foresters selecting primarily for growth rate. Similarly, in two replicated trials of eight northern provenances in Bangladesh, coppice shoots were produced on almost every stump when the trees were felled at five years of age. After two years the coppice grew much more rapidly than the seedlings in height, diameter, and biomass but followed closely the original ranking in performance of the original trees and provenances (J. Davidson, unpublished information).

6.4.5 Drought tolerance

There have been several physiological investigations which might help explain why one provenance is faster growing and more drought hardy than another.

In their study of root growth of *E. camaldulensis*, Awe *et al.* (1976) grew seedlings in tubes one metre long. They found that roots of the Lake Albacutya provenance reached the bottom of the tube more quickly than Katherine and other provenances. The capacity to produce a massive root system very rapidly would help the Lake Albacutya seedlings to reach water at greater depths in the soil, avoid drought, and continue to grow rapidly when shallower rooted plants were under severe moisture stress.

Stomatal conductance and CO_2 uptake for the Lake Albacutya provenance were shown by Moreshet (1981) to be greater than for the slower growing Katherine provenance in trees of a four-year-old trial in Israel, but only when water was available. In the hot dry summer both provenances ceased growing and their dormancy was indicated by very low evapotranspiration.

A demonstration of the superior adaptation to drought of the Lake Albacutya provenance was made by Grunwald and Karschon (1982) when they compared it with seedlings of the tropical Gibb River provenance from Western Australia during the dry season in Israel. The Lake Albacutya provenance reached lower water potentials with less reduction in water content while maintaining a higher growth rate than the Gibb River provenance.

While the results of these three studies show that plants of the Lake Albacutya provenance have physiological adaptations to drought under the conditions of these experiments, the results do not explain why the Katherine provenance is the most drought resistant of several tested, including Lake Albacutya, in an arid area of northern Nigeria (Jackson and Ojo 1973) and near the Kalahari Desert in Zimbabwe (Barrett and Carter 1970).

6.4.6 Frost tolerance

E. camaldulensis seedlings survive winter frosts to about –10°C in the higher reaches of streams in the south-east of Australia and about –7°C in the semi-desert areas near the centre of the continent. The lowest temperatures on record within the range of *E. camaldulensis* are –9°C at Yass, New South Wales, –7°C at Alice Springs, Northern Territory (Hall *et al.* 1981), and –8°C near Lake Albacutya, Victoria (O'Brien 1989). These temperatures were recorded in a meteorological screen 1 m above the ground; small seedlings close to the ground would have been subject to temperatures several degrees lower. In these areas severe radiation frosts follow a series of increasingly cold nights.

Seedlings of southern provenances were noted as being more resistant to winter frosts than those of northern provenances by Larsen (1967) and Pryor and Byrne (1969). Karschon (1968) found the Alice Springs provenance more resistant to frost than the Katherine or Murchison River provenances. Provenance differences were investigated in controlled environments on four provenances by Awe and Shepherd (1975). After hardening, to simulate winter conditions, the two southern provenances (Moree and Nathalia) were more resistant to damage in a frost chamber. However, without hardening, as in autumn or spring, the two northern provenances (Katherine and Agnew) were more resistant. This form of frost tolerance would be very useful in the many plantation areas where sudden frost follows a period of mild weather, too warm for hardening.

Further screening for frost tolerance would be desirable to find provenances and individuals with greater tolerance to either winter or spring frosts. It may be that simpler, indirect estimates, including leaf dry matter content as a percentage of fresh weight (Grunwald and Karschon 1977c) or leaf thickness would speed up the screening. Other tests for frost tolerance of eucalypt seedlings are described in Sections 8.4.2 and 12.4.3.

6.4.7 Salinity tolerance

E. camaldulensis has been highly regarded for some time as a tree with useful tolerance of salinity in the soil. Some populations were found to have greater tolerance than others (Karschon and Zohar 1975), but only in recent years have well controlled experiments been made to test for genetic variation in tolerance of salinity.

In a study of seedlings of 52 species of eucalypts in saline solution culture Blake (1981) found that *E. camaldulensis* was the second most resistant species. (The most resistant species in this experiment was *E. woodwardii.*) Sands (1981) examined the salt tolerance of three southern provenances and found that at a salt concentration of 400 mM the seedlings from Lake Albacutya were all healthy, though their growth was reduced, and the seedlings from Shepparton (300 km east) were nearly all dead; seedlings from Port Lincoln (600 km west) were intermediate.

Data in Table 6.3 summarize some of the detailed research of L.A.J. Thomson and show that variation between provenances can be considerable (Thomson *et al.* 1987). The salinity level in the root zone at the site of origin, which one would expect to be a factor important to the evolution of differences in NaCl tolerance in eucalypts, was not found to be closely correlated with NaCl tolerance in *E. camaldulensis*. Data indicate that localized selection for tolerance has taken place for populations of *E. camaldulensis* growing in more saline sites, but high levels of tolerance also exist in some populations growing in soils of low salt content, for example the DeGrey River (WA) population. While salt-tolerant populations and individuals may be present in non-saline environments, it is most effective to concentrate searches for the salt-tolerant genotypes in environments where salinity is likely to be exerting a strong selective force (Thomson *et al.* 1987).

Techniques developed by Thomson *et al.* (1987) were used to screen large numbers of seedlings of promising provenances of *E. camaldulensis* and *E. rudis* (a close relative from Western Australia). Hartney (1982) had developed tissue culture techniques for vegetatively propagating outstandingly salt-tolerant individuals. Large numbers of rooted plantlets were mass produced in a laboratory in Canberra and sent to Perth, Western Australia. There the plantlets were transferred from sterile culture tubes and grown in a nursery as planting stock for afforestation of salt-prone areas and worked-out bauxite mines (Hartney and Kabay 1984). Plantlets have also been sent to other parts of Australia, to Israel, and to Florida, U.S.A.

Similar techniques might be used for screening *E. camaldulensis* for tolerance of alkalinity since planting on calcareous soils often results in chlorosis of the leaves and poor growth. However, on some calcareous sites growth is satisfactory (Sesbou and Nepveu 1978). Also the natural occurrences of the species on some calcareous sites, for example near Port Lincoln, South Australia, suggests that lime-tolerant provenances may be found (Jacobs 1981).

Table 6.3

Ranking of provenances of E. camaldulensis *in order of decreasing tolerance of NaCl, in glasshouse trials (based on Thomson et al. 1987).*

		NaCl concentration (mM) causing mortality[1]		
Provenance	State[2]	Mean	Standard error	Homogeneous subsets[3]
DeGrey River	WA	636	11	a
Silverton	NSW	613	13	a b
Wiluna	WA	608	12	a b
Pentecost River	WA	587	11	b c
Emu Creek	Qld	565	12	c d
Quilpie	Qld	532	13	d e
Kangaroo Island	SA	522	12	e f
Lake Albacutya	Vic	510	12	e f
Whiteheads Creek	Vic	507	13	e f
Minlaton	SA	503	14	e f g
Douglas	Vic	499	13	e f g
Katherine	NT	490	13	f g
Moree	NSW	466	12	g

1 Seawater is equivalent to ca. 550 mM

2 NSW New South Wales, NT Northern Territory, Qld Queensland, SA South Australia, Vic Victoria, WA Western Australia

3 Provenances which do not share a common letter have means which are significantly different at the 5 per cent level

6.5 Base populations for selection

Variation within populations of *E. camaldulensis* has been noted by many investigators (Franclet 1956; Claudot 1963; Destremau *et al.* 1973; Karschon 1974; Sesbou and Nepveu 1978; Grunwald and Karschon 1983; and others). However, none of the programs for introduction and breeding of this species has yet taken detailed genetic study of variation within populations as a major objective. Without information on how much variation is likely to exist within the better populations, it is not possible to say with confidence how large an investment in selection and breeding is justified. However, present indications are that once good provenances have been found there are further opportunities for more genetic improvement (White 1986; Nikles 1987). For example, Bellefontaine (1979) has reported progress with *E. camaldulensis* seed orchards in Morocco where 39 kg of seed was collected in 1977 from a grafted seed orchard of plus trees of Lake Albacutya origin selected in Italy and Morocco, and presumably derived from the original 1964 collection from ten trees.

Another approach to assured seed supply with some improvement in genetic quality was used by Siddiqui *et al.* (1979) who heavily thinned a ten-year-old provenance trial to convert it to a seed production area. Only the very best trees of the best provenances on the 1.5 ha trial were retained. Large quantities of seed were harvested in subsequent years.

It is very easy to make rooted cuttings from coppice shoots of *E. camaldulensis*. It is prudent to have made extensive provenance and family assessments before starting a full-scale cuttings program. The benefits of selection in a population with a wide genetic base are described by White (1986) who achieved remarkable gains in growth with cuttings of *E. camaldulensis* in the Terai region of Nepal after thorough testing and selection among individual tree families from the Petford region; a splendid achievement in only six years.

Hybrids are readily made between *E. camaldulensis* and *E. grandis*, *E. tereticornis*, and other species of the subgenus *Symphyomyrtus* (Griffin *et al.* 1988). The 'eucalyptus hybrid' or 'Mysore gum' in India is predominantly *E. tereticornis*, and some mixture with *E. camaldulensis* (Pryor 1966; Boland 1981). Hybrids such as *E. grandis* × *camaldulensis* could perform well on sites which are a little too dry for *E. grandis*.

Thus, even if a country does not grow large stands of *E. camaldulensis* for wood production, consideration should be given to growing a number of provenance resource stands. Good individuals from such stands will then be available for selection for crossing with good trees of another species in an existing breeding program, say *E. grandis*. These hybrids can then be moved into a commercial scale vegetative propagation program. Provision should be made to manage such provenance resource stands as seed stands with a mild selection pressure each generation.

Seed crops in the natural stands of the most successful provenances are irregular and usually expensive to collect as most of the natural stands are small in area, in remote localities and vulnerable to damage, not least from seed collectors. On the recommendation of the FAO Panel of Experts on Forest Gene Resources, the CSIRO Australian Tree Seed Centre made additional collections of seed in 1977 at Lake Albacutya, Petford, Katherine, and Gibb River, from 25 or more well-separated trees at each place (Doran and Boland 1978). Further collections were organized by CSIRO in 1984 and 1985 from 178 trees in the Petford region (J.C. Doran personal communication) and 252 trees in the Lake Albacutya region (Thomson and Merwin 1987). Sufficient seed was collected to send one or two kilograms of each provenance to several forestry organizations in countries where planting of *E. camaldulensis* is increasing.

One example is in India where Boland (1981) recommended establishing several plantations of 10 to 50 ha of each provenance which would be converted to seed production areas by early and heavy thinning. Others are in Bangladesh where two areas of 10 ha each of Petford provenance (from 100 seed trees) were established in 1983 (Davidson and Das 1985), in Zimbabwe where several well-designed 'breeding seedling orchards' were established in

1985 (Barnes 1987)(Fig 21.2, p. 215), and in Thailand and China. They would serve as *ex situ* conservation stands, produce large quantities of much needed seed within ten years, and provide a reasonably broad genetic base for several generations of selection and breeding.

7

Eucalyptus deglupta

7.1 Natural forests

Eucalyptus deglupta usually is a large tree 35–60 m high, but occasionally up to 75 m, with diameters of 0.5–2 m or greater (Fig. 7.1). This is a tropical species (9°N to 11°S), the only eucalypt found in nature north of the equator and, with *E. urophylla*, one of only two eucalypt species not occurring in

Fig. 7.1. Part of an extensive riverain stand of *E. deglupta*, Pandi River, West New Britain Province, Papua New Guinea (photo J. Davidson).

Australia. The natural distribution (Fig. 7.2) is markedly discontinuous from Mindanao in the Philippines, through Sulawesi, Ceram and Irian Jaya in Indonesia, and parts of mainland Papua New Guinea and most of New Britain (Heather 1955; Davidson 1973b, 1978b; Turnbull 1974).

Best development of the species is to be found in riverain situations at altitudes mainly below 150 m in New Britain. Full altitudinal range is from sea level to about 1800 m. Throughout its natural range the climatic region in which *E. deglupta* occurs is the humid tropics (Fig. 4.2, p. 41). Mean annual rainfall is typically 2500–3500 mm but some areas receive up to about 5000 mm. Seasonal distribution varies according to local monsoon patterns. Except for a rainshadow in south-eastern Papua New Guinea, rainfall is seldom less than 100–150 mm per month in the driest part of the year.

In the lowlands mean monthly maximum temperature is around 30–32°C, while in the highlands the mean monthly maximum of the coolest month may be only 24°C.

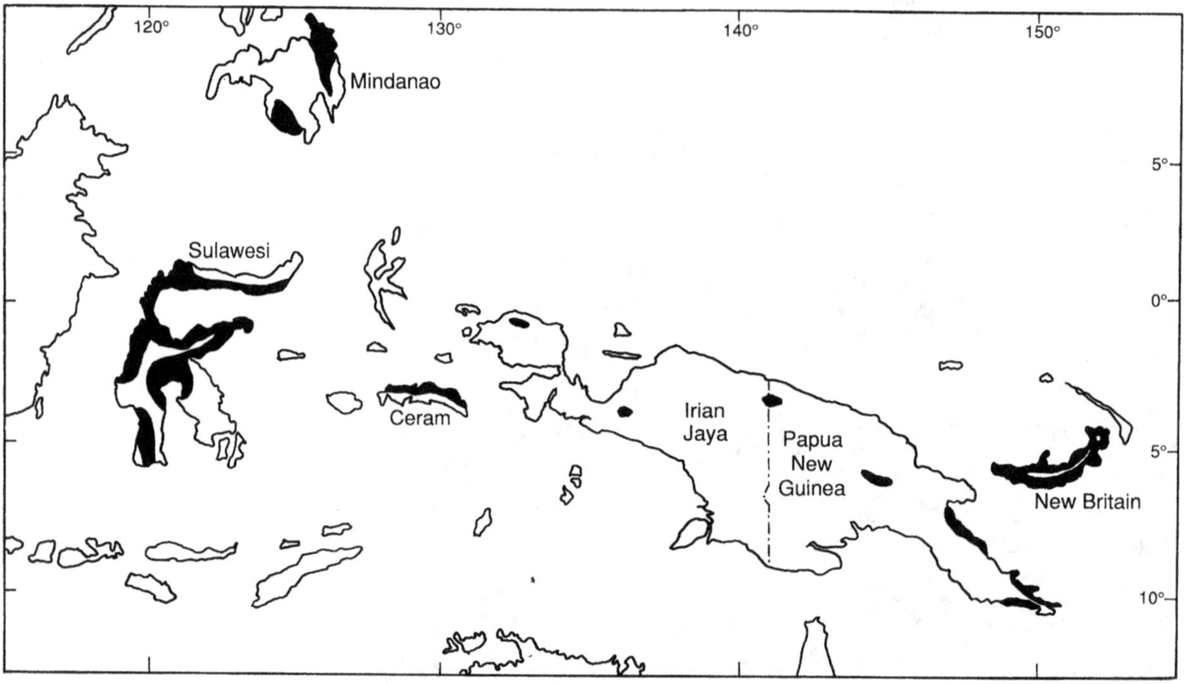

Fig. 7.2. Natural distribution of *E. deglupta* (based on Davidson 1972 and Cossalter 1977).

7.2 Plantations

E. deglupta is a species which has great potential for planting in wet tropical lowlands after harvesting of rainforest (Fig. 7.3). Growth rates are excellent on suitable sites (25–40 $m^3 ha^{-1} yr^{-1}$ over 15 years) and the wood and bark are good sources of fibre for pulp (Logan *et al.* 1987). Some major industrial projects are now under way in the tropics based, at least in part, on reforestation using this species in fast-growing plantations over short rotations of 10–12 years or less and a considerable database has been assembled on its performance (Whitmore 1978). Despite its spectacular growth rate and suitability for pulping, the initial enthusiasm for planting in many countries has been tempered somewhat by experience in plantations where the species was not well enough matched to the site. It is extremely site-sensitive, very susceptible to fire, does not coppice readily, and is susceptible to several pests and diseases.

Fig. 7.3. Plantation of *E. deglupta* aged 15 years at Keravat, East New Britain Province, Papua New Guinea. The mean height of the trees was 47 m and mean diameter at breast height 43 cm (photo J. Davidson).

Leading countries in plantation area are the Philippines and Indonesia (each with about 30 000 ha) followed by Brazil (about 10 000 ha). Other countries with significant areas (up to 4000 ha each) are Malaysia, Papua New Guinea, Ghana, Sri Lanka, Fiji, and Côte d'Ivoire. Total plantation area worldwide is approaching 100 000 ha.

On an ideal site, this eucalypt was recognized at one time as the world's fastest growing tree (Guinness Book of Records 1975). Dominant trees in plantations on rich, alluvial, well-drained river terraces at Gogol near Madang, Papua New Guinea planted at higher than usual densities (1.8 × 1.8 m and 2.4 × 2.4 m) have reached 38 m in height and 39 cm in diameter over bark at breast height at age three years (J. Davidson, unpublished data). Productivity of these young stands was 80–90 $m^3 ha^{-1} yr^{-1}$ measured inside one hectare plots, away from edge effects.

Paper Industries Corporation of the Philippines (PICOP) has extensive plantations of *E. deglupta* at its Bislig Bay locality in Mindanao. The goal there is eventually to have just under 20 000 ha of this eucalypt in plantations, about one-third of the total plantation area required to sustain the local pulpmill. Present planting rate varies from 1200 to 2000 ha of new planting each year but in addition several compartments are being planted to a second rotation of *E. deglupta*. In plantations with an initial stocking of about 625–680 stems ha^{-1} a yield of 25–30 $m^3 ha^{-1} yr^{-1}$ (overbark to 10 cm top diameter) can be expected on average sites over a ten year rotation. Much higher yields of 40–60 $m^3 ha^{-1} yr^{-1}$ (total volume over bark) can be obtained from higher stocking and above average sites in Mindanao.

The species is reported to be immune to the Diaporthe canker (*Cryphonectria cubensis*) (Hodges *et al.* 1979) and some other diseases which afflict other eucalypt species when transferred into the humid tropics.

Planting on poor sites, particularly under water-logged conditions, has led to heartwood staining and decay being reported from Papua New Guinea, the Solomon Islands and Philippines (Davidson 1973c; Harries *et al.* 1973). *E. deglupta* has been abandoned for commercial planting in Sabah, Malaysia, because growth slowed greatly after five years; now other species including *Gmelina arborea* and *Acacia mangium* are preferred there.

7.3 Genecology

E. deglupta is apparently the earliest eucalypt collected and described by botanists (as 'Arbor versicolor') predating the first collections in eastern Australia by more than 100 years. An illustration and description appeared in Rumpf (1743) of a specimen from Ceram probably collected in 1668. Though the specimen no longer exists, the location, description, and figure, especially in emphasizing the distinctive appearance of the bark and opposite arrangement of the adult foliage, leave little doubt that the specimen was of *E. deglupta* (Davidson 1978a).

This eucalypt was described validly by Carl Blume (1849) from a sterile specimen collected in 1821 from Sulawesi and found in Reinwardt's herbarium under the manuscript name of 'Populus deglubata'. Since then, six further synonyms have been published, four of them valid (Davidson 1978a). Botanists collecting separately, in isolation, at different times, and in different geographic areas, gave separate names to this eucalypt when found in western Mindanao (*E. multiflora* (Gray 1854)), northern Mindanao (*E. binacag* (Elmer 1914, 1915)), at Garaina in mainland Papua New Guinea (*E. schlechteri* (Diels 1922)), and on the island of New Britain (*E. naudiniana* (von Mueller 1886)) (Davidson 1978a).

The confusion associated with the name of this species was largely due to the inadequacy and variability of the early material collected and to lack of communication between herbaria. More recent collections made throughout the range of the species confirm that the previously described separate species fall within the rather wide range of variability shown by *E. deglupta*.

Examination of the herbarium material from natural forests and foliage specimens from provenance trials has shown that throughout lowland mainland New Guinea wild-growing *E. deglupta* (including the type of *E. schlechteri*) usually has broader and shorter leaves and smaller buds and fruit than samples from the island of New Britain (Fig. 7.4). However, on the Gazelle Peninsula at the eastern end of this island, where the type specimen of *E. naudiniana* was collected, the leaves tend to be broader like those of lowland mainland populations in size and shape. To the west along the north coast of the island of New Britain there is a trend towards smaller and narrower leaves with very pronounced drip tips.

Material from the highlands of New Guinea (e.g. Waria River, Papua New Guinea, and Wissel Lakes, Irian Jaya) has smaller and shorter leaves tending to

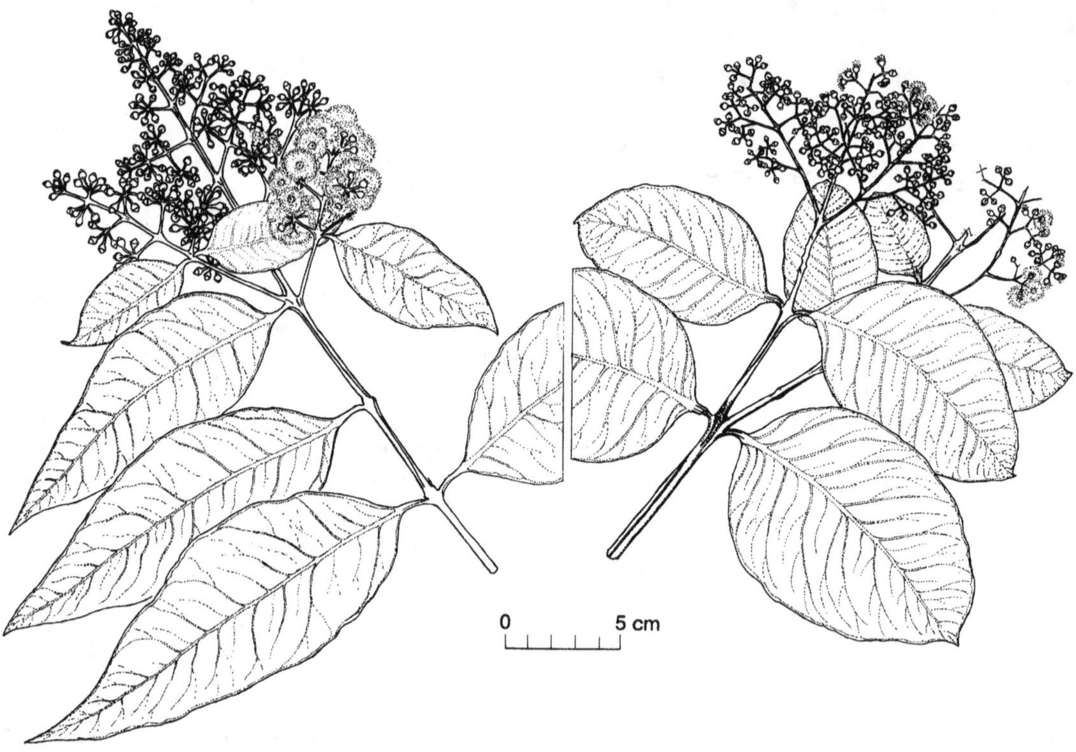

Fig. 7.4. Adult leaf shape of *E. deglupta* from New Britain (left) and Garaina, mainland Papua New Guinea (right) (from Davidson 1972; drawing by N.H.S. Howcroft).

orbicular, without pronounced drip tips. Examples from Ceram and Sulawesi (including the type of *E. deglupta* (Blume 1849)) seem to be similar to the lowland mainland New Guinea sources while those from Mindanao, Philippines (including *E. binacag* and the type of *E. multiflora*) tend towards the west New Britain types with smaller narrower leaves and pronounced drip tips.

Results of numerical studies on leaf, bud, and fruit characters lead to the conclusion that there are no distinct subgroupings which could be the basis of formal taxa such as subspecies. However, the variety is considered an appropriate category, should it be desired for practical purposes, to distinguish populations with the shorter broader leaves (*E. deglupta* var. *schlechteri*) from populations with longer, narrower leaves with pronounced drip tips (*E. deglupta* var. *multiflora*).

E. deglupta is the only species of eucalypt with a disjunct distribution separated by major sea and land barriers. At the same time its natural distribution ranges over an area greater than for any other eucalypt except *E. camaldulensis* (Chippendale and Wolf 1981).

7.4 Provenance variation

Given the wide variation in morphology, and the disjunct, wide, pantropic range of this species, separated by major sea and mountain barriers, it could be expected there would be a wide range of performance of different geographic sources when planted together on one site.

Provenance trials established in Papua New Guinea in 1969–70 were among the first such extensive trials of this species. Nine provenances were tested in Papua New Guinea in five trials at three sites: Keravat established in 1970–71 and 1971–72 seasons, Dami in 1970–71, and Gogol in 1970–71 and 1971–72 (Davidson 1983a).

There were very significant differences between provenances for both height and diameter growth and significant differences between sites for diameter growth only. There was no significant provenance–site interaction, at least up to age seven years.

The best provenances for growth rate were New Britain and Philippines while the New Guinea mainland provenances were not so good. Two periods were noted where deaths occurred. The first was between planting and 18 months of age, the second after three years of age. The first was probably a result of weed competition and/or adverse soil or climatic effects, the second was more likely from disease or from competition between trees. By age six years, there were very highly significant differences between provenances for survival.

West New Britain provenances and the Philippines provenance from south-east Bislig Bay, Mindanao showed best survival. The other Philippines provenance and the remainder showed a much lower survival rate. There were significant differences between Gogol and Keravat trial sites for survival but there was only a very slight provenance–site interaction.

There were no significant differences between provenances for straightness. Provenances with high scores were Raba Raba and Philippines (Cotabato) (Table 7.1). Differences between trial sites for straightness, however, were significant but there were no significant provenance–site interaction effects.

Table 7.1
Principal sites of seed collection of E. deglupta mentioned in Chapter 7 (after Davidson 1983a).

Location	Latitude °	Longitude °E	Altitude (m)
Mindanao, Phillipines			
Caliawan River			
(near Bislig Bay)	7 36N	126 07	510
Cotabato (Milbuk)	6 15N	124 18	390
New Britain, Papua New Guinea			
Gazelle Peninsula,			
Keravat (Rock Pile)	4 20S	152 00	5
Mainland, Papua New Guinea			
Mingende	5 58S	144 55	2500
Raba Raba	9 58S	149 50	150
Wilelo	5 35S	150 15	10

For fluting of the stem, differences between provenances were significant but there was no difference between sites. No differences between provenances or sites were found for degree of buttressing.

The provenance trials showed that glaucousness of trees in natural stands in the Philippines also developed on their progeny at each trial site. Leaves and twigs have a greyish powder on their surfaces, in contrast to the bright green leaves of the mainland Papua New Guinea and New Britain provenances (N.H.S Howcroft, personal communication).

Concerning leaf shape, there were significant differences between provenances for leaf width, length, and area. Raba Raba and Mingende provenances had the largest leaves, New Britain intermediate, and Sulawesi and Philippines the smallest leaves.

On assessing aerial rooting ability, significant differences were found between provenances in the number of aerial roots formed and in the height up the bole they were found. Mingende provenance produced the most, followed by Wilelo and Philippines (Cotabato) with an intermediate number, and then Raba Raba, Sulawesi, Philippines (Bislig), and Keravat with almost zero.

For wood volume production, the Philippines (Bislig) and New Britain provenances performed best and the mainland New Guinea provenances were inferior. There were significant differences between provenances for wood density at age two years. For weight of dry wood produced, Philippines (Bislig) performed well at Gogol and Keravat while the Keravat provenance performed better at Keravat than at the Gogol trial site. At age six, more wood had been produced at Gogol than at Keravat (57.0 and 32.8 t ha^{-1} respectively at an initial 2.4×2.4 m spacing).

These results obtained in Papua New Guinea by Davidson (1983a) have been supported by similar findings in other tropical countries where seeds of similar origins have been grown. For example in Puerto Rico, the Philippines (Bislig) and Keravat provenances performed well on two sites (Whitmore 1973, 1983; Whitmore and Macia 1975; Figueroa and Whitmore 1980). Seven provenances were established in Côte d'Ivoire in two separate trials (in 1974 and 1975). New Britain sources have performed well there but a mainland Papua New Guinea source, Ossima, performed poorly (Diabate 1983).

Provenance trials in east Kalimantan in Indonesia at three years of age also indicated the superiority of New Britain (Rock Pile) and Philippines (Bislig) sources in growth rate and straightness (Long and Dykstra 1983).

From tropical lowland plantings it would appear after the first decade of trials that Mindanao in the Philippines and west New Britain in Papua New Guinea contain the most productive provenance sources. However, well-designed trials testing seed collections from several localities in each of those two areas are necessary before a final selection is made of the best seed sources.

Within New Britain there is some evidence that the north-west coastal sources such as Wilelo are superior to the north-east coastal Gazelle Peninsula sources (Davidson 1983a). Further collections of seed have been made to confirm this.

No formal provenance trials have been carried out using sources within Mindanao, except for the two mentioned by Davidson (1983a) which differed in performance, the source near Bislig performing better than the source from Cotabato (Milbuk). Information is available at PICOP from several plots, unfortunately not arranged to a statistical design, which seem to indicate a considerable range of provenance performance (e.g. 8.6 to 13.8 cm in diameter and 7.9 to 14.3 m in height at age four years) following the ranking New Bataan, Cabadbaran, Bislig, Zamboanga, and Pasian in decreasing order.

Observations by J. Davidson in plantations in Mindanao indicated that a Papua New Guinea provenance was particularly susceptible to attack by a wood-boring buprestid beetle, *Agrilis* sp. Local Philippine provenances which appeared highly resistant at first were increasingly attacked. This borer poses a threat to *E. deglupta* plantations, and a challenge to *E. deglupta* breeders.

7.5 Base populations for selection

The base population for selection in Papua New Guinea was originally the Keravat plantations which were derived from three nearby Gazelle Peninsula localities, the Warangoi, Keravat, and Vudal Rivers. Plantations between 10 and 20 years of age totalling 303 ha were cruised and assessed in 1968. From this resource 150 trees were selected for an initial breeding population (one per 2 ha or about one in every 1000 trees).

Although early trials indicated only that the Gazelle source might be among the better provenances, out of necessity the selection and breeding program initially went forward on this population for more than a decade before the results of formal provenance trials became available to indicate that there could be better base populations available in west New Britain and in Mindanao.

In Mindanao, PICOP started out by making its selections from natural stands of four different Mindanao provenances (New Bataan, Bislig, Monkayo, and Pasian) not from their early plantations which were of Gazelle Peninsula, Papua New Guinea origin. Total selections reached 207 by 1980 for use in the first seed orchard, the majority (172) coming from the local Bislig source.

Areas totalling 150 ha in natural stands located in southern Sulawesi in Indonesia have been set aside as seed stands. Selection of superior trees has been carried out in 8–9 year-old plantations in East Kalimantan and in natural forests at over 50 separate locations in Indonesia. By 1984, 116 'plus trees' had been selected from these combined resources. Progeny from these trees are being evaluated separately to determine both provenance and family performance.

In countries other than Philippines, Indonesia and Papua New Guinea, the most widely distributed and commonly used seed has been obtained from Keravat on the Gazelle Peninsula of New Britain. Consequently most existing plantations in which selection might take place are either of uncertain origin, or from Keravat, or from land races of Keravat origin (such as one called Rock Pile in some reports).

7.6 Conservation

It has been recognized by FAO that some of the natural provenances of *E. deglupta* are in danger of extinction. There are three main reasons for the decline of the species in several areas (Davidson 1986b).

1. *E. deglupta* is intolerant of shade. Older stands are invaded by rainforest preventing regeneration of the species. This leads to pure rainforest and the elimination of *E. deglupta* from that site.
2. Mature stands are being heavily cut over for building, fence, and canoe timbers, and for firewood.
3. The species occurs, in many places, on fertile river flats and levee banks which are rapidly being cleared for food gardens (shifting agriculture) and other farming activities.

Considerable areas of *E. deglupta* are being logged on the islands of New Britain and Mindanao. The only good stand in Cotabato Province, Mindanao, has been endangered through logging activities of a local company. Near Ulamona, and elsewhere on the north coast of New Britain, near-pure stands of *E. deglupta* are being selectively logged for the finest stems. No other natural area is known where *E. deglupta* is growing as well as in riverain stands covering some 8000 ha along the north coast of east and west New Britain. This area represents an important gene resource worthy of conservation.

Clearing of stands of *E. deglupta* for agricultural purposes is especially prevalent in Mindanao and Indonesia. For example it is estimated that more than 60 per cent of the finest stand on the Caliawan River, an east-bank tributary of the Agusan River in Mindanao, has been destroyed ahead of agricultural activity. Destruction is continuing and, unless action is taken now to preserve this stand as a reserve for seed collection purposes, it will probably be nonexistent in a short time. This stand is particularly important as it is one of the largest riverain stands in Mindanao and has been used previously by the Bislig Bay Lumber Company as a seed source. Also, this particular provenance has been determined to be one of the better provenances for use in Papua New Guinea (Davidson 1983a) and elsewhere (e.g. Whitmore 1983).

7.6.1 Conservation *in situ*

Because of the particular mode of occurrence of *E. deglupta* (mainly along streams), conservation *in situ* as part of the viable natural ecosystem seems most desirable. This would be successful provided that:

(i) stands are relatively pure and extensive to minimize encroachment and replacement by rainforest in the seral community;

(ii) the area can be given full protection from natural and man-made hazards;

(iii) the genetic resources are available for collection and use both within and outside the country of origin;

(iv) necessary action to make the conservation a reality can be taken immediately, such as passing of legislation, marking of boundaries, public relations, and education.

The Government of the Philippines has declared an area of *E. deglupta* in Cotabato Province a forest reserve to protect it from logging activities. Bislig Bay Lumber Company and its associated company, PICOP, have reserved areas of *E. deglupta* for future seed collection. There is a similar need for preservation of stands on the islands of Sulawesi, Ceram, and New Britain.

For *E. deglupta* little is known about the minimum area needed to form a viable unit for conservation, either of the ecosystem or of the genetic resources, and much more research is needed. The area required to preserve a viable breeding population of *E. deglupta* alone need not be large. Such a reserve would result in the conservation *in situ* of a wide range of flora and fauna although the area may be inadequate to conserve the whole ecosystem. *E. deglupta* is a widespread species possessing great genetic diversity within its natural range, therefore a series of reserves representing a range of different provenances is preferable to a single large reserve in the centre of the range.

7.6.2 Conservation *ex situ*

Certain populations of *E. deglupta* are already endangered but conservation *in situ* is unlikely. Action to effect conservation of some of the stands *ex situ* has been taken, particularly of two Mindanao provenances (Caliawan River and Milbuk, Cotabato). Blocks of up to 10 ha of each provenance were established at three sites in Papua New Guinea. In these two examples, sufficient seed was collected for provenance trials, selection, and breeding as well as for conservation stands. Further, immediate action to conserve more provenances *ex situ* is required.

Seed of *E. deglupta* does not store well, unlike most other eucalypts. Viability is largely lost after two years even under refrigeration at +4°C. When stored at –15°C viability remained high for 7 years, then fell rapidly (Doran *et al.* 1987). Thus seed storage is of limited value for long-term conservation of the genetic resources of *E. deglupta.*

8

Eucalyptus delegatensis

8.1 Natural forests

Eucalyptus delegatensis and *E. nitens* are the only eucalypts which combine fast growth, cold tolerance, and production of high-quality sawn timber suitable for joinery and furniture. *E. delegatensis* is economically important in Tasmania, eastern Victoria, and southern New South Wales. In Tasmania and Victoria it has been the principal timber-producing species for many years, probably exceeded only by *E. obliqua* in total quantity of sawlogs and pulpwood harvested in those regions.

The wood of *E. delegatensis* is of high quality and similar to *E. obliqua* and *E. regnans*. In the Australian timber trade the name 'ash' is commonly used for all three species because of the distinctly porous annual rings which are similar to those of the northern hemisphere ash (*Fraxinus*). These species are sold in Europe and elsewhere under the name 'Tasmanian oak' because of the superficial resemblance to oak (*Quercus*). The wood is of light colour, low density, fissile, and it is one of the best eucalypt woods for pulp and paper. It is often available in wide clear boards in long lengths of select grade.

E. delegatensis forms forests over 40 m in height at higher altitudes than all other eucalypt species (Fig. 8.1). Occasional stands in the lower part of its altitudinal range are over 80 m tall, and are true

Fig. 8.1. Natural stand of *E. delegatensis* near Batlow, New South Wales (photo W.J. Egan).

giants, as their former name *E. gigantea* suggests. The altitudinal range is from 1100 to 1500 m in southern New South Wales at 35°S latitude, 850 to 1450 m in Victoria, up to 1220 m in northern Tasmania, and between 160 and about 900 m in southern Tasmania at 43°S. It is by far the most commercially important large eucalypt above 500 m altitude in Tasmania and above 1000 m on the Australian mainland (Fig. 8.2). At altitudes above the highest occurence of *E. delegatensis* one usually finds more open stands of shorter trees of *E. dalrympleana* and then woodland of snow gums such as *E. pauciflora* and *E. coccifera*.

Stands of *E. delegatensis* vary in size from one hectare to hundreds of hectares and are usually even-aged and of the one species, though several other species are sometimes associated (Boland and Dunn 1985). The sites where *E. delegatensis* grows in the northern part of its range are generally those combining deep, friable, moist soil on southern slopes and in gullies. However, it is found on all aspects in Tasmania where drought is much less frequent than on the mainland. There is snow each

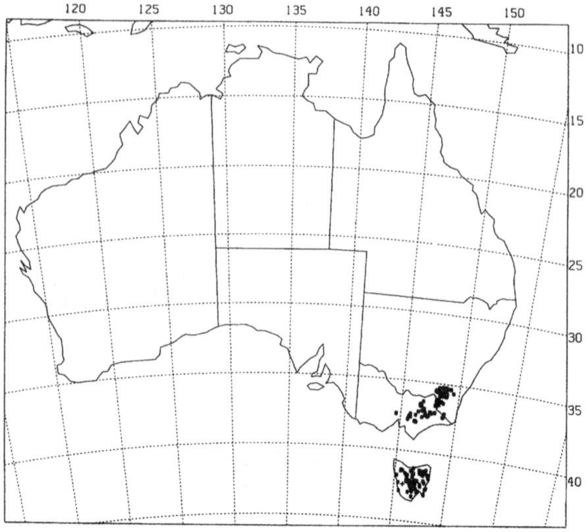

Fig. 8.2. Natural distribution of *E. delegatensis.*

winter in almost all *E. delegatensis* forests and the snow often lies on the ground for several weeks. Mean annual rainfall ranges from about 900 to 2500 mm.

8.2 Plantations

Although *E. delegatensis* is a major source of high-quality sawn timber and pulpwood from natural forests of the highlands of south-eastern Australia, it is not widely planted in Australia or abroad. Since about 1960 regeneration of several thousand hectares per year, in small coupes, has been effected by direct seeding with up to 0.5 kg of seed per hectare. In one area of southern New South Wales regeneration has been by natural seed fall under the group selection system (Hamilton and Cowley 1987). In the same period only a few hundred hectares have been planted with seedlings in Tasmania, Victoria, and New South Wales. Numerous small plantations totalling about 1000 ha have been established in New Zealand where *E. delegatensis* has been one of the main eucalypts planted for several decades. There is increasing interest in the species in the south of France where it grows well on acid soils but not on alkaline soils (Marien 1984).

8.3 Genecology

In 1977 and 1978 CSIRO made a thorough range-wide collection from 465 trees of *E. delegatensis* from 65 localities (Fig. 8.3). At each locality seed and botanical specimens were collected with a rifle from five or more trees at least 50 m apart: from 5 trees at 49 localities, 10 trees at 13 localities, and 30 trees at 3 localities (Boland *et al.* 1980b). The three large samples were for more detailed studies, for example the investigation by Moran and Brown (1980) of temporal variation in outcrossing rate within a population near Batlow in New South Wales.

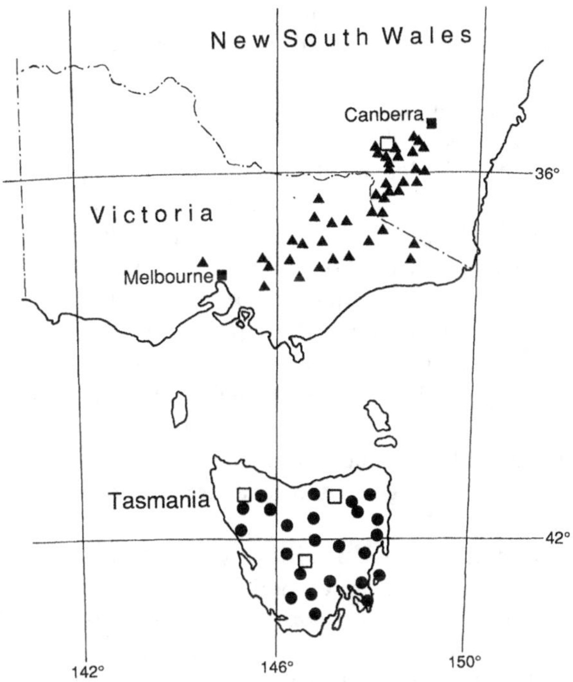

Fig. 8.3. Sites of *E. delegatensis* seed collections (triangles and circles) and provenance trials (squares) (after Boland 1985 and Moran *et al.* 1990).

Following analyses of 45 attributes of seedlings raised from these collections Boland (1985) concluded that the Tasmanian populations were sufficiently different from the mainland populations to warrant division into two subspecies: *E. delegatensis* subsp. *delegatensis* and *E. delegatensis* subsp. *tasmaniensis*. On the average the Tasmanian seedlings were more glaucous, had warty (verrucose) stems and leaves, had smaller rounder leaves, were slower growing, and were generally less tolerant of frost and snow than seedlings of mainland origin (Boland and Dunn 1985).

Further evidence of genetic differentiation between Tasmanian and mainland populations came from New Zealand provenance trials where there was a marked difference with respect to disease susceptibility and insect attack (Wilcox 1983). A common leaf-spot disease caused by *Mycosphaerella cryptica* was much more severe on mainland provenances, whilst defoliation by the chrysomelid beetle *Paropsis charybdis* was largely confined to Tasmanian provenances. Also, in three Australian trials insect defoliation was much greater in Tasmanian provenances than in mainland provenances at age 2–3 years (Ohmart *et al.* 1984).

8.4 Provenance variation

8.4.1 Growth rate

Two small trials were planted in Victoria in 1967 and 1973, comparing 10 and 14 Victorian provenances respectively (Abbott and Pederick 1984a). These trials gave preliminary indications that within the main area of the species in the mountain ranges of eastern Victoria there seemed to be only a limited amount of variation between the provenances tested. Altitude of seed sources within one area, Mount Stirling, did not have an effect on seedling growth rate or seed germination (Abbott and Pederick 1984b).

Between 1978 and 1980 eleven major trials were established in Australia (seven trials each of 64 provenances), New Zealand (three trials of 36 provenances), and France (one trial of 64 provenances). At three and five years there were large and statistically significant differences between provenances in tree height (Abbott and Pederick 1984a; Moran *et al.* 1990; M.D. Wilcox, personal communication). On the trial sites of highest quality the best provenances averaged 8 m at four years and the worst 5 m.

The early results indicated that provenances of *E. delegatensis* from the Central Highlands of eastern Victoria were among the leaders in all the trials, for example at age three years in the two trials summarized in Table 8.1. The most consistently fast growing provenances were from Royston (Table 8.2) and areas nearby about 100 km north-east of Melbourne. However, rankings changed in later measurements. At nine years in the trial near Batlow in southern New South Wales (latitude 35° 37'S) the growth of local provenances was best (P.K. Ades, personal communication). While most Tasmanian provenances grew slowly in all trials, several of the Tasmanian

Table 8.1
Early results (3 years) from two E. delegatensis *provenance trials planted in New Zealand in 1978, at Rotoaira (latitude 39°S, altitude 700 m) and Longwood (46° 11' S, 100 m) (from Wilcox 1983).*

Region	Mean height (m) 1981	
	Rotoaira	Longwood
Victoria	4.6	4.3
New South Wales	4.0	3.9
Tasmania	4.0	4.1
New Zealand	4.1	4.0

provenances were among the leaders in a trial in the South Island of New Zealand at latitude 46°S (Wilcox *et al.* 1985; King and Young 1992) and in Tasmania at 41° to 43°S (Raymond 1988b; Moran *et al.* 1990). Thus there are clear indications of provenance by planting site interactions in this species. (See Section 18.3.5 for further information on genotype–environment interaction.)

8.4.2 Frost tolerance

Twenty-nine Victorian seed sources of *E. delegatensis* from altitudes over the range 825 to 1200 m were examined in laboratory studies of frost resistance by Grose (1960). Twenty-four seedlings from each source were exposed to artificial frosts. Mean leaf damage of individual seed sources ranged from 10–70 per cent, and the regression of mean leaf damage against altitude of seed source was negative and statistically significant.

In the field trial near Batlow in southern New South Wales, observations were made of frost damage to the leaves of two-year-old saplings. The Tasmanian provenances suffered most damage and the Victorian provenances generally more than those from southern New South Wales. Tasmanian provenances from high altitude in the central plateau were much less damaged than those from low altitude near the east coast (Boland and Dunn 1985). The stems of many trees in this trial were bent or broken in a heavy snow storm in the third winter. There was severe damage in low altitude Tasmanian provenances and very little damage in the local NSW provenances (Moran *et al.* 1990).

In preliminary experiments by Abbott and Pederick (1984b) on altitudinal variation in frost resistance of *E. delegatensis*, seedlings were raised from two transects each of 500 m altitude near Mount Stirling, Victoria. When the partly hardened seedlings were exposed to –10°C for ten minutes there was less leaf damage for the seedlings originating from higher altitudes.

As part of a program of juvenile selection of eucalypts for frost tolerance, Marien (1979) found a strong correlation between tolerance of frost and altitude of seed source among 18 Tasmanian provenances of *E. delegatensis* (Fig. 8.4). These seedlings were subjected to natural frost of –12°C in an open nursery near Toulouse in the south of France. He also

Table 8.2
Sites of seed collection of E. delegatensis *mentioned in Chapter 8. (After Moran et al. 1990, where details are given of all the collection sites shown in Fig. 8.3)*

Location	State	Latitude °S	Longitude °E	Altitude (m)
Gungarlin River	NSW	36 17	148 30	1255
Mount Stirling	Vic	37 06	146 28	1300
Royston River Road	Vic	37 23	145 52	960
Mount Useful	Vic	37 40	146 31	1220
Ben Nevis	Tas	41 22	147 40	820
Ben Lomond	Tas	41 31	147 39	1220
Guildford	Tas	41 30	145 40	580
Bicheno	Tas	41 53	148 12	460
Maydena	Tas	42 33	146 34	650

NSW New South Wales, Tas Tasmania, Vic Victoria

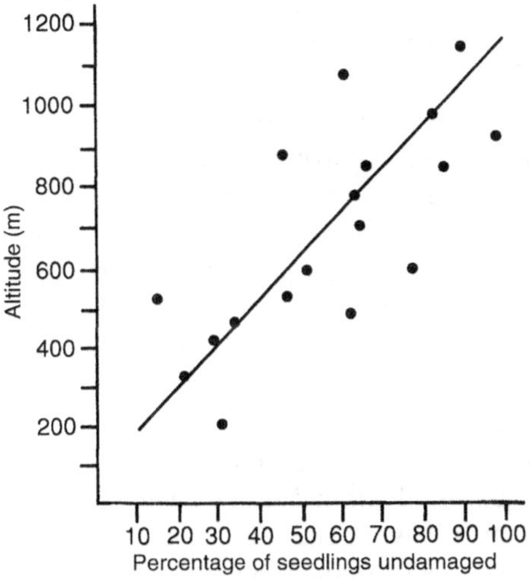

Fig. 8.4. Influence of altitude of seed source on frost tolerance among 18 Tasmanian provenances of *E. delegatensis*. Note that seedlings from higher altitude were more frost resistant (after Marien 1979).

found a negative correlation between altitude and height growth of one-year-old nursery seedlings of ten of these Tasmanian provenances (Fig 8.5). Thus there was, on a population basis, an adverse correlation of height growth and frost tolerance. However, there was a small proportion, less than four per cent, of very tall, frost-tolerant individuals in the frost-sensitive provenances. These individuals (correlation breakers) were selected for propagation by cuttings. Among the very frost-tolerant provenances there were no very fast-growing seedlings.

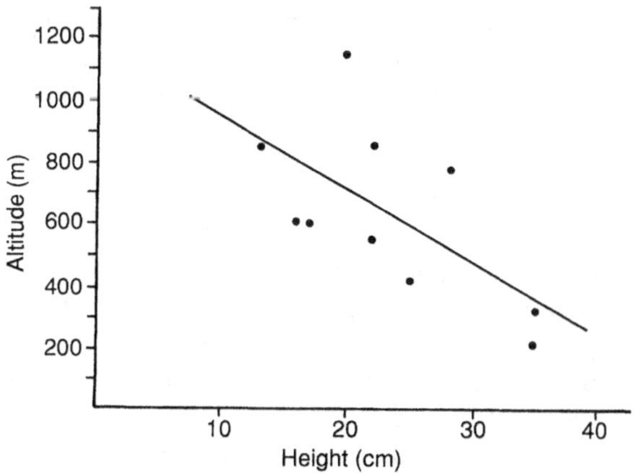

Fig. 8.5. Influence of altitude of seed source on height growth of one-year-old nursery seedlings of ten Tasmanian provenances of *E. delegatensis*. Note that seedlings from higher altitudes grew more slowly (after Marien 1979).

While investigating the reasons why *E. delegatensis* seedlings did not grow well above 800 m in northern Tasmania after natural or artificial regeneration, Webb *et al.* (1983) studied the effects of artificial frost on seedlings of five provenances. They found that leaf discs from hardened seedlings of four Tasmanian provenances were killed at –10°C while the one New South Wales provenance was undamaged. The ranking of the five provenances for frost tolerance was similar to that observed in the New South Wales provenance trial (Boland and Dunn 1985), with Gungarlin River, New South Wales, the most tolerant of frost and Bicheno, Tasmania, the least.

The technique of freezing excised leaf discs under controlled conditions to determine the lethal temperature (the frost minimum temperature capable of producing leaf death) was used by Hallam and Reid (1989). Four Tasmanian, two Victorian, and one New South Wales provenance of *E. delegatensis* were examined, using leaves taken from saplings growing in two large provenance trials in Tasmania. Hardening and dehardening of seedlings exposed to controlled night temperatures in controlled-environment cabinets was also studied. The provenance hardiness rankings obtained by Hallam and Reid for the field sapling material and the seedlings hardened in controlled environments were similar. Two high-altitude provenances from Tasmania, Ben Lomond and Ben Nevis, and the Gungarlin River provenance from New South Wales, were hardier than the Victorian and low-altitude Tasmanian provenances. While Hallam and Reid (1989) found Ben Lomond was the most hardy, Boland and Dunn (1985) ranked it among the less hardy provenances on the basis of their field observations of frost damage at the New South Wales trial site. At the colder of the two Tasmanian field sites, the difference in lethal temperature between the Ben Lomond provenance (most hardy) and the Bicheno provenance (least hardy, from a low altitude on Tasmania's east coast) was less than 0.5°C in early summer, when the lethal temperature was about –4°C. This difference increased to 2.7°C in midwinter, when the lethal temperature of the Ben Lomond provenance had dropped to –8.6°C. Seasonal and between-provenance differences in hardiness were less pronounced at the warmer Tasmanian site.

The frost resistance of several commercial eucalypt species growing in plantations in southern Tasmania was examined by Hallam *et al.* (1989) using the excised leaf-disc method described above. Of two Tasmanian *E. delegatensis* provenances studied, Guildford, at an altitude of 580 m in northern Tasmania, proved hardier than Maydena, altitude 650 m in southern Tasmania. The Guildford provenance hardened from a lethal temperature of –5.5°C in March to –8.7°C in August at the lowest plantation site, altitude 60 m, while the Maydena provenance hardened from –5.0°C to –7.4°C at this site. Lethal temperatures were higher at three other plantation sites which were higher in altitude but had generally higher night temperatures.

In summary, the studies on the frost resistance of *E. delegatensis* have established substantial variation in resistance among provenances, equivalent to differences of up to at least 2°C in winter survival temperatures. There is a general trend of increasing resistance with increasing altitude of seed source within both Victoria and Tasmania. There are indications that the high-altitude provenances from New South Wales may be hardiest, and the low-altitude provenances from eastern Tasmania the least hardy. A comprehensive survey of geographical variation in frost resistance in *E. delegatensis* remains to be undertaken. Patterns of variation in resistance within provenances also deserve further study.

8.4.3 Winter kill

Extensive field observations and experiments on regeneration of *E. delegatensis* in Victoria by Grose (1960) identified a form of winter damage different from normal frost damage produced by exposure to low minimum temperatures. Seedlings undamaged by severe frosts prior to the onset of snow cover would be damaged or killed on prolonged exposure to wet snow, a phenomenon termed 'winter kill'. Over a five-year period, mortality of first year seedlings during the winter months from winter kill averaged 50 per cent for seedlings on ashbed sites, making it a much more important source of mortality than frost damage.

Examination of leaf tissue in contact with wet snow for prolonged periods showed that affected leaves, which subsequently died, had water contents double that of unaffected leaves. The same symptoms could be produced in the laboratory by treatments which saturated the intercellular spaces in the leaves with water for several days. Damage in the field was most severe on southern slopes, where snow cover was more prolonged, and was worse in years when there was heavy snow. Seedlings which were larger and more glaucous, as was the case on ashbed sites, were least damaged, and seedlings growing under shrub cover, which as a result of shading were less glaucous, were most vulnerable. Glaucousness was thought to promote resistance to winter kill by reducing the infiltration of water into the leaves from melting snow in contact with them. Differences between provenances in resistance to winter kill have not been reported, although the work of Grose suggests that they could be of major importance in the establishment of *E. delegatensis* in a large part of its natural range, and there is considerable provenance variation in glaucousness as Boland and Dunn (1985) have shown.

8.5 Base populations for selection

The results from the two Victorian trials at four years supported the practice of preference for local seed collections for aerial sowing after natural forests in Victoria have been clear-felled for sawlogs and pulpwood. However, this was not the case in New South Wales or Tasmania where the Victorian Central Highlands provenances generally grew faster in trials than the local provenances. If a single provenance had to be chosen at the early stages of provenance testing it would probably have been the Royston region for its consistency in all trials on milder sites, though its advantage was small and not significantly different statistically from several other provenances (Wilcox 1983; Pederick 1986; Raymond 1988b).

Wilcox *et al.* (1985) noted that the local New Zealand seed sources of *E. delegatensis*, used for many years in plantation establishment, performed poorly relative to the better Australian provenances in provenance trials in New Zealand's South Island (Table 8.1, pp. 84). Also, two New Zealand seedlots generally had slower than average growth rate at three years in the four Australian provenance trials reported by Moran *et al.* (1990). After extensive collections in the Royston River region of Victoria, a 10 ha gene resource stand of the Royston provenance was planted in New Zealand in 1984 to replace the inferior New Zealand land race.

From the point of view of conservation of the genetic resources of natural *E. delegatensis* forests periodically felled for wood production, the early results of the Australian trials have highlighted the need for the establishment of seed zones. For example, among the Tasmanian provenances there were large and statistically significant differences between groups of provenances in leaf morphology, frost resistance, growth rate, and stem and branch quality. The provenances with the best growth rates in trials in Tasmania were from high rainfall areas of north-west Tasmania, whereas the high altitude and plateau provenances (for example Ben Lomond) had low growth rates, and provenances from low altitude were susceptible to frost. Local seed appears to be appropriate for these broad provenance zones. However, for maximum production in certain areas of Tasmania formerly carrying *E. delegatensis*, there may be a case for including 'exotic' provenances such as the Royston region in Victoria, in the base populations for selection.

In many cases, perhaps all, the overriding consideration in Australia will be political, to meet the community demand for the conservation of genetic integrity of local stands. Therefore, only locally collected seed will be permitted (as is already the case with *E. diversicolor* in Western Australia), even though distant provenances may be known to perform better than local provenances, as with Victorian *E. delegatensis* outgrowing locally collected seedlots in Tasmania.

9

Eucalyptus fastigata

9.1 Natural forests

Eucalyptus fastigata is confined to cool humid forests in a narrow zone running parallel to the New South Wales coast with a small extension into eastern Victoria. The species is found on moist, well-drained, fertile soils mainly on the eastern edges of tablelands and escarpments, and further inland in some places. The range of latitude is between 30°30' and 37°20'S (Fig 9.1, Table 9.1). The range of altitude is mainly between 700 and 1200 m but the species does occur below 300 m in moist gullies in the south (Fig. 9.2) and above 1400 m in the north. Frost and snow are common at higher altitudes throughout the range of the species. Rainfall between about 800 and 2000 mm is distributed fairly uniformly throughout the year in the south but there is a summer maximum in the north (Boland *et al.* 1984; Hall *et al.* 1981).

E. fastigata is closely related to but easily distinguished from its better known southern cousin *E. regnans*. The distribution of *E. fastigata* does not overlap with *E. regnans* geographically or morphologically though small stands of the two species are only a few kilometres apart near Bendoc in eastern Victoria close to the border of New South Wales.

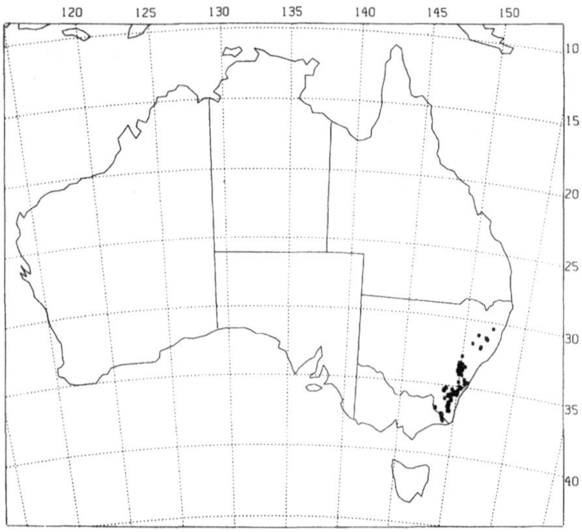

Fig. 9.1. Natural distribution of *E. fastigata*.

Table 9.1
Sites of seed collection of E. fastigata *mentioned in Chapter 9 (after Wilcox 1982c).*

Location	State	Latitude	Longitude	Altitude (m)
Barrington Tops	NSW	31 50	151 20	1370
Oberon	NSW	33 54	149 48	1220
Robertson	NSW	34 35	150 46	670–720
Tallaganda State Forest	NSW	35 29	149 30	970–1070
Badja	NSW	36 22	149 29	1000–1070
Brown Mountain	NSW	36 36	149 23	700–1200
Bondi State Forest	NSW	37 08	149 12	910
Bendoc	Vic	37 13	148 54	1070–1120

NSW New South Wales, Vic Victoria

Fig. 9.2. Natural stand of *E. fastigata* in the Monga State Forest, southern New South Wales (photo W.J. Egan).

In contrast to *E. regnans*, *E. fastigata* has brown fibrous bark over the whole trunk (hence the name brown barrel), has narrower leaves, has the ability to coppice, is somewhat more frost hardy, and generally does not grow as fast or reach the great size of *E. regnans*. Both these species are distinctive among the eucalypts in having their flower buds arranged in double umbels. Both species also produce similar high-quality wood of light colour and low density.

9.2 Plantations

In South Africa *E. fastigata* is being extensively planted in Natal and Transvaal between latitudes of 24° and 28°S and at altitudes of 1200 to 1800 m. Unpredictable and intermittent dieback has plagued the species when planted in the warm parts of the Transvaal mist belt and on heavy clay soils, but in the cooler regions of the Transvaal it has become a major species for the production of mining timber and pulp, occupying 31 000 ha in 1983 (Darrow 1984). The majority of eucalypt plantations in the eastern Transvaal highveld are of *E. fastigata*.

The species is well known in New Zealand where it has been grown since 1884 in scattered woodlots rather than in large plantations (Weston 1957). In the cooler areas of the North Island betweeen latitudes 38° and 40°S *E. fastigata* has been one of the four eucalypts favoured for plantations in recent times, the others being *E. delegatensis*, *E. nitens*, and *E. regnans* (Wilcox 1980). The area of *E. fastigata* plantations is approximately 2000 ha, mainly in the Kinleith Forest of NZFP Forests Ltd. Although the many existing plantings show that *E. fastigata* is already one of the healthiest and most adaptable eucalypts in New Zealand, greater frost hardiness would be an advantage. Also the species has an undesirable tendency for the bole to be short and the crown to break up early into heavy branches. Breeding work in New Zealand has concentrated on better branch and stem quality, greater frost tolerance, and faster growth to produce pulpwood and sawlogs (Wilcox 1980).

9.3 Provenance variation

9.3.1 Growth rate and frost tolerance

There is evidence of significant variation in frost tolerance among provenances of *E. fastigata*. This was first observed by Boden (1958b) when seedlings originating from 720 m, 1070 m, and 1170 m on a transect of Brown Mountain in southern New South Wales were exposed to natural frost in a Canberra nursery. Seedlings from the lowest altitude were severely damaged while those from the highest altitude were undamaged (Fig. 9.3).

In South Africa two of the Brown Mountain provenances from southern New South Wales (1170 m and 720 m) and one from the Australian Capital Territory (1070 m) were tested for cold tolerance and growth (Sherry and Pryor 1967). The two high-altitude provenances were less damaged by frost and cold winds than the low-altitude provenance. However, the low-altitude provenance had much faster height growth and by the second year had grown clear of the radiation frost layer where most mortality occurs. The effect of the radiation frost layer was confirmed because the smallest trees were the most severely damaged by frost. After 2.5 years the low altitude provenance had 20 per cent greater volume than the mid-altitude provenance and 75 per cent greater volume than the high altitude provenance (Sherry and Pryor 1967).

The early results of five South African trials planted in 1973 and 1980 (Roeder 1980a; Darrow 1984) showed no consistent differences between provenances, perhaps due to planting on relatively low, warm sites where all provenances suffered dieback, or to too few replications in the design as a result of shortage of seedlings. However, on a high-altitude site, at Jessievale in the eastern Transvaal highveld, the low altitude provenance from Robertson, New South Wales, was distinctly inferior in growth and survival (Darrow 1984). There was no evidence that the South African land race in these trials, a seedlot from Draycott, Natal, was superior to the newly imported, unselected wild populations from Australia.

The preliminary results of another South African trial planted in Natal in 1979 showed that on the

Fig. 9.3. Provenance variation in frost damage to *E. fastigata* from three localities on Brown Mountain in south-eastern New South Wales (Boden 1958). The severely damaged seedlings on the left were from the lowest altitude, 720 m. The undamaged seedlings on the right were from 1170 m and those in the middle from 1070 m (photo W. Pedersen).

frostiest of the three sites the Robertson provenance failed while four other southern New South Wales provenances from higher altitudes had reasonably good survival (Nixon 1984). Clearly there was a case for further provenance testing of this important species in South Africa. Additional seed collections for this purpose were made in Australia in 1984.

In New Zealand artificial frosts were imposed on seedlings from 18 seed sources of *E. fastigata*, 12 from the natural forests in New South Wales, five from New Zealand and one from South Africa (Wilcox *et al.* 1983a). Seedlings were hardened in the open nursery and taken to controlled environment chambers for frosting. They were scored as frost tolerant if less than 30 per cent of the foliage was damaged. Tests in autumn, winter, and spring all showed the northern-most provenances, Oberon (1220 m) and Barrington Tops (1370 m), to be most frost-tolerant. Most of the seedlings from these two provenances endured a winter frost of –9°C. The hardiest exotic seed source was from Draycott in Natal province, South Africa.

Two large, well-replicated field trials were established in New Zealand in 1979 each with 126 seedlots representing eight Australian provenances, and collections from plantations at Draycott, South Africa, and several localities in New Zealand (Wilcox 1982d). Of the 126 seedlots 115 were open-pollinated families of individual trees. The 51 New Zealand families were nearly all from plus-trees carefully selected for straight stems, good branching characteristics, and vigorous growth.

Frost damage and tree height were assessed after one year in one of the trials, and the tendency to early forking in both trials (Wilcox 1982d). The three hardiest seed sources, Oberon, Barrington Tops, and Draycott, were the slowest growing (Fig. 9.4). There was a fairly good correlation with results from natural frost (Wilcox 1982d) and from artificial frosts (Wilcox *et al.* 1983a). Two provenances from southern New South Wales, Bondi State Forest near Bombala, and Tallaganda State Forest near Rossi, were above average in a combination of all three traits of frost tolerance, growth rate, and form. These areas are therefore preferred as immediate commercial seed sources for planting in New Zealand in the near future. Several provenances (mainly but not exclusively from high altitudes in southern New South Wales) contained families and individuals of above average height, frost tolerance, and form, and therefore suitable for inclusion in a future breeding population.

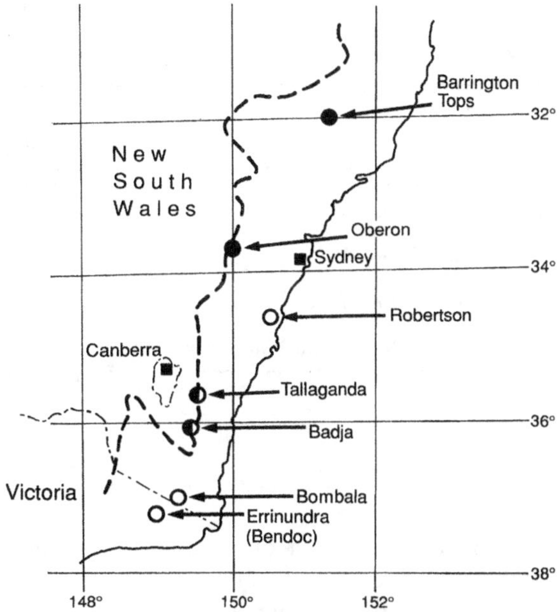

Fig. 9.4. Provenance variation in frost tolerance of *E. fastigata* seedlings subject to artificial frosts. Open circle least hardy, half circle moderately hardy, and closed circle most hardy The dashed line is the Great Dividing Range (after Wilcox 1982d; Wilcox *et al.* 1983a).

All the populations from New Zealand plantations and the provenance from Robertson, New South Wales, were comparatively frost-tender and heavily forked, and Robertson was amongst the slowest growing.

9.3.2 Leaf colour

In a study of leaf colour of seedings of *E. fastigata* Wilcox (1982c), found that both the Robertson provenance and all the New Zealand populations had a high proportion of seedlings with distinctly green cotyledons whereas all the other Australian provenances and the South African population had a high proportion of seedlings with a vivid purple colour on the underside of the cotyledons (Table 9.2).

The few seedlings with intermediate leaf colour were considered to be purple phenotypes in which the anthocyanin coloration was only weakly expressed.

Table 9.2
Variation in leaf colour among populations of E. fastigata *(from Wilcox 1982c).*

Seed source	Number of families	Mean frequency of cotyledon colour (per cent)		
		Purple	Intermediate	Green
Four New Zealand populations	42	15	3	82
Robertson, New South Wales	11	19	4	77
Draycott, Natal South Africa	6	85	5	10
Four high-altitude New South Wales provenances	46	78	9	13

9.4 Base populations for selection

Clearly the Robertson area, or a similar low altitude Australian occurrence not yet tested, is a prime suspect as the original source of the early New Zealand plantations of *E. fastigata*. None of these old plantations has much value as a commercial seed source or for future selection compared with several of the newly imported natural provenances, particularly for the colder areas where eucalypts are planted. Since 1982 most of the commercial seed of *E. fastigata* used in New Zealand has been imported from near Bendoc on the Errinundra Plateau in eastern Victoria, at the southern extremity of the species range. In warmer areas of New Zealand other faster growing species such as *E. saligna* would be preferred.

E. fastigata in South Africa is a case of a successful species with a narrow genetic base. It appears from seed records (Hagedorn 1984; Poynton 1979) that the present extensive plantations in South Africa originated mainly from a few trees in an early plantation and from their progeny. Only two importations of *E. fastigata* seed before 1923 to South Africa could be traced by Poynton (1979), in contrast to several importations of other eucalypts. Wilcox (1982c) has provided evidence from leaf colour that at least one South African plantation, at Draycott in Natal, originated from high altitude in Australia. The first seedlot was imported before 1904, and it seems unlikely that it was by chance from one of the best provenances in the wide natural range of this species, and from numerous well-separated trees of good quality.

Despite the recent good work in New Zealand and South Africa there is still no comprehensive, range-wide collection and testing of this species. Collections in the summer rainfall part of the range near latitude 30°S in the New England Tablelands of north-eastern New South Wales may have particular value in South Africa.

The South African and New Zealand studies on *E. fastigata* lend further support to the view that even such attractive land races as exist in these two countries are not a good enough base for current seed supply and future selection. Even after one or two generations of natural and silvicultural selection in the new environment these two land races, based on chance introductions, are not superior to newly introduced wild populations. Therefore there is likely to be considerable improvement in yield and quality by the simple procedures of provenance testing following by selection within a genetically broad sample of the best provenances, already started in both countries.

10

Eucalyptus globulus

10.1 Natural forests

One of the earliest of the eucalypts to be both validly named and brought into cultivation, *E. globulus* was formally described by Labillardière (1799) from a specimen he collected in south-eastern Tasmania in 1792 (Fig. 10.1). It is the best known of four closely related species (or subspecies) of blue gum, the others being *E. bicostata, E. pseudoglobulus,* and *E. maidenii* (Fig. 10.2).

The natural range of *E. globulus* is approximately 38°26'S to 43°30'S. Its principal occurrence is in eastern Tasmania in disjunct populations along the coast and up to 20 km inland in relatively dry open forest. Rainfall in some places is as low as 550 mm. The morphological affinity of trees in natural stands in south-eastern Tasmania with those in plantations in Europe (Orme 1978) suggests that the plantations originated from seed collections near the coast in that area.

E. globulus also extends 60 km inland at more than 500 m above sea level at latitude 43°S. There the rainfall is more than 1000 mm and the trees growing in closed forest are 70 to 80 m tall. It is also found in several very small stands along the wet and stormy west coast, but not along the north coast of Tasmania (Orme 1978). *E. globulus* also occurs on several of the Bass Strait Islands and in small areas in Victoria on the southern extremity of the mainland at Wilsons Promontory and on the coastal side of both the Otway Ranges (Fig. 10.3) and the Strzelecki Ranges (Figs. 10.2 and 10.4). Of the other three closely related species of blue gum, *E. bicostata* has an even greater natural range but *E. pseudoglobulus* and *E. maidenii* are more restricted (Fig. 10.4).

10.2 Plantations

E. globulus was the first of the eucalypts to become widely known outside Australia as an ornamental and plantation tree. This was at least partly because the unpalatability of the juvenile leaves to grazing animals conferred an advantage over other eucalypts where fencing was not practicable (Jacobs 1981). The total plantation area of *E. globulus* was more than 800 000 ha in 1973, when it may have been the most planted eucalypt species (Jacobs 1981). By 1990 its planted area was certainly much greater but it was exceeded by *E. grandis* and perhaps also by *E. camaldulensis* and *E. tereticornis.*

The plantations of *E. globulus* are in areas of mild climate free of severe frosts (not below about –6°C) and without severe drought. There are plantations of more than 300 000 ha in both Portugal and Spain (Table 1.1), and smaller areas in Bolivia, Chile, China, Colombia, Ethiopia, Peru, USA (California), and several other countries. The tallest trees in Europe are specimens of *E. globulus* more than 70 m tall in Portugal (Goes 1977, 1984), where growing conditions on parts of the central coast, about 40°N latitude, are close to ideal for the species (Fig. 10.5).

The wood is used largely for pulpwood, firewood, and mine timbers. Plantations are cut on a coppice rotation of about 8 to 12 years, usually two or three times. Some trees are allowed to grow to a large size and are used as sawn timber but because of growth stresses the output of high-quality boards is low. On the most favourable sites mean annual increments of more than 20 m^3 ha^{-1} are obtained, for example to

Table 10.1

Location of E. globulus *provenance collections in 1976 as shown in Fig. 10.4 and mean diameter of each provenance at seven years at the Scottsdale trial (Volker and Orme 1988).*

Species, provenance group, provenance location, state[1]			Number of seed trees	Latitude °S	Longitude °E	Altitude (m)	Diameter 7 yr Scottsdale (cm)
E. globulus							
Southern group (large fruit 2.5–3.5 cm)							
3	Leprena	T	10	43 31	146 55	20	12.3
4	Bruny Island	T	10	43 23	147 19	180	11.5
5	Channel	T	10	43 15	147 11	100	12.8
6	Geeveston	T	10	43 12	146 58	200	13.2
7	Denison	T	5	42 59	146 50	240	13.2
8	Uxbridge	T	5	42 47	146 50	500	11.7
9	Jericho	T	5	42 25	147 15	500	11.5
11	Taranna	T	10	43 02	147 53	120	12.6
Eastern group (large fruit 2.5–3.5 cm)							
10	Pepper Hill	T	5	41 36	147 47	560	11.5
12	Swansea	T	10	42 09	148 02	100	12.2
13	Rheban	T	10	42 40	147 53	80	12.5
14	Seymour	T	10	41 43	148 17	20	11.5
15	Scamander	T	10	41 27	148 13	50	11.2
16	St Helens	T	10	41 12	148 15	50	12.8
Bass Strait group (smaller fruit 1.2–1.8 cm)							
17	Flinders Is. nth	T	10	39 47	147 55	50	12.9
18	Flinders Is. sth	T	10	40 13	148 02	50	12.3
20	King Island	T	5	39 55	144 00	50	12.5
21	Otway Ranges	V	8	38 40	143 44	100	13.2
1	Macquarie Harbour	T	5	42 24	145 14	20	11.9
2	Henty River	T	5	42 01	145 17	50	11.7
E. pseudoglobulus							
26	Wibens Hill	V	5	37 35	148 44	250	12.2
27	Jeeralang	V	14	38 22	146 34	400	12.2
E. bicostata							
23	Mt Cole	V	10	37 17	143 17	700	9.7
24	Toombullup	V	5	36 53	146 09	840	9.9
25	Strathbogie	V	5	36 54	145 53	870	9.2
E. maidenii							
29	Tantanwangalo	N	5	36 46	149 38	380	10.2
30	Mt Dromedary	N	10	36 36	150 04	305	9.0
31	Nerrigundah	N	3	36 07	149 54	245	9.0

[1]Provenance groups of *E. globulus* from Orme (1988); provenance locations from Volker and Orme (1988); V Victoria, T Tasmania, N New South Wales.

Fig. 10.1. P.J. Redouté's drawing of the type specimen of *E. globulus* (Labillardière 1799) (photo CSIRO).

30 m^3 ha^{-1} in Ethiopia (Davidson 1989; Pohjonen and Pukkala 1990) but generally 10 to 15 m^3 ha^{-1} is regarded as an achievable yield for *E. globulus* over large areas (Jacobs 1981).

The other three species have not been planted on a large scale. However, their performance in trials in many countries suggests that *E. pseudoglobulus* is close to *E. globulus* in growth rate, and that the greater frost tolerance of some provenances of *E. bicostata* and *E. maidenii* compensates for their slower growth in some cases. All four species have potential to contribute to the *E. globulus* breeding population in future.

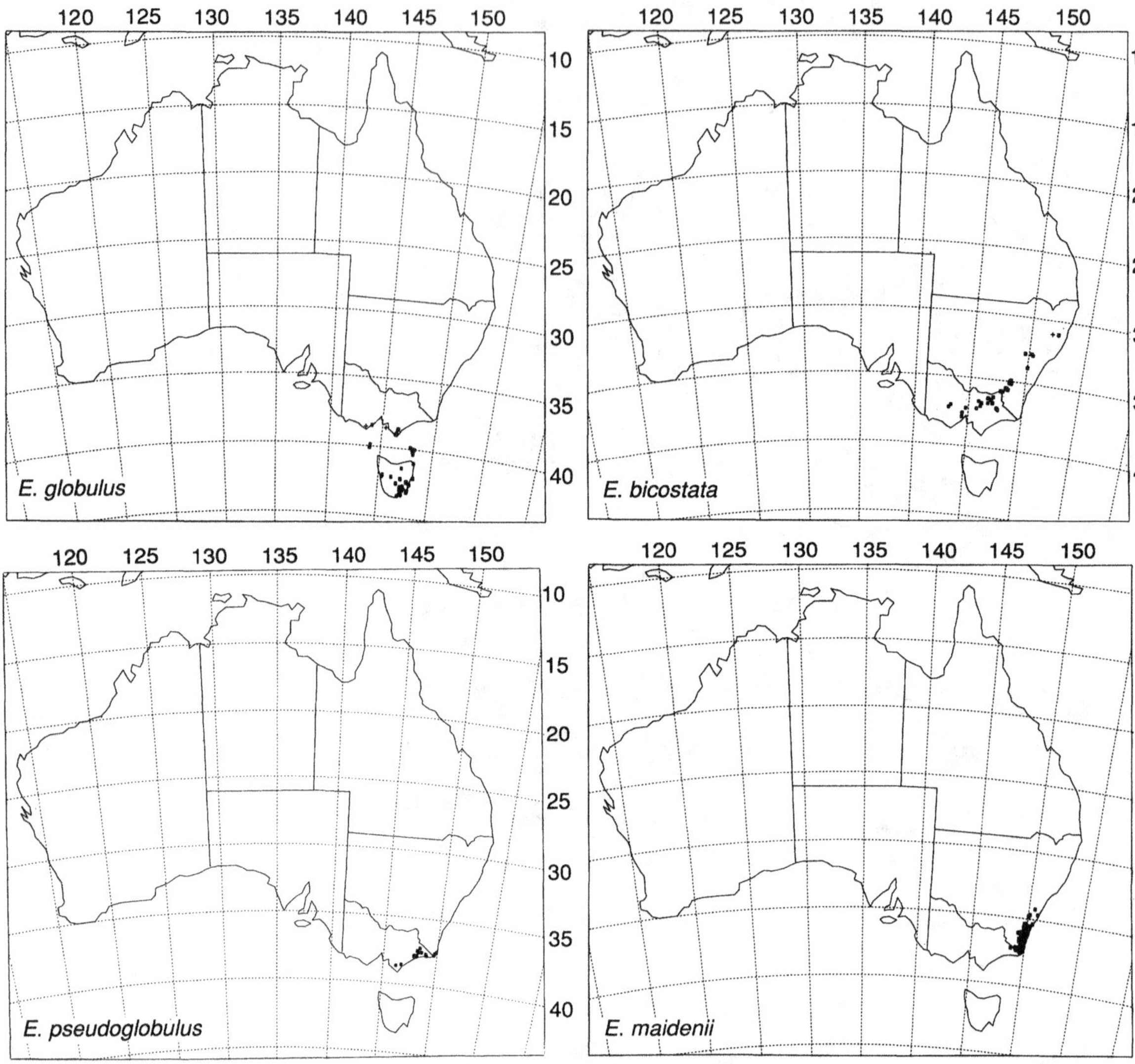

Fig. 10.2. Natural distribution of *E. globulus, E. bicostata, E. pseudoglobulus,* and *E. maidenii.*

10.3 Genecology

After a detailed study of 987 specimens of the 'blue gums' collected from 86 localities throughout south-eastern Australia, Kirkpatrick (1974, 1975) reported there were four main groups of populations distinctly different from each other and worthy of separate taxonomic status. The core areas of these taxa were separated by extensive areas carrying intermediate populations with no evident discontinuities in any of the 30 adult and juvenile characters studied. Only two characters were found to be particularly useful in distinguishing the four core populations: number of flowers in each inflorescence and the size of the fruits. On this evidence, Kirkpatrick (1974) considered that *E. globulus* should be divided into four subspecies: subsp. *globulus* (very large single fruits), subsp. *bicostata* (large fruits in threes), subsp. *pseudoglobulus* (small fruits in threes), and subsp. *maidenii* (small fruits in sevens). Although one, three, or seven are the common numbers there is some variation within each of the four taxa. For example, many of the Tasmanian

Fig. 10.3. Natural stand of *E. globulus* in the Otway Ranges, Victoria. One of the fine specimens sampled in the 1987 seed collections (photo CSIRO).

E. globulus populations sampled by Kirkpatrick (1975) included trees with three flowers per inflorescence.

Other taxonomists have agreed that there are four taxa but have sometimes used different names or treated Kirkpatrick's taxa as species. Pryor and Johnson (1971) listed four taxa at species level, *E. globulus, E. st-johnii* (invalidly according to Chippendale (1976)), *E. pseudoglobulus*, and *E. maidenii*, and stressed the affinity between the four by grouping them under the informal superspecies *Globulus*. Kirkpatrick's taxa were accepted as subspecies by Chippendale (1988), Chippendale and Wolf (1981), and Boland *et al.* (1984), but Orme (1978) and Brooker and Kleinig (1983) preferred the species names *E. globulus, E. bicostata, E. pseudoglobulus* (including *E. st-johnii*), and *E. maidenii*. In this chapter we use the four species names, *E. globulus, E. bicostata, E. pseudoglobulus,* and *E. maidenii,* with some justification from Chippendale (1988, p. 352) who noted that the four subspecies are 'sometimes treated as species'. We use the term 'blue gums' when referring to all four species. *E. globulus* is commonly called Tasmanian blue gum.

Seed collectors and taxonomic botanists usually wish to allot one of the four names to a particular tree or stand, some using species names and others subspecies. This proved especially difficult in the extensive seed collections of *E. globulus* in 1987 and 1988 in southern Victoria on the inland side of the Otway Ranges and Strzelecki Ranges where populations are found with some single-fruited and many three-fruited trees. Clearly such populations are not part of the core area of *E. globulus* botanically but may have equal silvicultural potential for plantations. Seed collectors have designated collections from intermediate populations in southern Victoria as *E. globulus* intergrading with *E. bicostata* or with *E. pseudoglobulus* (Gardiner and Crawford 1987, 1988).

When buds and fruits of the Jeeralang provenance collected in 1976 on the inland side of the Strzelecki Ranges (designated *E. pseudoglobulus* in Table 10.1) were examined by J.B. Kirkpatrick he noted that 'none of the specimens could be referred to typical subsp. *globulus* or subsp. *pseudoglobulus*' (letter to CSIRO, July 1976). 'These form part of the cline between subsp. *globulus* and subsp. *pseudoglobulus*...Some specimens have a passing resemblance to subsp. *bicostata*...They are best referred to *E. globulus sensu lato*'.

In his studies of seedlings in a greenhouse Kirkpatrick (1975) found the patterns of geographic variation in the juvenile characters were less distinct than in the adult phenotype. Nevertheless, lignotuber frequency was clearly lower in *E. globulus* than in the other three species. Also, there were indications of variation in frost and drought tolerance within the species.

10.4 Provenance variation

In 1976 the first range-wide provenance collection of Tasmanian *E. globulus* seed was made on the personal initiative of R.K. Orme (Fig. 10.4). This comprehensive and well-documented collection (Orme

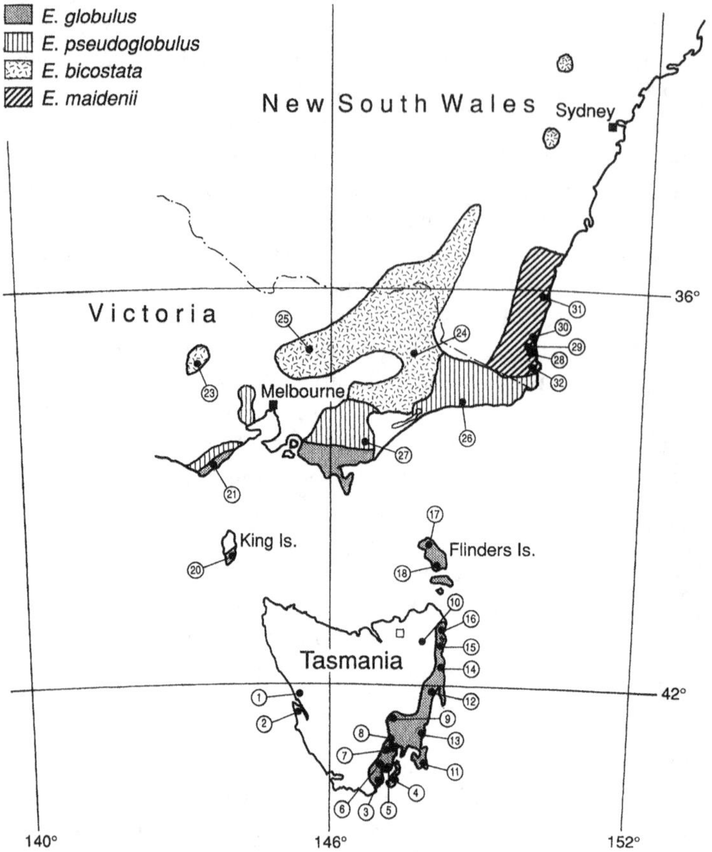

Fig. 10.4. Distribution of the four species of blue gum in south-eastern Australia showing location of seed collection sites (numbered circles) and provenance trial sites (squares) (after Volker and Orme 1988). The Geeveston trial is close to collection site 6 and Scottsdale is near site 10. This map does not distinguish between the core and intermediate populations recognized by Kirkpatrick (1974).

1978) was supplemented by collections from Victoria and New South Wales and resulted in the planting of provenance trials in several countries between 1977 and 1980 (Orme 1983). Further trials were planted in Western Australia and Portugal in 1985.

The need for provenance information on *E. globulus* was strongly put by R.C. Kellison of North Carolina State University in an unpublished report after a study tour in 1983 on the future of Portuguese forestry. 'It is incredulous that a species of such economic importance as *E. globulus* has never been provenance-tested. Despite the adaptability of that species to local conditions, the original seed sources probably came from a limited number of progenitors. The "land race" that has developed may or may not have been from the best provenance, and almost surely the genetic base is small and inbreeding is large.'

The reason that no range-wide provenance trials were planted before 1976 seems to have been partly due to a prevailing attitude that for *E. globulus*, and other species with a relatively small natural range, the natural origin is probably unimportant. However, there is a strong possibility that *E. globulus* has developed considerable genetic differentiation as yet not fully tested. It has a range of 600 km north to south and 300 km east to west in numerous genetically isolated natural stands, some of which differ markedly in soil and water relations.

10.4.1 Growth rate

Among the 20 or more blue gum provenance trials which used seed of Orme's 1976 collections and were planted between 1977 and 1985, the results of only two have been formally published (Volker and Orme 1988). Preliminary reports on the results of some of the other trials are available (Cannon 1982; Orme 1983; Butcher 1986) and further information has been obtained by correspondence with the investigators responsible for trials in Australia and

Fig. 10.5. Twenty-year-old *E. globulus* in a CELBI progeny test near Fouradoro, Portugal (photo K.G. Eldridge).

Portugal. As with provenance trials of other eucalypt species, there was the problem of too few families (five) in some provenances to estimate precisely the potential performance of the population which each provenance collection represents (Table 10.1). Further, as it was not possible to decide whether change in rank from site to site was due to experimental error or to provenance–environment interaction, the possibly important effect of interaction was, for the present, ignored. The reservations about making inferences from these trials are common to provenance trials of many other eucalypts.

Caution is needed before placing too much reliance on early results for ranking of particular provenances, especially when the statistical significance of the differences in growth between provenances is not high. To take an extreme example, at the Scottsdale trial the Seymour provenance dropped from 4th for height at 4 years to 19th for diameter at 7 years, out of the 21 provenances of *E. globulus*. In general, however, there were few large changes in rank of the provenances from 4 years (height) to 7 years (diameter) at the Scottsdale and Geeveston trials (Volker and Orme 1988).

No single provenance of *E. globulus* was consistently outstanding over the range of sites. However, the Geeveston, Channel, King Island, and Otways provenances were among the best, or at least above average, in several trials. Also the King Island provenance was notable for superior stem form in several trials, despite the five parent trees having poor stem form.

The provenances used in the Geeveston and Scottsdale trials are shown in Table 10.1 in groups designated by Orme (1983, 1988). Orme's Bass Strait group of *E. globulus* includes the Otway provenance in southern Victoria and the two provenances on the west coast of Tasmania. Diameter data for age seven years have been used to calculate the group averages shown in Table 10.2.

The difference between the three *E. globulus* groups is small and there is no clear evidence of a pattern of variation, except that the eastern group is somewhat inferior in growth rate (Table 10.2) and its flowering season is earlier (Section 19.2.2). *E. pseudoglobulus* was similar in growth rate to *E. globulus*, and was included in the Bass Strait group by Orme (1988). *E. bicostata* and *E. maidenii* are distinctly inferior in growth and this is also the case in every other trial, even at a cold dry, interior site in Portugal (M.H. Almeida, personal communication).

Where trials have included seedlots of land races from Portugal, Spain, California, or Colombia the land races have approximately the same average growth rate as the newly introduced, 'wild' Australian provenances. For example, in a trial in the central mountain area of Portugal at one year, 12 Tasmanian provenances and nine Portuguese and Spanish land races both had an average height of 1.7 m (M.H. Almeida, personal communication). In another trial at Furodouro near the coast in central Portugal planted in 1980, four Australian provenances from the Orme collection were compared with samples of the Portuguese and Californian land races. There was little difference between the six groups in diameter at eight years (Table10.3) and again no

Table 10.2
Diameter (dbhob cm) at seven years for groups of E. globulus *provenances at two trials in Tasmania (Volker and Orme 1988).*

Provenance group	No. provenances in group	Geeveston (cm)	Scottsdale (cm)
E. globulus			
Eastern Tasmania	6	9.6	11.9
Southern Tasmania	8	11.2	12.5
Bass Strait	6	11.1	12.4
E. pseudoglobulus	2	9.1	12.2
E. bicostata	3	8.1	9.6
E. maidenii	5	7.0	9.0

Table 10.3
Diameter (dbhob cm) at eight years in a E. globulus *provenance trial at Furadouro in Portugal in which family identity was retained (data from M. Brindbergs, Celbi).*

Seed source	No. of families	Diameter (cm)
King Island	4	12.4
Otways	4	13.1
Macquarie Harbour	4	12.4
Henty River	5	12.5
California	20	11.6
Portugal	bulk	12.4

superiority of the land races to the natural provenances.

A trial planted by APM Forests in 1971 on the northern side of the Strzelecki Ranges in Victoria contained 18 *E. globulus* provenances. The seed collections were made by T.G. Walduck. They were mainly from the Huon Valley region of south-eastern Tasmania, one from north-eastern Tasmania, and one from the Otway Ranges in Victoria. It was a well-designed trial with ten replications of ten-tree row plots. At 16 years there was substantial variation between provenances in stem volume and stem and branch form with the high-rainfall Huon region generally superior (Appleton *et al.* 1988). Walduck's Huon Valley collections were between Orme's Geeveston and Denison seed collection sites (see Table 10.1 and Fig. 10.4).

These observations are supported by early results from three smaller provenance trials. In Victoria two trials were established in 1976 with 10 seedlots, most of which were common to the main Tasmanian trials (L.A. Pederick, personal communication). One trial was on irrigated land in northern Victoria, and the other near Yarram at the eastern edge of the Strzelecki Ranges. At both sites at four years *E. bicostata* and *E. maidenii* grew relatively slowly. The two Victorian *E. globulus* seedlots from the Otway Ranges and from near Yarram grew as well as, or better than the Tasmanian collections from King Island, Bruny Island, and Uxbridge. The King Island provenance also performed very well in another small, well-replicated trial, planted in northern Tasmania in 1973 (Orme 1983). Three 'land race' seedlots from India, Italy, and Portugal all grew poorly.

The growth of *E. bicostata* and *E. maidenii* was inferior to *E. globulus* in every trial for which information has been obtained in Tasmania, Victoria, Western Australia, Colombia, and Portugal. However, *E. maidenii* is commonly planted at about 2000 m altitude in South China and in Kenya and Rwanda. In these two Central African countries *E. maidenii* is preferred to *E. globulus* and it has outperformed *E. globulus* in species trials (A. Kalinganire, personal communication).

The King Island provenance has shown a number of distinctive characteristics in provenance trials. It had the best height in some trials, though average in others, consistently good stem straightness and branch quality in all trials. Also it has distinctive bark colour being olive or dark yellow rather than the normal grey-green (Orme 1988; T.B. Butcher personal communication).

The 1976 seed collection on King Island was from only five trees, one of which had slow-growing progeny. This family ranked 69/73 in a 1985 trial in Western Australia (R. Mazanec, personal communication) and 15/18 in a similar trial planted in 1980 in Portugal (M. Brindbergs, personal communication). Further collections in 1987 and 1988 obtained small amounts of seed from another 46 trees most of which were open-grown and of poor form. However, early records (Campbell 1888) indicate that parts of King Island formerly had a dense forest of very tall trees of *E. globulus* which was destroyed by settlers' fires as the island was cleared for agriculture. Although most of the small remnant populations which regenerated after the fires are not tall or dense they are likely to retain genes of value for plantations. Further results of well-designed provenance trials are needed before deciding whether the King Island provenance of *E. globulus* is of special importance.

In summary, the state of knowledge of provenance variation in growth rate of the four species of blue gum after the first ten years of field testing was as follows.

1. No provenance of *E. globulus* was outstanding for consistently fast or slow growth in field trials, although Geeveston, Channel, King Island, and Otways were above average in several trials.
2. Among three provenance groups of *E. globulus* the Bass Strait and southern Tasmanian groups grew somewhat faster than the eastern Tasmanian group.
3. The two provenances from the west coast of Tasmania had growth rates and other characteristics similar to the Bass Strait group.
4. The growth of two *E. pseudoglobulus* provenances was similar to *E. globulus*.
5. The growth of *E. bicostata* and *E. maidenii* was consistently much less than for *E. globulus*.
6. Land races from plantations in several countries grew well, but not better than provenances from natural stands of *E. globulus*.

10.4.2 Wood properties

Provenance variation in wood properties is being investigated in Tasmania. In the first stage of the investigation wood samples from mature trees felled in natural forests for seed collection (Orme 1978) were pulped by the soda process (Turner *et al.* 1983).

A higher pulp yield was obtained from the west coast of Tasmania than from any other samples of *E. globulus*. The pulp yield of the relatively rare west coast trees (average 55 per cent) was similar to *E. regnans*, a species of high reputation for pulping. Trees from five localities in the northern part of the east coast, where *E. globulus* is abundant and accessible to seed collectors, had inferior pulp yield (average 44 per cent). There was a positive correlation between soda pulp yield of these parent trees and rainfall of their natural environment. Also high pulp yield was accompanied by somewhat higher strength in the resulting paper.

There was no correlation of pulp yield with wood basic density or tree age (Turner *et al.* 1983). From these data from naturally occurring trees it is not possible to decide whether the variation has a genetic component or is simply due to the trees having grown in very different environments. In the second stage of the investigation wood samples have been collected from eight-year-old trees in Tasmanian provenance trials. Until these results are available it is not possible to say whether the west coast trees will produce better pulpwood in plantations than other seed sources of *E. globulus*.

10.5 Base populations for selection

It must be emphasized that the observations on the pattern of provenance variation are only preliminary and may change with future measurements and interpretation of field trials. However, if the pattern of variation is confirmed it seems that all *E. globulus* provenances are suitable for inclusion in *E. globulus* breeding populations where selection aims for faster early growth. Land races from Portugal, California, and Colombia and some *E. pseudoglobulus* provenances also appear suitable for inclusion in breeding populations. For selection for other traits, such as improved stem straightness, branching, wood properties, or frost resistance, it may be that certain provenances will prove to be outstanding when current investigations are further advanced.

As was noted in Sections 1.1, 1.3, and 3.4 the genetic history of the large plantations of *E. globulus* in Portugal, Spain, and other countries is obscure. These results strengthen the case for new introductions for improvement of exotic plantations of *E. globulus*, since the land races tested in the current provenance trials were not superior in growth rate to several of the natural provenances. Natural provenances, as discussed in Section 19.6.2, are handicapped by reduced growth due to neighbourhood inbreeding effects in natural stands. Seed collected from plantations with a broad genetic base of promising Tasmanian provenances could be expected to grow faster than the original provenance collection, and therefore faster than the land race.

The possibility remains that distinct and repeatable differences between provenances will be demonstrated in future. Should there be natural provenances of *E. globulus* much superior to the currently used Portuguese, Spanish, Chilean, or other land races, genetic improvement of seed for plantations will be quickly and easily achieved by importing several kilograms of seed from a large number of trees of the superior provenances and making seed production areas. Valuable seed production areas might also be established by heavily thinning some of the existing provenance trials.

In 1987 and 1988 the Australian Tree Seed Centre of CSIRO organized collections from 612 trees of 82 provenances throughout the range of *E. globulus* (Gardiner and Crawford 1987, 1988). Had there been a clear indication that some provenances were better than others the collections would have been concentrated in those areas. Such was not the case and therefore a range-wide collection was made. These and other collections have been used to establish large provenance-progeny trials, for example 12 ha in 1988 and 60 ha in 1989 in Western Australia (Butcher 1988) and 16 ha in 1990 in Ethiopia (information from J. Davidson). Such trials will be the basis for selection, breeding, and vegetative propagation by several forestry organizations in Australia, Ethiopia, Portugal, and Spain. By the late 1990s the same trials can be expected to add considerably to the information on provenance variation in *E. globulus*.

11

Eucalyptus grandis and E. saligna

11.1 Natural forests

Eucalyptus grandis, as recognized by Pryor and Johnson (1971), has its southern limit of distribution at Minmi near Newcastle, New South Wales, at 32°52'S (Fig 11.1). Distribution from there is almost continuous up the New South Wales coast into southern Queensland to latitude 26°S. Further north the species is absent until about 18°S. Its distribution in the northern area is patchy with populations on the Eungella Tableland near Mackay, near Ingham, on the Atherton Tableland, and on the Windsor Tableland at the northern extreme of its range, 16°S (Burgess 1983).

Most of the natural forests of *E. grandis* in northern New South Wales and southern Queensland are on coastal lowlands and hills with an altitudinal range from sea level to about 600 m. Also it does not extend more than about 100 km from the sea. The altitudinal range in the tablelands of northern Queensland is from about 500 to 1100 m, but there is a population 450 km inland at Carnarvon, shown on the *E. saligna* map.

Booth *et al.* (1988) have developed a climatic profile for natural occurrences of *E. grandis*. They used the BIOCLIM computer model (see Section 4.3.3) to estimate climates of the locations from which 107 herbarium specimens of *E. grandis* had been collected. These authors found that the 12 bioclimatic indices which they examined varied widely over the 107 Australian locations for which climate was calculated by interpolation. For example, the range of mean annual temperature was from 14° to 22°C, mean daily minimum temperature of the coldest month from –1° to 11°C, mean daily maximum temperature of the warmest month from 25° to 31°C, and mean annual precipitation from 690 to 2480 mm. The summer maximum rainfall is more pronounced in the northern part of the species range.

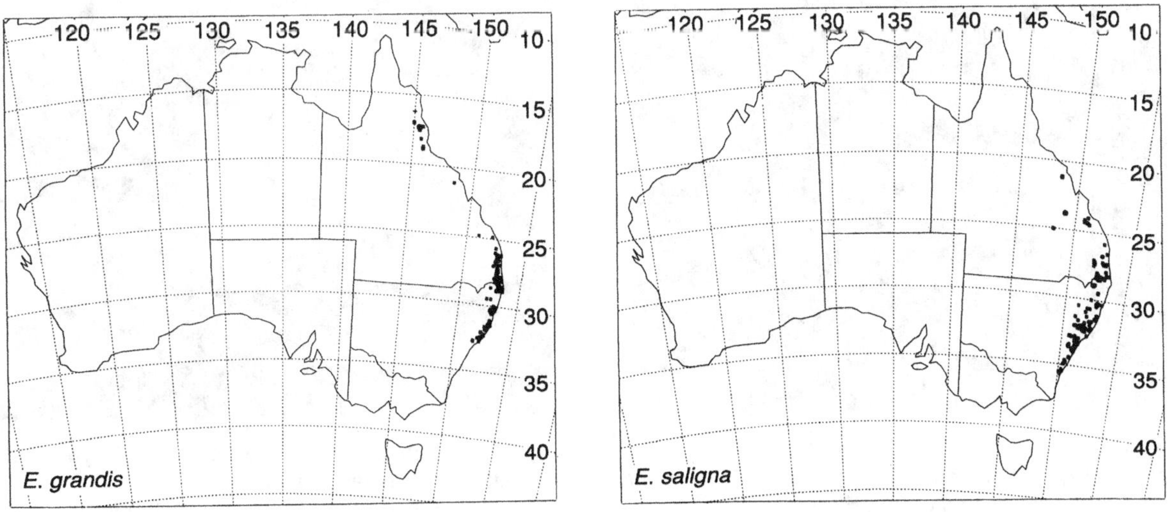

Fig. 11.1. Natural distribution of *E. grandis* and *E. saligna*.

Fig. 11.2. At the junction of rainforest with a tall open-forest of *E. grandis* near Atherton, north Queensland (photo G.L. Unwin).

Fig. 11.3. A pure stand of *E. grandis* with an understorey of rainforest species near Coffs Harbour, New South Wales (photo K.G. Eldridge).

Fig. 11.4. A large tree of *E. grandis* in a 40-year-old plantation at Westfalia Estates in the Eastern Transvaal, South Africa (photo K.G. Eldridge).

E. grandis grows on a range of soil types but generally occupies moderately fertile, lower valley slopes, often adjacent to rainforest (Fig. 11.2). However, it can occupy upper slopes and ridgetop sites, particularly on the western side of the Atherton Tableland near Herberton. It is not tolerant of flooding, despite its common name, flooded gum (Burgess 1983). *E. grandis* is often found growing in pure stands (Fig. 11.3) over an understorey of rainforest species, although it may also co-occur with *E. dunnii*, *E. pellita*, *E. pilularis*, *E. robusta*, *E. saligna*, or *E. tereticornis*.

E. grandis is a tall, rapidly growing eucalypt, reaching heights of up to 75 m, with smooth white or grey-white bark above a short rough basal stocking (Fig. 11.4). It belongs to the subgenus *Symphyomyrtus* (Pryor and Johnson 1971), and is very closely related to *E. saligna*, with which it was included until its valid description as a separate species in 1918 (Burgess 1983). The natural range of *E. saligna* (36°S to 24°S) overlaps with *E. grandis* in northern New South Wales and southern Queensland (Fig. 11.1). In a few stands the two species co-occur but generally *E. saligna* is found on less moist sites, upslope from the mesic lower valley slopes dominated by *E. grandis*. The two species are similar in appearance, being tall, smooth-barked trees, but *E. saligna* usually has heavier branching (Poynton 1979).

Despite their close relationship, there are no recorded hybrids of *E. grandis* and *E. saligna* either in their natural habitat or produced by hand pollination, although hybrids have been reported from plantations in South Africa and Florida (Griffin *et al.* 1988). Hybrids between *E. grandis* and a number of other species have been recorded. At the southern end of the range of *E. saligna* there are populations intermediate between *E. saligna* and *E. botryoides*, with rough bark covering much of the trunk (Brooker and Kleinig 1983).

11.2 Plantations

E. grandis has been favoured because of its ease of nursery care, rapid early growth, good form, and wood properties suitable for many uses (Hillis and Brown 1984). It will generally grow faster than any other tree species on good sites in subtropical and warm temperate climates (e.g. Barrett *et al.* 1975 for Zimbabwe) (Fig. 11.4). It is not suited to lowland regions of the humid tropics and is susceptible to disease in these environments (Evans 1982). Booth *et al.* (1989), using mapping based on the limits of climatic tolerance of *E. grandis* derived from its natural distribution, have correctly identified the areas in Africa where the species has succeeded in plantations (Section 4.3.3; Fig. 4.6).

Given a favourable site and good silvicultural treatment, volume yields in excess of 25 m^3 ha^{-1} yr^{-1} are often achieved (Poynton 1979) (Fig. 11.5). The timber of *E. grandis* is pale, relatively soft, and less dense than most of the eucalypts. It finds a wide range of uses including pulp for paper and rayon, charcoal, poles, and sawn timber for light construction (Jacobs 1981).

E. saligna has better coppicing ability than *E. grandis*. Its wood is denser and is suitable for general construction, flooring and pulp (Hillis and Brown 1984).

E. grandis is probably the most widely planted of the eucalypts for industrial wood production, with an estimated area of plantations of about 2 million hectares in 1987 (Burgess 1988). Most of these are in Brazil (over 1 million hectares) and South Africa (300 000 ha), but there are also large areas of plantations in Argentina, Australia, India, Uruguay, Zambia, Zimbabwe and other countries. Seedlots from the Coffs Harbour area have generally performed well (Burgess 1988).

E. grandis arrived in South Africa in the late 19th century, before it was distinguished from *E. saligna*, and there were mixtures of the two species in many of the planted stands (Poynton 1979). The *E. grandis* component of the mixture generally grew faster and seeded earlier than the *E. saligna*, so seed collections from the stands produced mixtures in which the proportion of *E. saligna* was reduced generation by generation. By now most of the trees in the plantation area of these two species are *E. grandis* (Burgess *et al.* 1985). A similar process is believed to have occurred in Zimbabwe, which appears to have received its first

Fig. 11.5. Pulpwood harvest from ten year old *E. grandis* in a plantation of Champion Papel e Celulose S.A. near Piracicaba, Brazil (photo K.G. Eldridge).

seed from South Africa (Barrett *et al.* 1975). Burgess *et al.* (1985) have concluded from studies of isozyme frequency patterns that the *E. grandis* currently being used in southern Africa probably had its original source in northern New South Wales or southern Queensland.

E. saligna is planted commercially in the North Island of New Zealand where it has performed better than *E. grandis* (Barr 1983). There are extensive plantations in many countries including Angola, Brazil, and Hawaii (Jacobs 1981).

11.3 Genecology

While *E. grandis* and *E. saligna* are easily distinguished in the southern part of their distributions, there are isolated populations in remote areas of central eastern Queensland, some more than 100 km from the coast at latitude 21°–25°S, which are difficult to assign to either species with certainty, being intermediate in their characteristics. A taxonomic investigation carried out by Burgess and Bell (1983) studied 24 populations taken from across the geographic ranges of the species, including 'core' populations clearly assignable to one of the two species and others which were difficult to assign to either (Table 11.1). They raised seedlings from all of the populations in a common environment, and measured a number of seedling phenotypic characters. Allozyme studies were carried out on bulked samples of germinants from each provenance. Adult characters were also determined from herbarium samples for some of the populations.

Analysis of the results using multivariate techniques showed that the core populations of the two species were distinctly different. There were some characters that distinguished the core *E. saligna* populations. Over 80 per cent of their seedlings were glaucous and had lignotubers; both these characters are completely absent in core *E. grandis* populations. Within the core *E. grandis* populations, there was no clear geographic pattern of variation in seedling characters. Seedlings of intermediate populations from central eastern Queensland were non-glaucous but had variable development of lignotubers. Their adult phenotypes, however, tended more definitely towards either *E. grandis* or *E. saligna*. Six of the fourteen allozyme loci

Table 11.1

Details of seed collections for provenance trials of E. grandis *and* E. saligna *at Coffs Harbour, northern New South Wales, and at Peachester, southern Queensland, from Burgess (1988); also some place names mentioned in the text.*

Location	State	CSIRO number	Latitude °S	Longitude °E	Altitude (m)	Distance from sea (km)	Number of trees
Atherton	Q		17 17	145 27	750		
Bellinger River	N		30 27	152 37	70	66	
Bellthorpe	Q	13022	26 49	152 45	140	40	10
*Blackdown Tableland	Q	13027	23 50	149 05	900	223	10
Buladelah	N		32 24	152 12	20		
*Carnarvon	Q	13263	24 57	148 03	1090	450	10
#Cascade	N		30 04	152 50	580	35	bulk
*Crediton	Q	12223	21 09	148 30	700	73	10
Danbulla	Q	12423	17 11	145 36	800	39	10
*Eungella	Q	13028	21 04	148 30	900	71	10
#Flat Rock	N	12974	35 20	150 16	45	20	9
#Gibraltar Range	N	13168	30 00	151 44	970	110	bulk
#Gladfield	Q	12226	28 13	152 05	500	124	bulk
Gympie	Q		26 13	152 40	100		
Herberton	Q	12970	17 21	145 27	1000	72	bulk
#Kenilworth	Q	13033	26 40	152 33	600	56	3
*Kroombit	Q	13026	24 23	151 00	860	101	10
Lorne	N	13017	31 37	152 43	40	12	10
Minmi	N		32 52	151 39	30	19	
Nulla Five Day	N	13018	30 48	152 49	150	20	10
Orara East	N	13019	30 13	153 02	135	11	10
Orara West	N	7823	30 10	153 00	18	18	12
Paluma	Q	13025	19 01	146 08	920	29	10
Pandanus	Q	12409	17 42	145 28	940	67	10
Peachester	Q		26 49	152 53	100		
Pomona	Q	12798	26 22	152 52	140	21	2
Ravenshoe	Q		17 36	145 29	913		
#Stewarts Brook	N	11045	32 00	151 19	1040	115	bulk
Tuckers Knob	N		30 22	153 00	110	17	
Wallingat	N	13016	32 22	152 28	80	30	9
Whian Whian	N	13021	28 31	153 22	475	18	8
Windsor Tableland	Q		16 15	144 59	1100		
#Wollemi Creek	N	7786	32 55	150 33	300	115	10
Woolgoolga	N		30 06	153 12	20		
Woondum	Q	13023	26 17	152 48	100	27	10
Zambia		10630			seed orchard		bulk

N New South Wales, Q Queensland

E. saligna, * = *E. grandis–E. saligna* complex, remainder are *E. grandis*.

The first four letters of the location are used to identify provenances in Table 11.2.

examined were also effective in discriminating between the three groups of populations. Although their affinities were not clear, it was considered that the isolated populations of central eastern Queensland might have potential in breeding programs.

11.4 Provenance variation

11.4.1 Growth rate

Early provenance trials of *E. grandis* suffered from being based on inadequate collections of seed, which did not cover the full geographic range of the species and sampled only one or a very few trees per provenance (e.g. Barrett *et al.* 1975; Roeder 1980b). The impression gained from these reports is that the results are inconclusive; there were either no significant differences between provenances, or, if significant differences were found, there was no obvious geographic pattern of variation (e.g. Pires *et al.* 1983; Rakotomanampison 1983; Darrow 1983c). In 1980, international trials of *E. grandis* were commenced through the auspices of IUFRO in 13 countries, using more comprehensive provenance collections made by the CSIRO Australian Tree Seed Centre (Pederick 1983).

There are three published reports of well-designed provenance trials based on appropriate sampling of natural populations of *E. grandis*, in Zimbabwe (Matheson and Mullin 1987), Hawaii (Skolmen 1986), and Australia (Burgess 1988).

Trials of *E. grandis* and *E. tereticornis* in Zimbabwe have been reported by Matheson and Mullin (1987). The trials are part of a local breeding program to improve the productivity, stem form, coppicing ability and frost resistance of *E. grandis.* Twenty four *E. grandis* provenances, including three Zimbabwean seedlots, were included in trials at three sites. The Australian provenances covered the full geographic range of the species. Provenances from Buladelah and Coffs Harbour in New South Wales grew fastest, and also scored very well on a selection index combining form and growth, although some from the Atherton Tableland in northern Queensland grew nearly as fast. The southern Queensland provenances performed poorly. There were significant differences among the four Atherton Tableland provenances separated by less than 25 km, showing that local variation can be important in *E. grandis.* A similar pattern occurred among northern New South Wales provenances, with the Woolgoolga provenance performing poorly in comparison with that from nearby Coffs Harbour. The authors suggested that such differences might be due either to differing selective forces, as there were marked soil and topographic differences among the Atherton Tableland sites, or to differing levels of inbreeding in the sampled populations. The three Zimbabwean seedlots performed significantly worse than the best wild Australian provenances, showing that any improvement in the local land races from the original seed introductions to southern Africa had not yet equalled the gains obtainable by selecting the natural populations most suited to the environment. Similarly, in an Australian trial Ades and Burgess (1983) found that *E. grandis* seed from a South African seed orchard produced slower growth than open-pollinated seed from several Australian plus trees.

E. grandis and *E. saligna* have been tested at four different sites in Hawaii (Skolmen 1986). Eleven seedlots of each species were tested, all of them from Australia except for one Hawaiian land race of *E. saligna.* Seven of the *E. grandis* seedlots were represented by nine or more parent trees. The Australian seedlots were taken from sites across the full latitudinal and altitudinal range of *E. grandis,* although the southern extreme of the *E. saligna* distribution was not sampled. Analysis of heights and diameters at four and five years of age showed that most *E. grandis* provenances grew faster than most *E. saligna* provenances. There were significant differences between individual provenances for both species. Provenances of *E. grandis* from near Buladelah and Coffs Harbour, New South Wales, and from Gympie in southern Queensland and Ravenshoe in northern Queensland did well at all four test sites, although there were some differences in relative performance between sites. The morphologically intermediate populations from central Queensland were ranked in the middle or towards the bottom for growth. Several southern *E. saligna* provenances were the worst performers at all of the sites. Although the trials were well replicated, with eleven replicates of four-tree row plots, the use of row plots meant that growth differences between provenances would tend to be exaggerated with increasing age because of between-row competition. The lower density of *E. grandis* wood (367 kg m^{-3} average, compared with 413 kg m^{-3} for *E. saligna*) reduced somewhat the advantage of *E. grandis* over *E. saligna* in terms of biomass production. Coppicing ability was about the same for both species, with an overall mean of 65 per cent of five year-old trees successfully coppicing after clear felling.

Australian provenance studies on *E. grandis* have been reported by Burgess (1988). Two trials were planted in 1982 on sites originally carrying *E. grandis* forest, one in the Coffs Harbour area and one at Peachester in south-eastern Queensland. These two trials included the same provenances as in the trials in Zimbabwe and Hawaii. They included twelve provenances of *E. grandis*, seven of *E. saligna*, five provenances from intermediate populations, and a seedlot from a Zambian seed orchard. Most of these provenances were represented by ten parent trees.

In an earlier set of trials, planted in 1972, 20 *E. grandis* provenances collected over a smaller latitudinal range in northern New South Wales and southern Queensland, including eight covering a range of altitudes in the Coffs Harbour area at latitude 30°S, were planted at two sites in the Coffs Harbour region. These trials used five parent trees from each collection site.

The provenance trials of *E. grandis* and *E. saligna* planted in 1982 demonstrated significant between-provenance differences in tree height and diameter at 3–4 years of age. Table 11.2 shows the mean diameters of the provenances, and the statistical significance of differences between them. It was notable that at Peachester, *E. saligna* performed well, having five of its provenances in the best nine for

Table 11.2

Provenance variation in E. grandis *and* E. saligna *in diameter (dbhob cm) at 4 years at Peachester and at 3 years at Coffs Harbour. Provenance names are explained in Table 11.1 (from Burgess 1988).*

Peachester age 4 years			Coffs Harbour age 3 years		
Casc#	13.15	a	Bell	6.38	a
Woon	12.89	a b	Whia	6.35	a
Gibr#	12.81	a b c	Woon	6.30	a b
Kroo*	12.79	a b c	Wall	6.17	a b c
Keni#	12.68	a b c d	Pand	6.13	a b c d
Bell	12.45	a b c d e	Danb	5.97	a b c d e
Woll#	12.43	a b c d e	Herb	5.94	a b c d e
Glad#	12.23	a b c d e f	Palu	5.91	a b c d e f
Danb	12.16	b c d e f g	Casc#	5.72	a b c d e f g
Null	12.11	b c d e f g	Pomo	5.64	b c d e f.g
Pomo	12.08	b c d e f g	Cred*	5.58	c d e f g h
Palu	11.93	c d e f g h	Kroo*	5.49	c d e f g h
Flat#	11.79	d e f g h	Lorn	5.43	d e f g h
Whia	11.67	e f g h	Keni#	5.39	e f g h
Cred*	11.57	e f g h	Orar E	5.23	f g h i
Eung*	11.43	f g h	Eung*	5.17	g h i
Pand	11.42	f g h	Gibr#	4.92	h i
Herb	11.34	f g h	Stew#	4.91	h i
Blac*	11.33	f g h	Flat#	4.58	i j
Orar E	11.32	f g h	Null	4.46	i j
Lorn	11.21	g h i	Gibr#	4.23	j k
Wall	11.02	h i	Woll#	3.94	j k
Carn*	10.38	i j	Zamb	3.73	k
Zamb	9.81	j k	Blac*	3.61	k
Stew#	9.42	k	Carn*	3.61	k

Means included in the same bracket do not differ significantly at the 5% probability level, Duncan's multiple range test

= *E. saligna*, * = *E. grandis–E. saligna* complex, remainder are *E. grandis*.

diameter growth, and three in the best six for height growth. This is contrary to the generally accepted view that *E. saligna* grows slower than *E. grandis*. At Coffs Harbour *E. saligna* did not perform well, the intermediate populations ranked in the lower half of the provenances, and the Zambian seed orchard near the bottom. The best *E. grandis* provenances at both sites came from a wide latitudinal range, including north and south Queensland and New South Wales. The Orara East provenance (CSIRO seedlot 13019) had significantly slower growth than the best provenances. This is of interest because Orara East is a commonly-used seed collection location in the Coffs Harbour area, indicating that some of the base populations sent to other countries may well have been sub-optimal. However, CSIRO seedlot 7823, collected only 6 km away from seedlot 13019 and at a similar altitude and topographic location, has performed well in provenance trials in other countries. The marked difference in performance of these two seedlots of *E. grandis* from within one small geographic area contrasts with the apparent absence of a broad clear overall pattern of geographic variation in this species.

In the two earlier Australian trials planted in 1972 there were significant differences in volume production between *E. grandis* provenances for both the altocline (the 8 provenances representing a range of altitudes in the Coffs Harbour area) and the latocline (representing a range of latitude). Since all the altocline provenances were collected from within an area of about 49 × 36 km, this result supports the conclusion of Matheson and Mullin (1987), that there is significant variation between provenance collections at a local level in the performance of this species, but no obvious local geographic pattern. Regressions of latitude and altitude against volume were not significant, and there was no suggestion of any general trend of productivity with provenance latitude or altitude. It should be noted that the altocline did not effectively sample the altitudinal range of *E. grandis* at the latitude selected: only one site higher than 200 m (Cascade, 580 m) was sampled. There was some indication that provenances located further from the sea grew less rapidly. Burgess (1988) compared the performance of particular *E. grandis* seedlots in trials in a number of countries. Certain seedlots such as Tuckers Knob performed well at most sites, others such as Orara East and Orara West were variable, and others such as Bellinger River and Minmi performed badly at most sites. The Minmi seedlot was collected from a number of isolated relict trees in an open field and its poor performance may well be due to a high degree of inbreeding.

Burgess concluded that selection of genetic material of *E. grandis* for provenance trials and breeding programs should concentrate on exploiting variation at a local level rather than search for the 'best' general area, as certain local provenances from across the broad geographic range have all performed well in the carefully conducted trials reviewed above. A rewarding strategy might be to plant a wide range of provenances and select the best individual trees for a breeding program regardless of their provenance. This seems a reasonable conclusion as far as most characteristics are concerned, although in the case of frost there is some evidence that inland, high-altitude provenances are best, and for resistance to fungal attack, provenances from north Queensland appear best (see Sections 11.4.2 and 11.4.3).

In New Zealand, where *E. saligna* is the preferred eucalypt species for the north of the North Island, a trial planted in 1976 included eight Australian provenances of *E. saligna* (Wilcox *et al.* 1983b). The *E. saligna* provenance from Barrengary Mountain (34°44'S) above Kangaroo Valley near Nowra in southern New South Wales had the best diameter growth at each of three test sites at age three years. Provenances from Bateman's Bay (35°42'S) in southern New South Wales and Coffs Harbour in northern New South Wales also grew faster than the New Zealand seedlots, and the Coffs Harbour provenance had the best score for stem form. The commonly planted New Zealand land race of *E. saligna*, 'Bartlett's' strain derived from the Silverdale plantation near Auckland, was much inferior in growth to the newly imported seed from natural forests; its average ranking was 21 out of 25 seedlots. Selection for superior timber quality in the land race seems to have been at the expense of growth rate (Wilcox 1980; Barr 1983). *E. grandis* provenances included in these New Zealand trials did not grow as fast as the better *E. saligna* seedlots, and were more susceptible to damage by the eucalyptus tortoise beetle (a chrysomellid, *Paropsis charybdis*).

11.4.2 Frost tolerance

Frost is one of the environmental factors limiting the latitudinal and altitudinal range of *E. grandis* in overseas plantations, and there are many records of severe frost damage in southern Africa, South America, and south-eastern USA. Planting criteria based on climatic conditions have been developed in Natal, South Africa. There, experience indicates that mean temperature of the coldest month should be no lower than 11°C and mean daily minimum temperature of the coldest month should be no lower than 4°C for optimum growth of the species: at sites with lower minima other eucalypt species will be more productive (Schönau 1984).

E. grandis and *E. saligna* have considerable capacity to harden so as to resist frost damage; Burgess (1984) has recorded the survival of young saplings of both species grown far outside their natural range at Canberra, Australian Capital Territory. The saplings were undamaged by a series of winter frosts, in which screen temperatures fell to about –8°C. These plants received several weeks of hardening brought about by night temperatures lower than 5°C prior to the most severe frosts.

Differential survival of nine *E. grandis* and five *E. saligna* provenances was observed by Mullin and Barnes (1977) in a provenance trial at a high altitude site in Zimbabwe. Many seedlings were damaged or killed by severe frosts six months after planting. *E. saligna* provenances survived better than *E. grandis*. The *E. saligna* provenance from near Kangaroo Valley, with 50 per cent survival, was superior to all *E. grandis* provenances. Two *E. saligna* provenances from northern New South Wales proved significantly less hardy, with percentage survivals of 24 and 18. Within *E. grandis* the higher altitude provenances, from northern Queensland, southern Queensland and New South Wales survived significantly better than low-altitude provenances from New South Wales. The percentage of seedlings surviving in individual *E. grandis* provenances ranged from 7 to 28 per cent.

Intermediate *E. grandis-E. saligna* from the Blackdown Tableland in central eastern Queensland survived frosts much better than other *E. grandis* populations in Brazil (report cited in Burgess and Bell 1983).

A breeding program has been underway since 1961 in Florida, USA, to decrease the cold damage suffered by *E. grandis*, which was grown for pulpwood (Franklin and Meskimen 1984). Severe damage and death, even in saplings and adult trees, are produced by freezing conditions associated with circulation of cold air-masses from the north of the continent in some winters. These advective freezes often follow a period of warm weather, so that there is little or no chance for frost resistance to develop since hardening requires low night temperatures. Both the 'freeze tolerance', the degree of damage scored immediately after a freeze, and the 'freeze resilience', a measure of how well the tree recovers its form and vigour during the subsequent growing season, have been used by Franklin and Meskimen as selection criteria in their breeding program, which also aims at improving volume production, form, and coppicing ability. They noted that rapid growth decreased the likelihood for damage from a radiation frost on a still night, as taller saplings raised their crowns above the coldest air near the ground surface. *E. grandis* was chosen as the principal species for improvement because of its rapid growth and ability to survive mild freezes, although it was the least hardy species examined (comparing plants of the same height).

Franklin and Meskimen's breeding approach for *E. grandis* has been to work with the best trees from as many natural provenances and other sources as could be conveniently obtained, rather than only one or two of the 'best' provenances, so as to maintain genetic diversity and retain as much promising breeding stock as possible. Seedlots from Australia, South Africa and other countries were screened in the breeding program. It was found that Australian and South African seedlots performed best on average of the first generation material for volume production and freeze tolerance. Two generations of selection had improved volume production by 84 per cent over the average of the first generation families, and increased the cold tolerance score by 15 per cent. (Trees were ranked on a 6 unit scale, 5 undamaged, 0 total crown kill). However, in subsequent trials that included the IUFRO seed collection (Pederick 1983) which was used for the 1982 provenance trials described in Section 11.4.1, the Herberton provenance from the Atherton Tableland had superior cold tolerance to the improved Florida material and only slightly lower height growth after one year (Franklin and Meskimen 1984). Provenance collections in Australia have not been made with a view to selecting the most cold-tolerant populations: seed collected from trees growing close to valley bottoms in high altitude locations, subject to the most severe frost regimes, might have proved even more cold-tolerant than the Herberton provenance.

Individuals which were judged to be hybrids between *E. grandis* and other species, *E. robusta*, *E. camaldulensis* and *E. tereticornis*, were found by Franklin and Meskimen to be more freeze-tolerant than pure *E. grandis* collected from the same *E. grandis* mother trees. The authors considered that field-verified laboratory testing of frost resistance would have greatly improved the speed of their breeding program, as some generations were not exposed to damaging field frosts capable of identifying cold-hardy material.

11.4.3 Disease

Campinhos and Ikemori (1978) found that *E. grandis* provenances from the Atherton region developed fewer cankers from the fungus *Cryphonectria cubensis* than did the more southern seedlots they tested. The superior resistance of the northern provenances was confirmed in a subsequent report (Campinhos *et al.* 1983). This difference may have evolved because of the higher summer rainfall and humidity in the northern part of the species' range.

A field trial in Belo Oriente, in the Brazilian state of Minas Gerais (Dianese *et al.* 1984b) tested susceptibility to the rust fungus *Puccinia psidii* (guava rust), which causes significant losses and growth reduction to a number of eucalypt species in three states of Brazil. A total of 23 seedlots from 13 eucalypt species was tested. Four replicates of 25-tree plots of four month-old seedlings of each seedlot were planted out in a cleared area within a *E. grandis* plantation that had proved susceptible to the fungus. The *E. grandis* land race from South Africa which had been used for the original plantation was markedly more susceptible to attack than were all other seedlots, being the only seedlot rated as highly susceptible. Ninety-nine per cent of its seedlings were infected, while percentages of infected seedlings for the other three *E. grandis* provenances ranged from 38 to 48 per cent, these provenances being rated as susceptible.

Dianese *et al.* (1984a) studied the response of a number of eucalypt seedlots to a disease of unknown origin, 'Mal de Rio Doce', which affects some 30 000 ha of eucalypt plantations in the Brazilian state of Minas Gerais. This disease is now the most damaging disease of eucalypts in Brazil. The study was carried out at the same location as the study on *Puccinia psidii* referred to above, and used a similar experimental design and the same range of species and seedlots. Cloned individuals obtained by vegetative propagation of a single resistant *E. grandis* tree from an Australian seedlot proved very resistant in the trial. The difference in disease symptoms (dead branches per plant and percentage of infected plants) between the resistant clone and the South African seedlot was statistically significant. The other two seedlots of *E. grandis* had intermediate damage symptoms but had grown faster than the resistant clone. The authors concluded that although some provenances of other species, *E. camaldulensis* and *E. pellita*, should be planted on the basis of their good growth and resistance to this and other diseases, selection of tolerant *E. grandis* clones also should be considered.

Many provenances of *E. grandis* have been included in provenance trials in Kerala state, India (Deo *et al.* 1986). Pink disease, caused by the fungus *Corticium salmonicolor*, causes cankers in the woody shoot tissues, and has severely affected pulpwood plantations of *E. tereticornis* in Kerala. It was noted that *E. grandis* plantings in the area were much less affected, so trials were established at several sites to identify provenances of *E. grandis* that were both resistant and fast-growing. Insufficient time had elapsed for the tests to be conclusive, but all provenances had so far remained free of infection and growth rates were promising.

11.5 Base populations for selection

For such a commercially important species, the base populations of *E. grandis* used for the first stages of breeding programs in a number of countries have been inadequate. However, the international program of provenance trials organized through IUFRO (Pederick 1983) has distributed a rangewide provenance collection to most of the countries where *E. grandis* is commercially important. Given the irregular patterns of significant local variation in performance reviewed above, further worthwhile gains can be expected from the recent introduction of additional Australian collections to these countries. Of particular interest are the little-tested low-altitude tropical populations (for example southwest of Cairns at an altitude of 450 m) and the high-altitude populations, up to 650 m, in the Coffs Harbour district, which may be expected to be of value for disease resistance and frost resistance respectively.

Advanced breeding programs for *E. grandis* are in progress in several countries. At Aracruz, in Brazil, where breeding has concentrated on vegetative propagation of outstanding clones, more than 300 seedlots of *E. grandis* have been introduced since 1977. These progenies of individual trees, mainly from the Atherton area (Doran 1978), are the basis of a program of recurrent selection providing select parents to the Aracruz program of interspecific crosses, mainly with *E. urophylla*. The hybrids will be more resistant to the fungus *Cryphonectria cubensis,* and produce wood of higher density, than pure *E. grandis* (Campinhos and Ikemori 1989).

The breeding program in Florida (Franklin and Meskimen 1984) has been reviewed in the frost section, 11.4.2. Breeding in South Africa (Van Wyk 1985b) has aimed to improve stem form, log-end splitting, and other wood characteristics as well as to improve volume production. This program has made use of controlled crosses of plus trees selected in the South African land race. Efforts have been made to expand the base population using seed collected in natural forests in Australia. Breeding objectives now include improved clonal performance, so that genetic gains can be quickly transferred to commercial use through vegetative propagation. Interspecific hybrids are being made of *E. grandis* with *E. camaldulensis*, *E. tereticornis*, and *E. urophylla* to combine desired characteristics of the parents such as adaptation to drought and to increase wood density (Van Wyk 1985a).

A more detailed investigation of natural variation of *E. grandis* in the Coffs Harbour area commenced in 1990 (I.P. Burgess personal communication). This involves a very intensive sampling of 19 sites between sea level and 600 m, and resampling of the same trees in subsequent years. Isozyme studies will provide information on the spatial and year-to-year variation in outcrossing rates, and field trials will assess the effect of outcrossing rates on progeny performance. Such information should resolve some of the yet-unanswered questions about the nature and causes of local variation in the species.

Conservation status of the remaining natural forests of *E. grandis* and *E. saligna* is reasonably good, but many local areas have been completely or partly cleared for farmland, or for eucalypt plantations some of which did not use the local seed source and have therefore led to the loss or contamination of local provenances.

12

Eucalyptus nitens

12.1 Natural forests

This fast-growing and relatively uncommon species of the higher parts of the mountains of south-eastern Australia is remarkable for extreme provenance variations over a limited environmental range. Its disjunct distribution is in a large number of small populations from 38°S in central and eastern Victoria to 30°S in northern New South Wales (Fig. 12.1).

The range of altitude of *E. nitens* is 670 to 1280 m in Victoria in the two main regions of occurrence, and up to almost 1600 m in northern New South Wales where the species is found near Ebor and Barrington Tops. In southern New South Wales there are three small stands along the Great Dividing Range. The largest single stand is on the Errinundra Plateau in eastern Victoria close to the New South Wales border. The largest total area of the species is the sum of many small stands in three areas in the eastern part of the Central Highlands of Victoria, designated by Pederick (1979) as the Toorongo, Rubicon, and Macalister provenances (Fig. 12.2, Table 12.1).

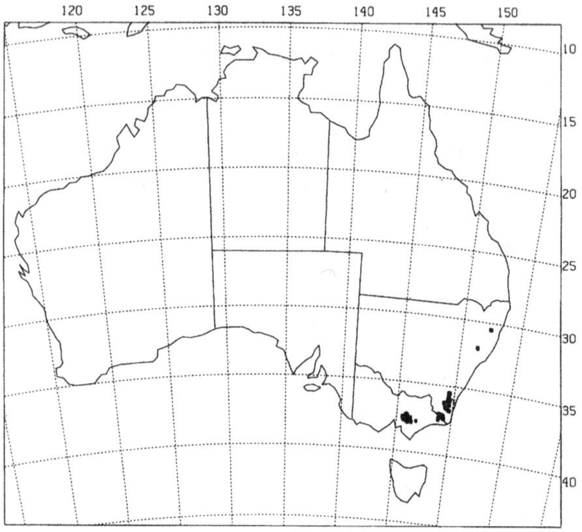

Fig. 12.1. Natural distribution of *E. nitens.*

Fig. 12.2. Detail of the natural distribution of *E. nitens* showing the largest total area of the species composed of the Rubicon, Toorongo, and Macalister provenances, and the largest single stand on the Errinundra Plateau (after Pederick 1979).

Table 12.1

Provenance regions of E. nitens *mentioned in Chapter 12 and representative sites of seed collection (after Pederick 1977, 1979).*

Location	Latitude °S	Longitude °E	Altitude (m)
Rubicon provenance region			
Royston Road	37 26	145 54	1000
Toorongo provenance region			
Mount Toorongo	37 48	146 07	1050
Tanjil Bren	37 47	146 12	880
Christmas Creek	37 53	146 17	790
Mount Saint Gwinear	37 50	146 21	1150
Macalister provenance region			
Connors Plain	37 32	146 28	1260
Errinundra provenance region			
Gunmark Range	37 17	148 52	1050
Southern New South Wales provenance region			
Nimmitabel (Glenbog)	36 38	149 24	1220
Northern New South Wales provenance region			
Barrington Tops	32 00	151 30	1520
Ebor	30 24	152 29	1460

The natural forests of *E. nitens* are all in localities of high rainfall, almost free of drought, with fairly deep, fertile soils, and with frequent frosts (Fig. 12.3). However, it is not an environment exclusive to this species. In the south the high altitude stands of *E. regnans* and some of the more sheltered stands of *E. delegatensis* are on sites which appear to be the same as those on which *E. nitens* grows nearby. Similarly in New South Wales, the higher altitude stands of *E. fastigata* occupy many sites apparently suitable for *E. nitens*. Chance, in the form of availability of seed, or lack of it, at the time of natural regeneration after wildfire, evidently plays a role in determining which of several eucalypt species occupies each particular piece of the mosaic of similar environments over the countryside.

E. delegatensis, E. fastigata, and *E. regnans* are members of the ash group of eucalypts (series *Obliquae* of Pryor and Johnson 1971) while *E. nitens* is one of the southern blue gums (series *Viminales*, subseries *Globulinae* including *E. globulus* and *E. cypellocarpa*). However, because the wood of *E. nitens* is similar to wood of members of the ash group in having low density, light colour, and few kino veins, the Australian sawmilling industry markets timber of *E. nitens* with the ash group.

12.2 Plantations

From its brief history as a plantation species *E. nitens* has a reputation for frost hardiness and extreme early vigour over a range of moist, cool sites. Also, if young trees are harvested in spring, they have some limited coppicing ability, unlike *E. regnans* and *E. delegatensis* which do not coppice. Enthusiasm for planting *E. nitens* is tempered by shortage of seed, some concern about persistent branches, and by loss of growth from defoliating insects. New plantings of *E. nitens* in New Zealand decreased to almost zero in the late 1970s because of defoliations by a chrysomelid beetle, *Paropsis*

Fig. 12.3. *E. nitens* on the Errinundra Plateau, Victoria (photo K.G. Eldridge).

charybdis (Wilcox 1980). In Tasmania defoliation by *P. charybdis* and damage by coreid shoot-wilt bugs and other insects also reduce growth somewhat, but not enough to inhibit planting for pulpwood (Elliott and de Little 1984). The annual planting area of *E. nitens* in northern Tasmania has exceeded 2000 ha since 1985 (Tibbits 1986), much of it replacing the slower-growing natural forests of *E. delegatensis* after clearfelling for pulpwood and sawlogs. In South Africa plantations are being extended in the Eastern Transvaal highveld for mine timbers and pulpwood.

12.3 Genecology

A distinct pattern of variation in juvenile leaf shape of *E. nitens* was observed by Pederick (1976a, 1977, 1979). Shepherd *et al.* (1976) found that leaf shape, but not leaf area, was a stable character not varying with the temperature at which the seedling was grown. Seedlings of the Errinundra provenance region (Fig. 12.2) did not contain stilbene, a fluorescent polyphenol (Pederick and Lennox 1979), which was present in almost all seedlings of other provenances. In field trials Pederick (1979) found that the Errinundra provenance could also be distinguished from the others by its change, after only one year, from juvenile foliage (glaucous, opposite leaves) to the adult form (green, alternate, petiolate leaves (Fig. 12.4) and by its adult leaves having a denticulate margin with gland-like structures. The other provenances continued to produce juvenile foliage for at least two years and in some

Fig. 12.4. Provenance variation in a three-year-old trial of *E. nitens* near Powelltown, Victoria. The juvenile foliage on the row of Rubicon plants (left) contrasts with the adult foliage on the row of Errinundra plants (right) (from Pederick 1979) (photo L.A. Pederick).

cases up to five years after planting. Of the Victorian provenances, those in which the juvenile condition persisted longest originated at highest altitude.

There was a discontinuity in juvenile and adult leaf characters of *E. nitens* between eastern Victoria (the Errinundra provenance region) and further west in Victoria (the Toorongo, Rubicon, and Macalister provenances). However, from eastern Victoria to northern New South Wales there was continuous (clinal) variation in juvenile leaf shape from narrow lanceolate to broad lanceolate.

The Errinundra-like characters were also found in the progeny of a proportion of the trees in three separate stands on the south-western edge of the Toorongo provenance area in Victoria, Mount Toorongo, Tanjil Bren, and Christmas Creek.

Varietal status was proposed by Pederick (1979) for the Errinundra form, together with the three Errinundra-like stands in the Toorongo area, as *E. nitens* var. *errinundra,* while the rest of the species was to be called *E. nitens* var. *nitens*. There is a proposal to make the Errinundra form a separate species, *E. denticulata* (Cook and Ladiges 1991).

12.4 Provenance variation

12.4.1 Growth and form

Provenance trials representing almost the whole natural range of *E. nitens* were established in Victoria at latitude 37° to 38°S in 1968, 1972 and 1975 (Pederick 1977, 1979). The 1968 trial gave the first evidence of the striking difference in appearance of three-year-old trees of the eastern and western varieties in Victoria (Fig. 12.4). The 1972 trial on two sites contained three bulk seedlots and 53 open-pollinated familes representing six provenance regions: Rubicon, Toorongo, Macalister, Errinundra, and southern and northern New South Wales (Table 12.1, Fig. 12.2). The 1975 trial planted on one site concentrated on the Toorongo provenance with five trees from each of eight locations around the perimeter of the area shown as Toorongo in Fig. 12.2.

Early growth rate of *E. nitens* appeared to be strongly controlled by foliage type and the age at which adult foliage appeared, since the fast-growing provenances at five years were those with the persistent form of juvenile foliage (Pederick 1979). The ranking at three and five years was maintained at 12 years (Pederick 1985). However, prediction of future performance from height in the first year, when all provenances have juvenile leaves, is not reliable due to the combination of different times of transition to adult foliage and the slower growth rates with adult foliage. Indeed the Errinundra provenance was among the tallest at the end of the first year in trials in Victoria and California but its ranking for height growth declined in the following years (Pederick 1977; Ledig 1983).

In the Victorian provenance trials and in similar trials in Tasmania (D. de Little personal communication) and New Zealand (King and Wilcox 1988) the three central Victoria provenances Toorongo, Rubicon, and Macalister were the most vigorous after five years, followed by New South Wales south and north, and lastly Errinundra. The growth rate advantage of a persistent juvenile stage was very clear, but this was accompanied by heavier and more persistent branches.

The Toorongo provenance of *E. nitens* has been the most consistent in growth rate, and has produced slightly better branching and straighter stems than the Rubicon and Macalister provenances in Victorian trials (Pederick 1985). Exceptions are the collections from Mount Toorongo, Tanjil Bren, and Christmas Creek which include individuals with Errinundra-like characteristics and slower growth.

Several provenance trials of *E. nitens* have been conducted in South Africa and Zimbabwe between latitudes 18° and 30°S in the summer rainfall zone. All the trials were at high altitude sites where other fast-growing eucalypts are damaged by frost but *E. nitens* is generally not damaged.

In all the trials in South Africa and Zimbabwe the New South Wales provenances grew faster than those from Victoria, in contrast to the results in Victoria, Tasmania, and New Zealand.

The six provenance regions of *E. nitens*, Toorongo, Rubicon, Macalister, Errinundra, southern New South Wales, and northern New South Wales represented by 27 seed sources were tested in two trials in South Africa at 29°S. For the first two years the Victorian provenances from Toorongo, Rubicon and Macalister showed the best height growth but by age five Rubicon and Macalister were the worst (Nixon and Hagedorn 1983). At age five the most vigorous provenances were those from New South Wales with little difference between the northern and southern sources. A local land race collected at Jessievale State Forest was similiar to the New South Wales sources in growth, morphology, and frost resistance.

Trials involving seven New South Wales and five Victorian provenances were established in 1980 at two sites in the Eastern Transvaal highveld. After three years at the Jessievale trial site (26°S, altitude 1733 m), the provenances from New South Wales had slightly better rates of survival than the Victorian provenances, but the difference was not significant. However, the New South Wales provenances were significantly better than those from Victoria for height, diameter, mean tree volume and volume per hectare (Darrow 1983a). At four years, the mean height of the New South Wales provenances was 14 per cent greater, the diameter 20 per cent greater and the mean tree volume 40 per cent greater than the mean for the Victorian provenances (Purnell and Lundquist 1986). At the Elandshoogte site (25°S, 1700 m) the advantage of the New South Wales provenances was more pronounced, the corresponding figures being 44 per cent for height, 20 per cent for diameter, and 173 per cent for tree volume. Ranking of some individual provenances was not consistent between the two sites: for example Tallaganda, from southern New South Wales, ranked first for all growth traits at Elandshoogte but below the trial mean at Jessievale.

Nine provenances of *E. nitens* were established in replicated trials on two sites at Nyangui at latitude 18°S in Zimbabwe. Quaile and Mullin (1984) have reported the superior performance of the New South Wales provenances on both sites. On the high altitude site at 2135 m the New South Wales provenances averaged 23.5 m and the Victoria provenances 20.5 m at 13.5 years. On the lower site at 1880 m the corresponding heights were 29.9 and 24.9 m. This is in agreement with the results of the South African trials.

The difference in results between the southern African trials at latitudes of 18° to 26°S and those from Victoria, Tasmania, and New Zealand at 38° to 42°S are apparently due to the northern provenances being adapted to a rainfall pattern with a summer maximum, and the southern provenances to a winter maximum. In the South African trials, at least part of the reason for the poorer performance of the Victorian provenances is that they are more susceptible to fungal attack, as described in Section 12.4.4. There may also be a more direct effect of the summer rainfall climate favouring the growth rates of the northern provenances. The Bendoc provenance from the Errinundra plateau (Fig. 12.2), which was not affected by fungal disease because of its early development of resistant adult foliage, still performed poorly relative to New South Wales provenances (Purnell and Lundquist 1986). However, it was also a poor performer in most of the trials in the winter rainfall zones.

Australian meteorological data (Hall *et al.* 1981) indicate that Dorrigo near Ebor in northern New South Wales receives about 66 per cent of its annual rainfall in the six summer months November to April; further south at Bombala in southern New South Wales and Bendoc in the Errinundra area of Victoria rainfall is uniformly distributed with 50 per cent in summer; and 200 km to the west the Toorongo and Rubicon provenances fall in the winter rainfall zone, Tanjil Bren and Warburton receiving 42 per cent of their rainfall in summer. Temperatures at Nyangui, Zimbabwe (Quaile and Mullin 1984) are also more comparable with those recorded for northern New South Wales than for the Victorian parts of the range of *E. nitens*.

12.4.2 Wood properties

In contrast to the striking variation in leaf characters and growth rate, the geographic variation in wood properties sampled over the whole range of *E. nitens* was negligible both from a five-year old Victorian provenance trial and from old trees in natural forest (Nicholls and Pederick 1979). Nor did the detailed studies of several wood characteristics in older trees in the Victorian trials reveal provenance differences of any consequence (McKimm 1985a,b; McKimm and Ilic 1987). However, in the two provenance trials in Zimbabwe, Quaile and Mullin (1984) found that the New South Wales provenances had slightly lower basic density than those from Victoria. In South African trials Hagedorn (1983) found no geographic trend in basic density, but distinctly less spirality of grain of the New South Wales provenances and of the local Jessievale seedlot. Spirality of grain of trees of the Errinundra and Macalister provenances at eight years was so extreme that only a small proportion of trees met the standards for poles.

12.4.3 Frost tolerance

An indication of considerable provenance variation in frost tolerance was obtained from a field trial at a cold site in the south eastern Transvaal by Nixon and Hagedorn (1983). They found that 46 per cent of 18 month old trees of the Errinundra provenance were severely damaged by frost, while only 2 to 3 per cent of the trees from northern and southern New South Wales and a local South African seed source were severely damaged. The Toorongo and Rubicon provenances were intermediate in their damage levels. Although this trial was well replicated, each provenance was represented by only two to eight families.

Twelve provenances of *E. nitens* were included in a provenance trial of 15 eucalypt species in Natal province, South Africa, reported by Nixon (1984). Damage caused by natural frosts in the first winter was assessed on a subjective scale of 0–4 and expressed as a percentage of the maximum possible score. Two provenances from northern New South Wales proved most resistant, with a mean damage of 30 per cent. Six provenances from southern New South Wales averaged 54 per cent damage, while four Victorian provenances averaged 66 per cent. The Toorongo provenance received most damage, as was the case at the Transvaal trial site. Again, several of the provenances were poorly represented, with four or fewer families.

A comprehensive survey of genetic variation in the frost hardiness of *E. nitens* was undertaken by Tibbits and Reid (1987), who examined 99 individual-tree seedlots and two bulked provenance collections sampled across the species geographic and altitudinal range. The six broad geographic provenances defined by Pederick (1979), namely Rubicon, Toorongo, Macalister, and Errinundra from Victoria and the provenances from southern and northern New South Wales, were each sampled so as to give for most of these provenances five individual-tree seedlots from each of two, three, or four altitudes covering the altitudinal range of the provenance. Within each broad geographic provenance the difference between the highest and lowest altitudes sampled ranged from 190 m for northern New South Wales to 550 m for Rubicon.

One experiment used seedlings from all 101 seedlots raised in a glasshouse and then exposed for one week to a 12°C day/8°C night temperature regime in a controlled environment chamber. These seedlings did not harden appreciably and all provenances were severely damaged by artificial frosts of –4.1°C. Over the range –3.2 to –3.8°C, the Errinundra and southern New South Wales provenances were significantly less hardy than the others, sustaining more leaf damage and higher mortality and having poorer stem recovery from damage than the other four provenances.

Seedlings hardened outdoors in the winter in Hobart, Tasmania, were tested for frost hardiness using a technique in which excised leaf discs were subjected to frost in a freezing chamber. Damage was subsequently evaluated by the extent of leaching of cell electrolytes from the discs, and lethal temperatures causing death of the leaf tissue estimated for each seedling. Seven seedlings were tested from each of three families from both high and low altitude populations for all six provenances. Analysis of variance of lethal temperatures, which averaged –5.4°C, showed that differences between provenances accounted for 75 per cent of the variance, and between-population differences (i.e. between low and high altitude populations within a provenance region) accounted for only 8 per cent of the variance. The balance was accounted for by differences between families within altitudinal populations (3 per cent) and among seedlings within families (13 per cent). As was the case for unhardened seedlings, Errinundra and southern New South Wales were the least frost-hardy provenance regions, having significantly higher lethal temperatures than all other provenances except Toorongo. The difference in lethal temperature between high and low altitude populations was greatest for southern New South Wales, where it was 0.6°C; only in this provenance was it significant. However, it should be noted that the difference between the highest and lowest altitude families tested was less than 300 m in four of the six provenances, the exceptions being Macalister (390 m) and southern New South Wales (310 m).

The leaf-disc method was also used by Tibbits and Reid (1987) to follow seasonal hardiness changes of eight trees from each of eight families in a three-year-old *E. nitens* plantation. Lethal temperatures were above –4°C in late summer, and fell to below –10°C for the hardiest families in midwinter, when there was a 2.3°C difference between the most hardy family, from the low altitude Rubicon population, and the least hardy family, from the low altitude southern New South Wales population. Highly significant changes from season to season in the rankings of individual families were observed over a one year period of field hardiness monitoring.

Field trials were conducted at two frost-prone sites in northern Tasmania, Hampshire, and Racecourse Plains. Each trial included 97 *E. nitens* families in a randomized complete block design, with single-tree plots replicated 24 times. Seedlings were planted out in summer and assessed in autumn, before the onset of frost damage, and again in July after frequent frosts down to –7°C at Hampshire and –10°C at Racecourse Plains. There was only slight damage at Hampshire, and no significant differences among provenances or altitudinal populations. At Racecourse Plains, foliar damage from frost averaged 30 per cent and there were highly significant differences between the provenances, which accounted for 73 per cent of the variance. Southern New South Wales and Errinundra received most damage, as they did in the artificial frost trials. However, if only the high-altitude populations from each provenance were compared, differences between provenances were not significant. Similarly, considering only the hardiest family from each provenance, differences between provenances were also not significant. Within provenances, higher altitude seed sources generally received significantly less damage than did those from lower altitudes, and there were significant differences between individual families within provenances. Foliar damage in the hardiest families in the trial averaged less than 10 per cent, while damage in the least hardy families exceeded 50 per cent.

The results obtained by Tibbits and Reid (1987) are noteworthy because similar patterns of intraspecific variation in frost resistance were obtained using unhardened, partly hardened and fully hardened seedlings, and using both artificial frosting techniques and exposure to field frosts. This shows that screening of provenances and families within a species for frost resistance using artificial frosting techniques can be a reliable guide to field hardiness. Such screening, using young seedlings, can greatly advance selection programs as it produces an immediate result which is not affected by year to year variation in frost climate that may reduce or eliminate the information obtained by field trials.

12.4.4 Disease

A disease causing leaf spotting and premature defoliation of the juvenile foliage of *E. nitens* has been a major cause of the poor performance of Victorian provenances in the South African provenance trials (Nixon and Hagedorn 1983; Darrow 1983a; Purnell and Lundquist 1986). The fungus *Mycosphaerella nubilosa* has been identified as the cause of the disease (Lundquist 1983; Purnell and Lundquist 1986). In the trial at Jessievale State Forest, defoliation caused by the fungus was scored by visual assessment using nine damage classes ranging from 1 (0–3 per cent defoliation) to 9 (76–100 per cent defoliation). At age three years, mean provenance scores ranged from 1.8, or about 5 per cent defoliation (Ebor provenance from Northern New South Wales) to 6.4, or about 50 per cent defoliation (Mount St Gwinear, Toorongo provenance, Victoria). The Victorian means were significantly different from those from New South Wales provenances. The range from a minimum provenance score of 5.6 to a maximum of 7.3 was not as pronounced at age four years, but the difference between Victoria and New South Wales was still significant. Within provenance variation was large, with some trees from Victorian provenances and many from New South Wales showing good resistance. Breeding for disease resistance is seen as the only practical method for control, and the disease may become more significant as the area planted to *E. nitens* increases. Further investigations are planned to determine the heritability of resistance (Purnell and Lundquist 1986).

12.5 Base populations for selection

E. nitens has a reputation for small seed crops both in natural forest and plantations, and consequently commercial seed supply is difficult and expensive.

The initial introduction of *E. nitens* into South Africa provides a clear example of the deleterious effects of introduction and propagation from a narrow genetic base (R.C. Purnell personal communication; Purnell 1986). One seedlot collected from an unknown, but probably small, number of parent trees near Nimmitabel in southern New South Wales was received from Australia in 1930 and planted out at Jessievale State Forest. Seed collected after 11 years from this stand was used to establish seed stands on De Hoek State Forest, and seed collected from these trees was in turn used to establish seed stands at Nelshoogte State Forest. Seed from Nelshoogte produces very poor quality trees, many being stunted and slow-growing. Although the plantings at Jessievale and De Hoek were substantial (3000 and 2350 trees respectively), because of the late flowering and small seed crops of the species it is believed that only a few trees contributed to the seed collected from all three forests, so that three generations of inbreeding, subsequent to a restricted initial introduction, is the likely cause of the poor performance of the Nelshoogte seed.

For South African breeding programs, Purnell and Lundquist (1986) concluded that selection of provenances should be based on growth rate, stem form, wood quality, and disease resistance. Although the South African trials indicated that New South Wales provenances were more suitable than those from Victoria, they were unable to firmly recommend specific New South Wales provenances because of the small differences between them, and uncertainties caused by including only one or a very few parent trees in some of the provenances tested. Fresh collections were made in 1985 from the New South Wales provenances, sampling 10 to 20 trees per population and keeping the seedlots from individual trees separate, so as to provide an adequate base population for future testing and breeding.

The forestry subsidiary of an Australian company, Associated Pulp and Paper Mills (APPM), started plantations of *E. nitens* in north-west Tasmania on a pilot scale in 1974. By 1985 the Tasmanian annual planting of 2000 ha made *E. nitens* Australia's principal plantation eucalypt in terms of current annual planting (*E. diversicolor* in Western Australia was second). Almost all the seed for more than ten years of *E. nitens* planting by APPM was supplied by one seed collector in Victoria from the Toorongo provenance. He climbed selected trees with spurs and sawed off seed-bearing branches, an unusual and dangerous method of collection. Each seedlot supplied to Tasmania was from 10 to 20 large trees and usually less than 0.5 kg per tree. The trees were at least 100 m apart and no tree was climbed twice (K.W. Johnson personal communication). The plantation records of APPM show which seedlot is planted in each compartment. When the Company selected 200 plus trees for seed orchards and for vegetative propagation they were able to minimize inbreeding in their breeding population by limiting the number of plus trees selected from each seed collection (de Little and Tibbits 1988).

The combination of many small collections from the best provenance and good plantation records has provided Tasmanian foresters with an excellent genetic base for future selection of this species.

Logging in the natural *E. nitens* forests of the central Victorian and northern New South Wales provenances has almost stopped as most of the remaining stands in these areas are now included in national parks or equivalent reserves. Much of the natural *E. nitens* forest in southern New South Wales has been harvested for sawlogs in recent years, often with inadequate regeneration. At Errinundra, where harvesting of sawlogs may continue for some years it is government policy to regenerate with local seed, despite the evidence that Toorongo seed is likely to produce more wood (Pederick 1979).

13

Eucalyptus obliqua

13.1 Natural forests

Eucalyptus obliqua, the type species of the genus (Fig. 1.2, p. 6), is one of the most widespread and economically important eucalypts of south-eastern Australia. The natural range extends from northern New South Wales to southern Tasmania between latitudes 29° and 43°S and westward as far as Kangaroo Island, South Australia (Fig. 13.1). It is known as messmate stringybark and is an attractive component of the landscape in much of the moister hill country in the part of the continent where most Australian people live.

The main occurrences of *E. obliqua* are in the central and southern parts of Victoria and in Tasmania where it is common except on the west coast. In New South Wales it occurs on small areas of the northern and southern tablelands. Occurrences in South Australia are in the Mount Lofty Ranges near Adelaide, Kangaroo Island, and in the extreme south-east of the State (Boland *et al.* 1984). In the northern parts of its range it occurs at altitudes of 500–1200 m and descends almost to sea level in some southern locations (Green 1971). The upper altitudinal limit is about 750 m in Tasmania, 1000 m in Victoria, and 1200 m in northern New South Wales (Boland *et al.* 1984). *E. obliqua* occurs on a wide variety of soil types, with rainfall usually in the range 750–1250 mm without any extreme dry period, and in some places with moderate frosts and snowfalls.

On the most favourable sites *E. obliqua* is a tall straight tree of more than 80 m with rainforest species in the understorey. In some coastal areas it is a bushy forked tree of no more than 10 m with heathland species as the understorey (Fig. 13.2). In most of its range *E. obliqua* grows to about 30–40 m and produces good quality, general-purpose sawn timber and pulpwood (Fig. 13.3). It is also much used in rural areas for poles, posts, and firewood. Regeneration after logging, several thousand hectares per year, is mainly by natural seedfall, and coppice from stumps and lignotubers. In wetter areas

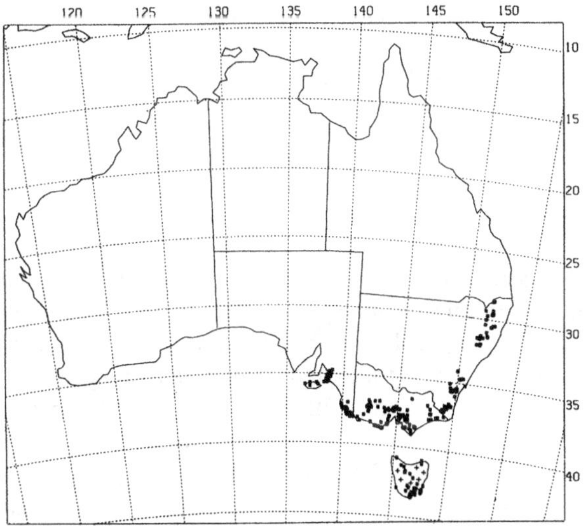

Fig. 13.1. Natural distribution of *E. obliqua*.

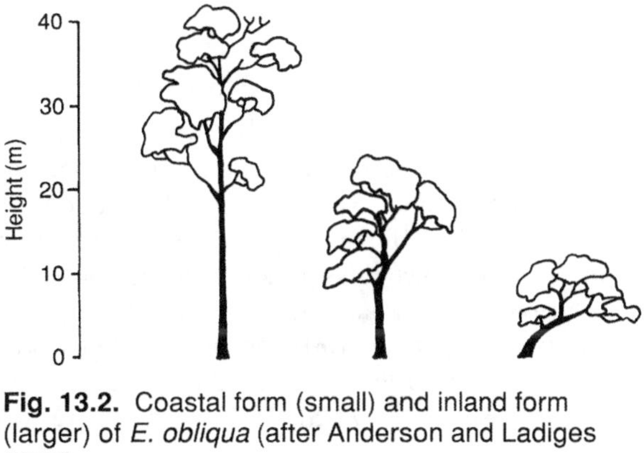

Fig. 13.2. Coastal form (small) and inland form (larger) of *E. obliqua* (after Anderson and Ladiges 1978).

Fig. 13.3. Natural forest of *E. obliqua* near Smithton, Tasmania (photo CSIRO).

regeneration is by direct seeding, or occasionally by planting seedlings, following clearfelling and burning the logging slash. Seed is usually obtained locally, acknowledging the prevailing concern to maintain these managed forests in as natural a condition as possible.

13.2 Plantations

E. obliqua is not a major plantation species anywhere in the world, in contrast to its great importance for wood production in Victoria and Tasmania. The reason may be its somewhat slower growth and greater sensitivity to frost compared with the usual alternative eucalypt species for moist and cool planting sites: *E. delegatensis*, *E. globulus*, *E. nitens*, *E. regnans*, and *E. viminalis*.

13.3 Genecology

With such extremes in the natural environment of *E. obliqua* one would expect genetic selection to have led to distinctly different ecotypes. Several research projects have shown this species is indeed one of the most geographically variable of the eucalypts.

A study of genetic variation in *E. obliqua* was commenced in 1966 by the Australian Forest Research Institute (now CSIRO Division of Forestry) as a guide to both taxonomic revision and practical tree breeding in this species (Green 1971). A widespread geographic sampling was carried out by collecting open-pollinated seed, botanical specimens and parental data at 22 provenance localities separated by at least one degree of latitude as shown in Figure 13.4 and Table 13.1. Not all major areas of occurrence were sampled, for example in north-east-

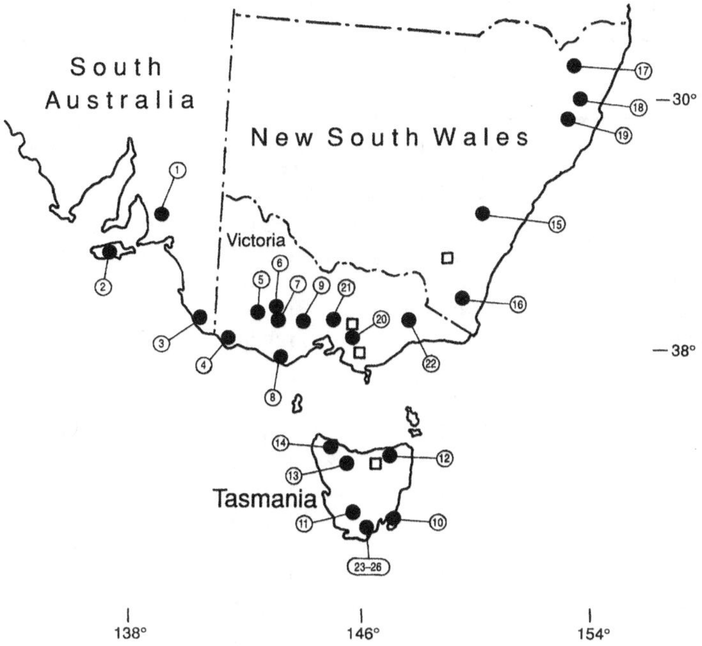

Fig. 13.4. Sites of *E. obliqua* seed collections (circles) and provenance trials (squares) (after Green 1971).

ern Victoria, the Strzelecki Ranges and Wilson's Promentary in south-eastern Victoria, and the type locality of the species on Bruny Island in south-eastern Tasmania. Local variation among populations within one provenance was sampled by one additional collection by K.C. Felton in 1968 in four stands of greatly differing growth rate located within 2 km of each other at Strathblane in south-eastern Tasmania.

Variation among provenances in morphology of fruit and adult leaves and among parent trees within provenances was statistically significant for several characteristics. In particular, the Furner provenance from the south-east of South Australia flowered several years earlier and had much larger fruit and seed than any other provenance, giving some support for the designation *E. obliqua* var. *megacarpa* Blakely for *E. obliqua* near Millicent, 30 km south of Furner. The Millicent trees have even larger fruit than those at Furner (A.C. Matheson, personal communication). The other four varieties proposed by Blakely (1934) were not supported by Green's studies.

When seedlings were raised in a greenhouse from these collections Green (1971) found that lignotubers were most frequent in the four provenances from South Australia and south-western Victoria. This observation was confirmed in field trials established in 1970, where the South Australian provenances were easily identified by the small size of the seedlings, bushy form, large lignotubers, and narrow yellowish-green leaves. In these trials the New South Wales provenances were also distinctive in the first few years because of their broad, thick, dark green leaves which were nearly horizontal. In contrast, leaves on seedlings of Tasmanian and Victorian origin were almost vertical. This striking difference in orientation of seedling leaves was noted by Barber (1965) (Fig. 13.5) who observed seedlings in the field. Only by bringing the different populations together in the one environment of a provenance trial was it possible to confirm that leaf orientation in *E. obliqua* is strongly inherited and not an environmentally determined character: seedlings of northern New South Wales origin did not produce pendulous, vertical leaves when grown in southern Victoria or Tasmania (K.G. Eldridge, observation).

Both continuous (clinal) and discontinuous (ecotypic) variation were noted by Green (1971). Subsequently Anderson and Ladiges (1978, 1982) and Anderson (1982a,b) made an intensive study which clearly established the existence of populations adapted to particular soil types (edaphic ecotypes) in

Table 13.1

Sites of E. obliqua *seed collection and provenance trials mentioned in Chapter 13. The numbers indicate these locations on the map in Fig. 13.4 (after Brown et al. 1976).*

	Location	State	Latitude °S	Longitude °E	Altitude (m)
17	Gibraltar Range	NSW	29 38	152 08	1000
18	Styx River	NSW	30 35	152 12	520
19	Yarrowitch	NSW	31 21	152 01	1160
15	Mount Werong	NSW	34 06	149 54	1160
16	Brown Mountain	NSW	36 37	149 29	370
	Bendoc	Vic	37 10	148 55	900
22	Swifts Creek	Vic	37 08	147 54	730
20	Powelltown	Vic	37 53	145 51	370
21	Broadford	Vic	37 19	145 12	450
9	Daylesford	Vic	37 26	144 12	700
6	Redbank	Vic	36 50	143 20	460
7	Mount Cole	Vic	37 16	143 12	700
5	Halls Gap	Vic	37 09	142 28	590
8	Lavers Hill (Otway)	Vic	38 40	143 21	270
4	Drik Drik	Vic	38 04	141 24	100
	Creswick[t]	Vic	37 26	143 54	540
	Macedon[t]	Vic	37 25	144 35	550
	Narbethong[t]	Vic	37 30	145 39	370
	Silver Creek[t]	Vic	38 16	146 16	150
3	Furner	SA	37 25	140 19	30
1	Mount Pleasant	SA	34 45	139 02	520
2	Parndana, Kangaroo Is.	SA	35 47	137 16	150
13	Nietta	Tas	41 22	146 04	300
14	Mawbanna	Tas	40 55	145 21	90
12	Forester	Tas	41 04	147 40	90
11	Maydena	Tas	42 43	146 32	610
10	Murdunna	Tas	42 57	147 53	150
23	Strathblane 50 m*	Tas	43 06	146 26	160
24	Strathblane 34 m*	Tas	43 06	146 26	85
25	Strathblane 29 m*	Tas	43 06	146 26	79
26	Strathblane 18 m*	Tas	43 06	146 26	73
	Lisle[t]	Tas	41 13	147 19	300

* Site index estimated as mean height of predominant trees at 50 years

[t] Trial sites

NSW New South Wales, SA South Australia, Tas Tasmania, Vic Victoria

E. obliqua. The seedlings they studied in experiments in a greenhouse and in field trials were raised from samples from three sites of contrasting soil type in the Otway district of southern Victoria. They collected seed from 15 trees at each site. One collection site had tall trees, high rainfall, and fertile acidic loam (pH 5.2). The other two sites 30 km away had smaller trees, lower rainfall, and less fertile sandy soil, one neutral (pH 6.5) the other alkaline (pH 8.5) (Fig. 13.2).

Germination of the provenance from acidic loam was severely reduced at high concentrations of calcium, whereas the provenance from the most alkaline soil maintained high germinability over a wide

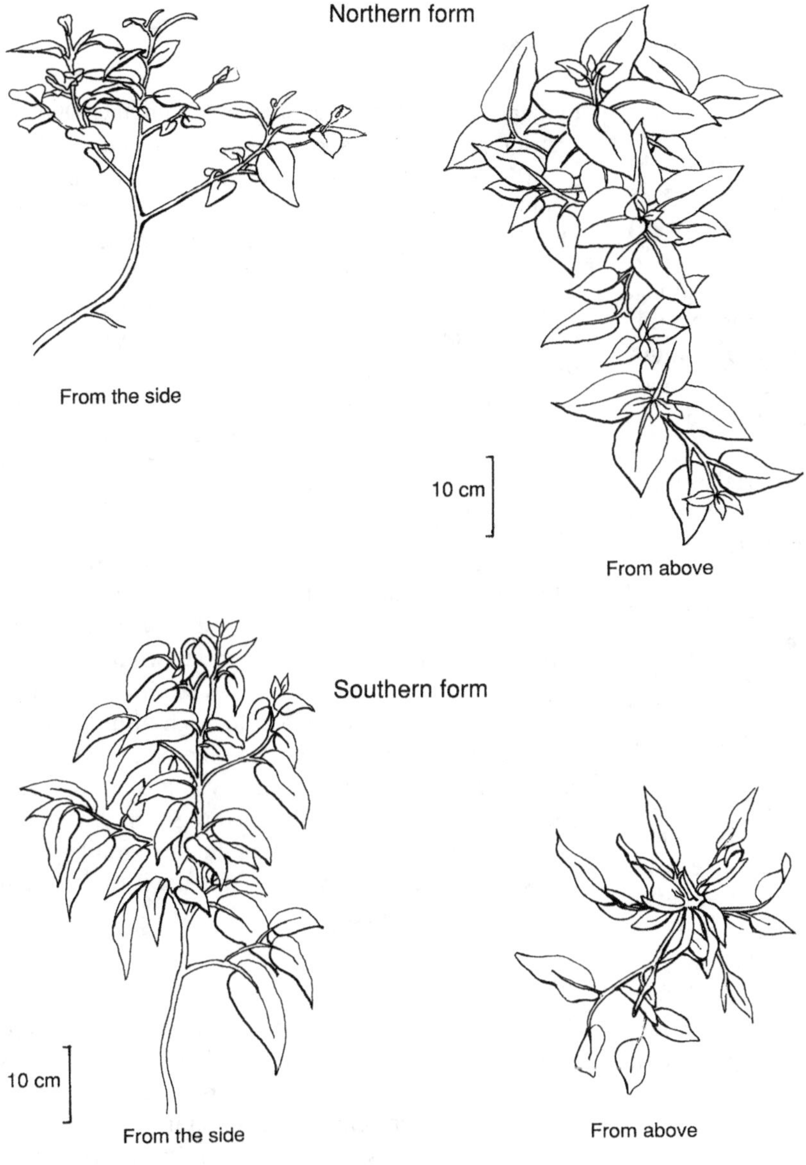

Fig. 13.5. Contrast in leaf orientation of seedlings of the northern form of *E. obliqua* (flat leaves, New South Wales) and the southern form (pendulous leaves, Tasmania) (redrawn from Barber 1965).

range of calcium concentrations (Anderson 1982b). When grown in the greenhouse on calcareous sands of pH 6.5 and 8.0, seedlings of the provenance from acidic loam showed marked reduction in growth and were severely chlorotic with red and necrotic patches on the leaves. The two provenances from neutral and alkaline sands were less chlorotic and grew almost as well on alkaline soil as on acidic soil (Anderson and Ladiges 1978). Seedlings of the three provenances were planted in trials at each of the three collection sites and studied for three years. The provenance from acidic loam again showed severe chlorosis, slow growth and low survival when grown on more alkaline calcareous sands, but was the fastest-growing population on fertile acidic loam (Anderson and Ladiges 1982).

13.4 Provenance variation

13.4.1 Growth rate

The first provenance study of *E. obliqua* aiming to find the best place to collect seed for plantations was established as a small trial near Creswick, Victoria, in 1959 (Pederick 1974). The eight seedlots were established in 12-tree plots in a trial with two replications. At 13 years the Powelltown provenance from central Victoria grew faster than provenances from Mount Cole in western Victoria and Bendoc in eastern Victoria. The local Creswick provenance was the slowest growing (Pederick 1976a).

Three long term cooperative field trials of *E. obliqua* were established in Australia in 1970 and 1971 (Brown *et al.* 1976). They were based on the collections of 22 provenances made by Green (1971), mentioned previously, and were the first range-wide provenance trials of the eucalypts in which all families were identified. The design was a cubic lattice of 216 families (6 × 6 × 6) with 6 replicates of 8-tree plots. Each trial contained more than 10 000 trees and occupied 5 ha. These trials have shown clearly that all the Tasmanian provenances grew well at each of the three trial sites and that the growth of the Lavers Hill provenance from the Otways district in Victoria was equal to or better than any of the Tasmanian provenances. The three South Australian provenances were the slowest growing (Brown *et al.* 1976; Matheson *et al.* 1986), so slow that by 18 years most trees of South Australian origin were either severely suppressed or dead. Correlations between seedling performance in the greenhouse and tree growth in the field were generally good (Brown *et al.* 1976). Lavers Hill, which was identified as the fastest growing provenance in the greenhouse, was also the best in the field trials. Similarly, Furner was the slowest growing in both cases.

Local variation, as tested by the four adjacent provenances from Strathblane, Tasmania, was small and not correlated with predominant height (mean height of predominant trees at 50 years) in the four parent stands, which were 50, 34, 29, and 18 m respectively. This finding is in contrast to the striking evidence for edaphic ecotypes within the Otways district referred to above (Anderson and Ladiges 1978). Evidently, genetic variation between local populations cannot be presumed to be present, or absent, until appropriate experiments have been carried out.

While there was statistically significant genotype–environment interaction in growth and survival over the three sites, the magnitude of the interactions was small (Brown *et al.* 1976). Not much growth would be lost by planting at any one site those provenances identified to be best at another site. Also the local provenance was not the best for growth at any of the three sites, all within the region of the natural range where the species grows well. Outside this region on sites which are harsher due to frost, drought or other factors, a generalization that the Lavers Hill and Tasmanian provenances are best would not be appropriate. For example, the 216 families were also planted in a small trial on a cold, high-altitude site at Bendora near Canberra, outside the natural range of *E. obliqua*. The provenances which survived and grew well were from northern New South Wales, not those which were successful on the three milder sites. Rankings at Bendora were not statistically correlated with glasshouse heights.

13.4.2 Wood properties

A study of variation in wood density was carried out when one of the trials, at Silver Creek, Victoria, was seven years old. Discs of wood from thinnings of 14 provenances were examined in detail (Nicholls and Matheson 1980). There were significant differences between provenances for bark thickness, proportion of heartwood, incidence of kino rings, and within-ring wood density. However, there was no geographic pattern to the variation in these characters.

When the trial at Lisle in northern Tasmania was thinned at age eight years samples were collected for a study of pulping characteristics (Matheson *et al.* 1986). Provenances which produced the fastest growing trees not only produced the strongest pulp but yielded more pulp per unit mass of wood. These provenances also were cheaper to pulp as they required a lower soda charge. Genetic correlations indicated that selection of fast growing provenances would lead to a correlated increase in pulp strength without the necessity for the expensive procedure of screening provenances for pulp strength. Variation between provenances in wood density was not statistically significant and, as in the study by Nicholls and Matheson (1980), there was no indication of a geographic pattern.

13.5 Naturally occurring hybrids

Several studies have been made of the potential of naturally occurring hybrids of *E. obliqua* and *E. regnans* as seed sources for plantations. These two closely related species overlap in both geographic distribution and flowering time (Ashton 1981). The common names 'gum topped messmate' and 'gum top' describe the greater proportion of smooth deciduous bark on the upper part of the main stem of these trees compared with the usual *E. obliqua* characteristic of thick, rough bark up to the highest branches. 'Gum top' is found over considerable areas in south-eastern Tasmania and in parts of the Otway district in Victoria where it is called 'Otway messmate'. Elsewhere trees with bark characteristics intermediate between *E. obliqua* and *E. regnans* are in smaller numbers, usually at the junction of stands of the two species (Ashton 1958, 1981; Ashton and Sandiford 1988; Ashton and Williams 1973). The seedling characteristics of F_1 hybrids made by artificial cross pollination by Ashton (1984) were similar to seedlings raised from suspected hybrids sampled in natural stands.

In trials established in Victoria at Macedon in 1955 and at Creswick in 1959, the Otway messmate seed source grew distinctly taller than other provenances of *E. obliqua* (Pederick 1974, 1976a) (Fig. 13.6). Another trial established in 1959 in Victoria, at Silver Creek, compared Otway messmate with three other suspected hybrid populations, and *E. obliqua* and *E. regnans* (Griffin and Eldridge 1980). At age 15 years the four intermediate populations had produced stands with intermediate values for tree size, height of the stocking of rough bark, bark thickness, and lignotuber frequency. However, the growth of the Otway messmate was equal to *E. regnans*. The wood characteristics of *E. regnans*, *E. obliqua* and intermediates were not significantly different in this trial (Nicholls and Griffin 1978).

Fig. 13.6. Provenance variation in *E. obliqua* in a 17-year-old trial at Macedon, Victoria. Trees on the right are Otway messmate and those on the left are of a local seed source near Macedon (photo L.A. Pederick).

The collection site of the Otway messmate used in this trial was close to the Lavers Hill collection site for the three large trials described earlier. It is possible the presence of *E. regnans* genes in the Lavers Hill provenance may have contributed to its good performance in all these trials.

13.6 Base populations for selection

In the three trials with 22 provenances of *E. obliqua* (Brown *et al.* 1976) the combination of provenance trial and progeny test provided an opportunity to make a quick start in a breeding program, though at a high cost for testing. Immediately the better provenances were identified selection could have commenced to choose the best trees in the best families of those provenances. Several years would have been saved compared with a two-stage strategy for testing and utilizing provenance variation (Griffin 1983) in which a provenance trial, without family identity, is carried out, and the progeny tests are established several years later with only the better provenances. Small quantities of seed of all families of the J.W. Green collection are still in store at CSIRO Division of Forestry, Canberra. In fact, no breeding program has yet developed for *E. obliqua* in Australia because there is no planting program.

In New Zealand, where there are several excellent small plantations of *E. obliqua* and some good reports of timber quality from farm trees (Wilcox 1980), 62 families selected from those in the three large Australian trials were established as a base population for a possible future breeding program. They were from five provenances which had the best growth at an early stage of the Australian trials, an example of the two-stage strategy suggested by Griffin (1983).

Conservation of the genetic resources of *E. obliqua in situ* could be considered to be reasonably adequate due to the widespread range of the species, its inclusion in many national parks and equivalent reserves, and its regeneration by natural seedfall in many logging operations in forest reserves. This is despite the loss of much of the natural forest by clearing for agriculture. However, caution is advisable in planning new seed collections of the better provenances for a breeding program as many of the present 'natural' forests are the result of either a century of disgenic selective logging or, worse, the recent use of non-local seed for regeneration by broadcasting seed after logging.

14

Eucalyptus regnans

14.1 Natural forests

Eucalyptus regnans is the second tallest tree species in the world, only exceeded in height by the coast redwoods of California (Ashton 1975b). The natural forests are in Victoria and Tasmania in discontinuous areas between latitudes 37° and 43°S (Fig. 14.1). In Victoria it occurs at altitudes between 200 and 1100 m and is restricted to mountainous areas south of the Great Dividing Range in the eastern part of the state, but there is also an occurrence in the Otway Ranges south-west of Melbourne. In Tasmania it occurs between sea level and 700 m principally in the north-east, and in the south-east in the valleys of the Huon and Derwent Rivers (Griffin *et al.* 1982; Boland *et al.* 1984).

E. regnans is an ecologically demanding species, being intolerant of drought and severe cold and requiring a moderately fertile soil with good drainage. Fog and mist are a feature of *E. regnans* forests. The natural occurrence, and successful plantation sites, are restricted to areas with suitable soils and with rainfall of more than 1000 mm evenly distributed throughout the year. Forests of *E. regnans* are pure, even-aged stands of 70 to 80 m mature height, and occasionally more than 95 m (Ashton 1975b). Smaller trees of other genera form a rainforest understorey. Natural regeneration follows in great profusion from seed shed from old trees killed by infrequent but devastating wildfire.

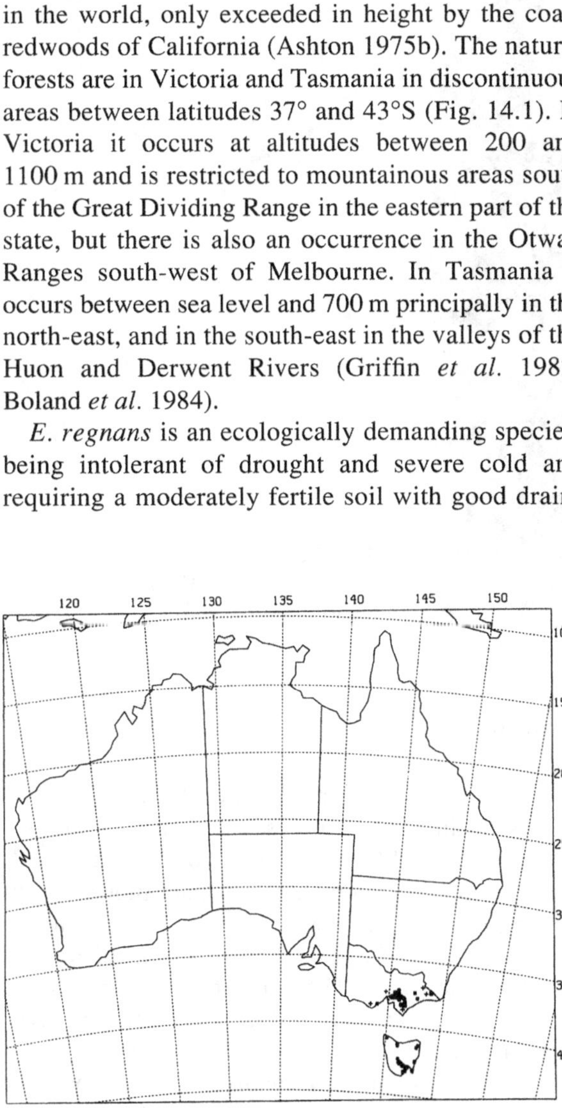

Fig. 14.1. Natural distribution of *E. regnans*.

Naturally occurring hybrids of *E. regnans* and *E. obliqua*, which are common at junctions of stands of the two species, were considered in Section 13.5.

The natural forests of *E. regnans* are well known for rapid height growth and large standing volumes (Fig. 14.2). Probably the largest standing volume recorded was in an old forest of 500 ha in the Toorongo Plateau in Victoria where the total stem volume averaged 1900 $m^3 ha^{-1}$ (Ferguson 1948). Growth rates in young natural stands on the better sites in Victoria are commonly about 15–20 $m^3 ha^{-1} yr^{-1}$ (Borough *et al.* 1984). The sawmilling and pulp and paper industries of Victoria and Tasmania depend mainly on *E. regnans* and two other species of the ash group, *E. delegatensis* and *E. obliqua* (series *Obliquae*, Pryor and Johnson (1971)). The wood of *E. regnans* is of relatively low density, light coloured, and straight-grained. Large logs attract a high price for joinery, furniture and sliced veneer (Fig. 14.3). It has a reputation as one of the best of the eucalypts for pulp yield and quality for all types of mechanical and chemical pulp (Higgins 1984). The harvested forests are generally regenerated by burning the logging debris and broadcasting local seed of the same species; large quantities of seed are used as the rate of broadcast sowing is about 1 kg ha^{-1} (Pederick 1990).

Fig. 14.2. Natural forest of *E. regnans* near Powelltown, Victoria (photo CSIRO).

14.2 Plantations

In Victoria *E. regnans* has been planted intermittently at the rate of a few hundred hectares per year since 1960, a total of more than 10 000 ha. The area of plantation in Tasmania is much smaller. Plantations established in Victoria between 1960 and 1965 on sites where the species formerly grew naturally had a mean annual increment of about 20 $m^3 ha^{-1} yr^{-1}$ at 20 years (Mann 1988). More recent plantations of *E. regnans* with improvements in site selection, establishment practices, and orchard seed are producing more than 30 $m^3 ha^{-1} yr^{-1}$.

Plantations in other countries have been very successful in the few places where climate and soil are close to ideal (Guinness Book of Records 1975). For example, in Kenya a one hectare plot at South Kinangop had a mean annual increment of 56 $m^3 ha^{-1} yr^{-1}$ at 42 years and predominant height of 70 m (Dyson 1975). The tallest tree was then 82 m, perhaps on its way to exceeding the tallest *E. regnans* in Australia.

In New Zealand *E. regnans* is one of the four most promising eucalypts for planting on colder

Fig. 14.3. Felling a large *E. regnans* near Marysville, Victoria (photo News and Information Bureau).

sites, together with *E. fastigata*, *E. delegatensis*, and *E. nitens* (Wilcox 1980) and is being planted for pulpwood. One company in the North Island, New Zealand Forest Products, had a total of more than 4000 ha of *E. regnans* plantation in 1986 (Hayward 1987). Elsewhere it has grown well in trials in sheltered temperate localities in Portugal and South Africa and at high altitude in the tropics in India, Sri Lanka, and Zimbabwe (Streets 1962; Jacobs 1981). In its natural range even the 'poorest' stands have a top height of 50 m. It is either excellent or absent. Since it is a king among trees as its name implies, *E. regnans* 'must be treated like a king if it is to give of its best wherever it is planted' (Jacobs, undated).

14.3 Provenance variation

14.3.1 Growth rate

The potential for fast growth led to investigations of seed sources to ascertain if use could be made of any adaptive variation within the species. Field experiments were established in Victoria in 1960 and 1963 to study the genetic variation along an altitudinal transect of a natural forest of *E. regnans* (Eldridge 1964, 1972).

Seedlings were grown from 64 trees sampled from seven populations on the slopes of Mount Erica, Victoria. The populations were distributed along a transect of 5 km increasing in altitude from 370 to 1000 m at latitude 37° 55'S. The two planting sites at 500 and 420 m in the Strzelecki Ranges 60 km south of Mount Erica were typical of altitudes of many forest sites in Victoria suitable for planting the species. The two large trials to study altitudinal variation were well-designed statistically and were probably the first combined provenance and progeny trials of eucalypts. Results obtained over more than a decade indicated the presence of significant genetic variation both between and within populations (Eldridge 1972). The variation between populations was clinal for several parameters of growth and form. Higher altitude seed sources produced slower growing trees with straighter stems. At age seven years the average height of provenances was 20 cm less for each increase of 100 m in altitude of seed source. Variation in height between populations was significant, and there was variation between families within populations. The statistical significance of variation between families was greater for diameter than for height, and was more apparent at ten years than at six years.

The physiological basis of the genetic variation with altitude was examined by growing seedlings from the extremes of altitude on Mount Erica, 270 m and 1100 m, under controlled environments. The high-altitude population grew more slowly than the low altitude population at sub-optimal temperatures, developed greater resistance to artificial frost, and responded a little to photoperiod. At low temperature (9°C by day and 4°C at night) the high-altitude population almost stopped growing under short days (8 hours) and became very resistant to frost. Under long days (12 hours) it grew more and was not so resistant to frost. By contrast, the low-altitude population showed no response to photoperiod (Eldridge 1968). It was concluded that seedlings of high-altitude populations of *E. regnans* are better adapted to survive severe winter conditions.

Wood samples from the two well-replicated trials were examined by Doran (1974) who found highly significant differences in wood density within populations but not between them. Doran (1975) also studied the occurrence of kino veins in the two provenance trials and found that this characteristic was more influenced by environmental than genetic factors.

Another trial planted in 1960 compared 20 Victorian and one Tasmanian (Florentine Valley) seedlots with 49-tree plots but only two replications. Although there were no significant differences between seedlots, there was an early trend to slower growth for the higher altitude Mount Erica seedlots (Eldridge 1964).

Preliminary trials by the Forests Commission of Victoria suggested that the magnitude of variation between Victorian provenances of *E. regnans* may not be as great or of such importance for breeding as variation between trees within provenances (Pederick 1976a). In subsequent trials of the progeny of selected trees of three Central Victorian provenances established in 1977 (47 families) and 1979 (29 families) the Neerim provenance grew a little slower than the Toolangi and Powelltown provenances (Pederick 1988).

In 1974, the Forest Research Institute (now CSIRO Division of Forestry) in cooperation with two State Forest Services and five private companies commenced a rangewide provenance study of *E. regnans* in Victoria and Tasmania (Griffin *et al.* 1982). Seed was collected at 26 localities in Victoria and 23 in Tasmania from an average of 14 trees per locality (Table 14.1, Fig. 14.4). With special collections from

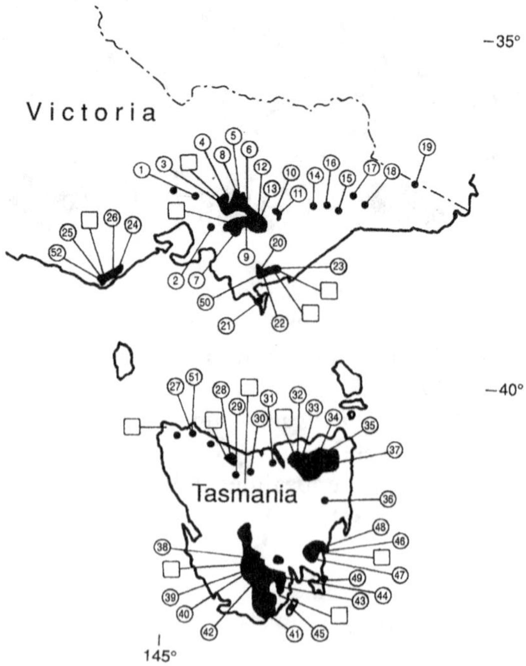

Fig. 14.4. Sites of seed collection (circles) and provenance trials (squares) of *E. regnans*, and the natural distribution of the species (shaded) (after Griffin *et al.* 1982).

Table 14.1

Sites of seed collection of E. regnans *for provenance trials mentioned in Chapter 14. The numbers indicate these locations on the map in Fig. 14.4 (after Griffin et al. 1982).*

Location	Latitude °S	Longitude °E	Altitude (m)	Distance from ocean (km)	Number of trees sampled
Victoria					
1 Mt Disappointment	37 25	145 12	640	52	11
2 Kallista	37 55	145 23	245–365	28	13
3 Toolangi	37 31	145 31	720–780	61	16
4 Narbethong	37 31	145 38	600–800	70	55
5 Armstrong's Creek	37 36	145 49	640–760	83	11
6 Braham's Road (Powelltown)	37 47	145 49	760–820	56	14
7 Tarago	37 56	145 55	360–600	51	16
8 Rubicon	37 20	145 56	550–560	100	13
9 Mt Toorongo Track	37 48	146 07	1000	84	14
10 Barkly River	37 27	146 29	650–750	112	11
11 Green Hills (Mt Useful)	37 41	146 32	910–990	95	15
12 Mt Erica	37 55	146 22	490	95	10
13 Mt Erica	37 54	146 24	990	95	7
14 Valencia Creek	37 33	147 00	975	78	13
15 Quarry Creek	37 26	147 39	700–800	66	50
16 Dargo	37 25	147 12	1000–1200	80	13
17 Ah Chou Creek (Buchan)	37 14	148 03	600–800	66	10
18 Yalmy River	37 21	148 27	825	48	12
19 Bendoc	37 18	148 58	1040–1080	64	10
20 Narracan	38 15	146 13	300	46	14
21 Wilson's Promontory	39 06	146 22	370–400	3	11
22 Mirboo East	38 31	146 23	300–360	24	18
23 Traralgon Creek	38 26	146 31	550–600	45	50
24 Carisbrook	38 36	143 45	365–490	8	11
25 Upper Ford River	38 39	143 29	335–395	12	14
26 *E. obliqua* - Lavers Hill	38 36	143 39	400	15	bulk sample
50 Gunyah	38 31	148 23	300–360	24	4
Tasmania					
27 New Haven	41 01	145 22	240	25	3
28 Ferndene	41 09	146 01	180–220	10	11
29 Lorinna	41 35	146 05	450–500	46	11
30 Christmas Hills	41 28	146 36	335	34	11
31 Beaconsfield	41 18	146 45	335–465	21	15
32 Lisle	41 15	147 22	580–600	29	15
33 Ben Nevis	41 25	147 33	250–300	44	10
34 Forester	41 05	147 46	200	22	12
35 Dan's Valley	41 23	147 53	440–460	39	15
36 Royal George	41 56	147 56	620	30	11
37 Gould's Country	41 13	148 09	150	11	13
38 Florentine Valley	42 29	146 27	360–380	80	15
39 Styx	42 49	146 36	340–520	56	15
40 Moogara	42 47	146 54	600–700	28	11
41 Strathblane	43 22	146 58	200–280	5	12
42 Stoneyhurst Creek	42 40	147 04	450–460	39	10
43 Kaoota	43 01	147 09	480–500	8	12

Table 14.1
(Continued)

	Location	Latitude °S	Longitude °E	Altitude (m)	Distance from ocean (km)	Number of trees sampled
44	Fern Tree	42 55	147 15	320–500	6	7
45	Bruny Island	43 24	147 16	260–280	4	12
46	Levendale	42 31	147 32	150	25	13
47	Nugent	42 42	147 49	340–400	11	13
48	Swan Port	42 24	147 51	420–480	12	11
49	Murdunna	42 58	147 57	350	5	12

50 trees at each of three Victorian localities, the total number of seed trees was 760, a thorough sampling of the whole species.

The range-wide provenance study was the first stage of a two-stage breeding project (Griffin 1983, 1989a). In stage one the 49 provenances were planted in 1977 in well-replicated trials without identification of families. The trials were planted on 12 sites, five in Victoria and seven in Tasmania. The objective was to determine geographic variation patterns and any major provenance–site interactions. Combined provenance-progeny trials of all 760 families were considered impractical with the limited resources available. Stage two was planted in 1985 with a larger sample of families from two regions of provenance which performed best in stage one (Section 14.4).

Nursery growth was a poor indicator of field growth potential, but as judged by similarity of provenance rankings over sites, genetic variation in growth rate was clearly expressed one year after field establishment. The fastest growing seedlots at one, five, and ten years over all 12 sites were from the Strzelecki Ranges and lower altitude Central Highlands of Victoria (Griffin *et al.* 1982, and data from A.R. Griffin) (Table 14.1, Fig. 14.5).

The range-wide provenance trials of *E. regnans* established in 1977 have shown there is considerably less geographic variation in growth rate and morphology than in three eucalypts of similar geographic range which have also been subject to thorough and comprehensive provenance trials: *E. delegatensis*, *E. nitens*, and *E. obliqua*, Chapters 8, 12, 13. There was no indication of a difference between Victorian provenances as a group and Tasmanian provenances as a group, which is so evident in *E. delegatensis* (Chapter 8) and *E. obliqua* (Chapter 13). Nor was

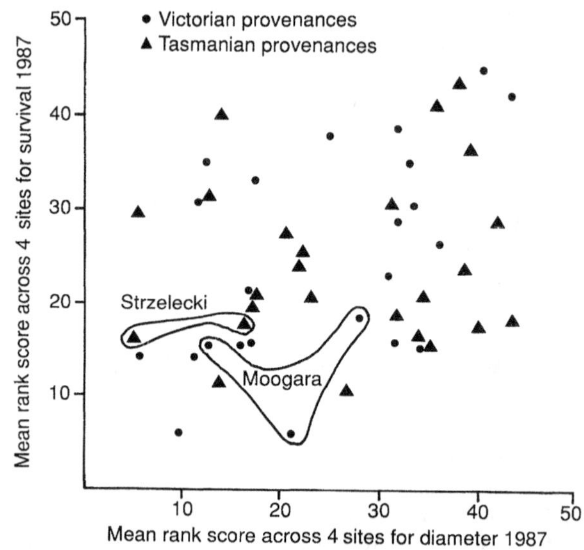

Fig. 14.5. Survival and diameter rankings at age 10 years derived from four *E. regnans* provenance trials in Tasmania. Tasmanian provenances are designated with a triangle and Victorian provenances with a circle. At each trial site the provenances were ranked from 1 (best) to 49 (worst) for both traits. The mean rank score is the mean rank over the four sites. The enclosed areas show the provenance regions which were re-sampled to provide many additional parent trees for stage two of the breeding program (after Volker and Raymond 1989).

there evidence of major genotype–environment interaction in growth over the 12 sites. This is not surprising considering the small range of environmental conditions on the very favourable sites which constitute the natural range of the species, and that *E. regnans* formerly grew naturally on all 12 test sites. However, some provenances were distinctly better adapted than others in frost tolerance.

14.3.2 Frost tolerance

Early studies showed differences in frost tolerance between natural populations of *E. regnans* with the high altitude provenances in a particular area more tolerant of frost than those from low altitude (Ashton 1958; Eldridge 1968, 1971).

At one of the planting sites of the range-wide trials, at Maydena in the Florentine Valley, Tasmania, frost damage was severe on most of the provenances in winter 1978 (Griffin *et al.* 1982). The most frost-tolerant provenances were from the highest altitude in Victoria and from the interior south-east and central Tasmania. After the record frosts of winter 1983 (Davidson and Reid 1985) the ranking of provenances for frost damage to six year old trees was almost the same as in the first year.

Both artificial and natural frosts were used in New Zealand studies of variation in frost tolerance of *E. regnans*. Seedlings hardened in the open were subject to advective frosts in controlled environment rooms (Rook *et al.* 1980). The seedlings tested were from 38 provenances of *E. regnans* from its natural range, including 25 of the 49 in the Australian trials, and from New Zealand stands. Similar material was planted out in 1977 at four sites in New Zealand and was frosted naturally in winter. Under both natural and artificial frosts provenances varied in a similar pattern in their frost tolerance. The range in frost tolerance among provenances was about 2°C in autumn, 2.5°C in winter and 1.5°C in spring. Most frost tolerant provenances of *E. regnans* were from interior, upland sites in southern central Tasmania (e.g. Moogara, Styx River, and Florentine Valley) and from high altitudes in Victoria (over 900 m) (Fig. 14.6). Intermediate hardiness was shown by provenances from Victorian stands at altitudes between 500 and 800 m and by coastal south-central and south-eastern Tasmanian provenances. Southern Victorian provenances from the Otway Ranges and Strzelecki Ranges, and northern Tasmanian provenances were the least frost tolerant. Seedlings from the hardiest provenances would be expected to survive frost of about –5.5°C in spring (early October) without severe damage on sites where such low temperatures were preceded by night temperatures low enough to harden the plants.

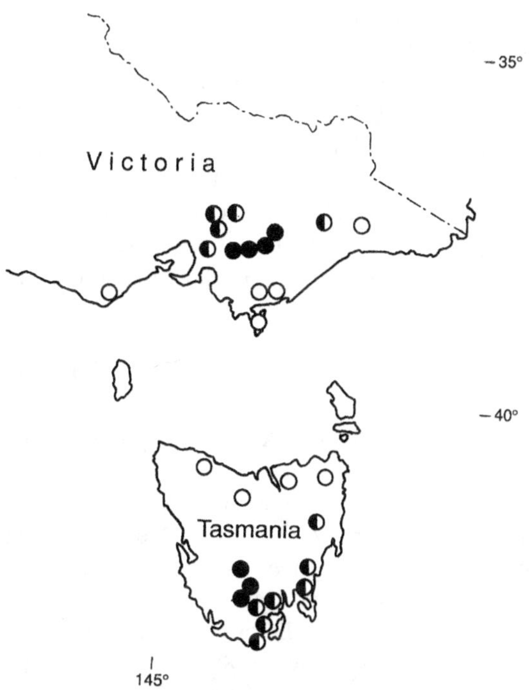

Fig. 14.6. Provenance variation in frost tolerance of *E. regnans*. Data from three New Zealand field trials and from artificial frosts. Open circle least hardy, half circle moderately hardy, and closed circle most hardy (after Rook *et al.* 1980).

Open-pollinated families in field trials in the North Island of New Zealand experienced severe frosts in the winter of 1978, enabling Wilcox *et al.* (1980) to make an early assessment of genetic variation in the incidence and severity of frost damage in 144 families from 36 provenances. The range of frost damage within families was 0–50 per cent. Some families appear to combine excellent growth rate with good frost hardiness, as was observed by Ashton (1958) who found very hardy seedlings colonizing the edge of a frost hollow in the middle of a fast-growing and not particularly frost-hardy population of *E. regnans* in central Victoria.

Frost tolerance is evidently a highly repeatable trait, with good agreement between results from several experiments: five field trials (Rook *et al.* 1980; Wilcox *et al.* 1980; Griffin *et al.* 1982) and three controlled environment studies (Ashton 1958; Eldridge 1968, 1971; Rook *et al.* 1980).

In the Australian trials a group of southern Tasmanian provenances (Moogara, Styx, and Florentine) combined reasonably fast growth (averaged over 12 planting sites) with high frost tolerance at the Maydena trial where they ranked 2, 5, and 10 out of 49 (Griffin *et al.* 1982). The two provenances with fastest growth at 4.5 years, from Mirboo East and Traralgon Creek both at low altitude in the Strzelecki Ranges in south-eastern Victoria, ranked 39 and 34 out of 49 for frost tolerance at age one year at the Maydena trial. However, at 10 years the Strzelecki provenances had good survival on average over all 12 sites (Fig. 14.5), despite early frost damage on some sites.

14.4 Base populations for selection

In stage two of the Australian project further individual-tree seed collections were made in 1983 and 1984 to supplement the seed in store from the earlier collections, making a total of 370 open-pollinated families of the provenances chosen for a combination of frost tolerance and fast growth at 4.5 years (Figs. 14.5 and 14.6). The provenances chosen were from the Strzelecki Ranges (38° 30'S) in Victoria and near Moogara (42° 50'S) in Tasmania.

Four progeny tests were established in northern Tasmania in 1985 with individual family identity maintained (Hand *et al.* 1989). Each test has 20 or 25 replications of single-tree plots and will be suitable for selection of plus trees for breeding populations and for conversion to seed orchards after thinning. One test contains only the fast-growing Victorian provenances, one contains only the frost-tolerant Tasmanian provenances, and the other two contain all 370 families. The test of Tasmanian families is on a site near Burnie only 3 km from the coast where it is hoped the mild temperatures will promote abundant seed production. The fastest growing part of this test was thinned at 4 years using a selection index based on growth rate and resistance to defoliation by insects, and the remainder was to be thinned later.

In New Zealand the initial strategy for rapid genetic improvement of *E. regnans* combined the two Australian stages for two reasons. First, there was an immediate need for a local supply of seed of 'a good all-round provenance with wide adaptability', for example from a stand at Tokoroa which originated from seed imported from southern Tasmania (Wilcox 1982a). Second, there were trees of plus-tree standard in a large number of the 36 provenances under test, though a higher frequency in the better provenances. Using early results from the provenance trials and from progeny tests of 141 open-pollinated families of selected trees of various provenances, clonal and seedling seed orchards have been established with the best material regardless of origin (Wilcox 1982a,b).

Conservation of *E. regnans* forests is a lively political issue in Australia (Eldridge 1990). Although there is only a very small area where the former grandeur of slender forest giants can still be seen, a considerable proportion of the original natural gene resources remains intact, perhaps one third. This is because most of the present forests in Victoria are from natural regeneration following the widespread bushfires of 1939, and in Tasmania there are extensive areas which regenerated naturally in the early years of harvesting. However, much of the forest resulting from direct seeding after harvesting since 1960, or from planting seedlings, cannot be regarded as a natural gene resource as the seedlots used for a particular area were not always from a large number of local seed trees but sometimes from only one tree, from a distant provenance, from a mixture of provenances, or even from a mixture of species when *E. regnans* seed was in short supply.

15

Eucalyptus tereticornis

15.1 Natural forests

Eucalyptus tereticornis (forest red gum) is a woodland and open forest tree (Fig. 15.1). It is closely related to *E. camaldulensis* and the wood properties and growth characteristics of the two species are similar.

This eucalypt has a very extensive natural distribution in a long narrow strip from southern Papua New Guinea to Victoria, Australia, a range of more than 30° of latitude (Fig. 15.2, Table 15.1). Climates vary from monsoonal with markedly wet and dry seasons at 9°S in Papua New Guinea, to a predominantly summer rainfall with a very dry winter in Queensland, an equal distribution of rainfall between winter and summer in southern New South Wales, and a dry summer and cold, wet winter at 38°S in eastern Victoria. Altitudinal range is from sea level to about 1000 m with a wide variation in annual rainfall from 500–1500 mm.

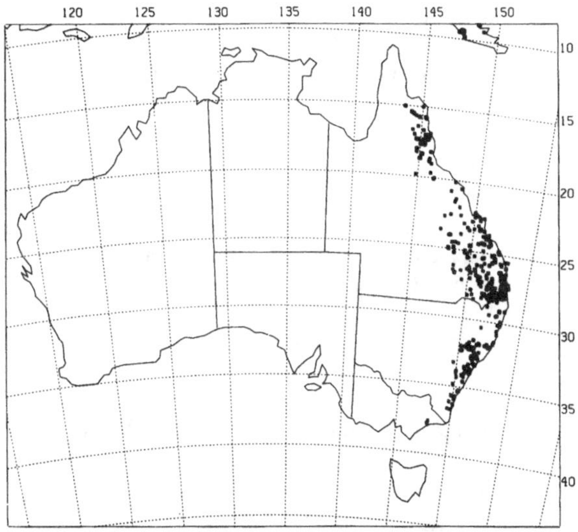

Fig. 15.2. Natural distribution of *E. tereticornis.*

Fig. 15.1. Natural stand of *E. tereticornis* near Helenvale, north Queensland (photo CSIRO).

Table 15.1
Principal sites of seed collection of E. tereticornis *mentioned in Chapter 15 (from CSIRO records).*

Location		Latitude °S	Longitude °E	Altitude (m)
Bulolo (plantation)	PNG	7 10	146 40	640
Dobodura near Popondetta	PNG	8 55	148 30	50
Sogeri near Port Morseby	PNG	9 30	147 26	580
Morehead River c/t	Q	15 02	143 40	60
Laura	Q	15 26	144 19	105
Cooktown	Q	15 41	145 10	120
Kennedy River c/t	Q	15 05	144 19	90
Lakeland Downs	Q	15 45	144 41	170
Helenvale	Q	15 43	145 14	160
Palmer River	Q	16 07	144 47	365
Herberton (Watsonville)	Q	17 22	145 19	790
Mount Garnet	Q	18 30	144 45	875

c/t population with characteristics intermediate between *E. camaldulensis* and *E. tereticornis*; PNG Papua New Guinea, Q Queensland

E. tereticornis is mainly found along the east coast within 100 km of the sea, while *E. camaldulensis* occurs mainly in the dry inland of the Australian continent, west of the Great Dividing Range. The natural occurrences of the two species overlap (Section 6.3). Regions with trees intermediate between *E. tereticornis* and *E. camaldulensis* include extensive areas of north-eastern Queensland (Doran and Burgess, 1992; Figs. 15.3, 15.4), and a small area in eastern Victoria.

Soil conditions seem to impose limits on the natural occurrence of the species because it is not found on heavy clays or dry, shallow soils, prefering fertile alluvia, sandy loams, and gravel terraces.

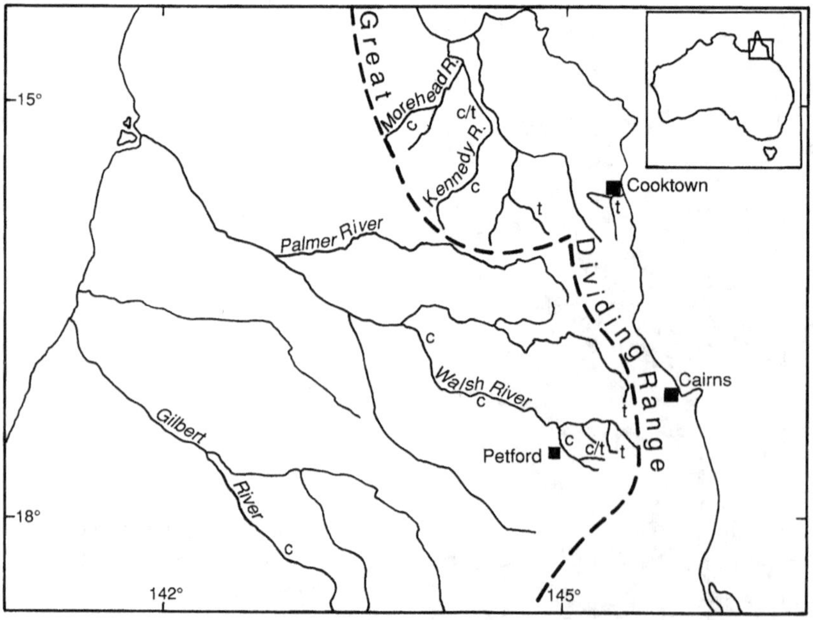

Fig. 15.3. Detail of distribution of *E. tereticornis* in north Queensland showing collection sites with trees intermediate (c/t) between *E. tereticornis* (t) and *E. camaldulensis* (c) (after Doran and Burgess 1992).

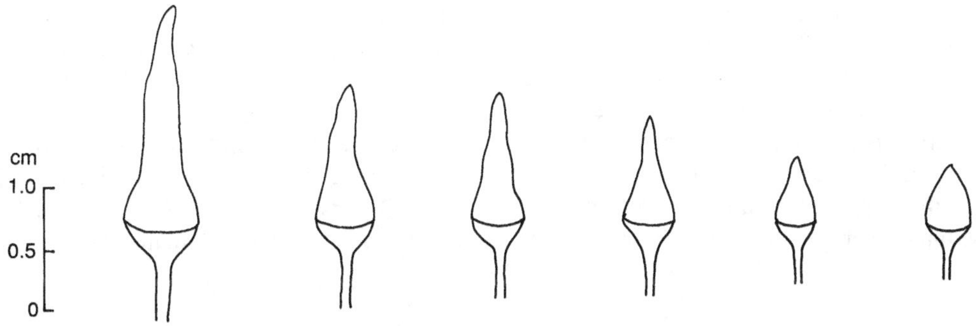

Fig. 15.4. Flower buds of *E. tereticornis* (left), *E. camaldulensis* (right), and intermediates from the Walsh River area of north Queensland (drawing S.M. House).

15.2 Plantations

E. tereticornis was one of the earliest eucalypts exported from Australia. It was introduced into several tropical and subtropical countries from the late nineteenth century, for example Pakistan in 1867, Ethiopia 1895, Zimbabwe 1900, Philippines 1910, Uganda 1912, and India 1919 (Zacharin 1978; Jacobs 1981).

Many of these early introductions were derived from very few original seed trees. For example the 'Mysore Hybrid' in India is believed to be derived from one small stand in the Nandi Hills (Pryor 1966; Chaturvedi 1976). The forms known variously as 'eucalyptus hybrid', 'eucalypt hybrid', and 'Mysore gum' represent about half of the eucalypts planted in many parts of India (Jacobs 1981). Although the word hybrid is often used to describe these populations, Pryor (1966) and other botanists consider they are mainly *E. tereticornis* with only occasional evidence of hybridization with other species, notably *E. robusta*. Boland (1981) suggested that *E. camaldulensis* is also involved in the mix known as 'eucalyptus hybrid'.

A land race of *E. tereticornis,* designated *E.* '12ABL', which appears to have originated from seed collected from a single tree (Martin 1971) has been planted widely in West Africa and southern China. The name is abridged from 'tree number 12 at the forest station Ambila-Lemaitso' in Madagascar (J. Esteve and C. Cossalter, personal communication). Another local land race of *E. tereticornis* or possible hybrid is *Eucalyptus* 'C' from Zanzibar, also probably derived from very few trees. It has been planted in several East African countries (Jacobs 1981).

E. tereticornis has been planted in many countries with results that already indicate it is a species well suited to a variety of sites and localities. Its wood is used for fuel, charcoal, poles, posts, mining timber, paper, hardboard, and particleboard (Hillis and Brown 1984).

India leads in planting of this species with large-scale plantations and countless woodlots of the 'Mysore gum' and 'eucalyptus hybrid' approaching a total of 500 000 hectares. About 250 000 hectares of *E. tereticornis* have been planted in Brazil and a large area in Angola. Planting in other countries has been less extensive and mostly more recent. Substantial areas (39 000 ha) have been planted to *E. tereticornis* and *E. camaldulensis* in Uruguay but separate figures for each species are not available. Some 16 000 ha in Vietnam and 15 000 ha in Argentina have been planted (Jacobs 1981). Almost 4000 ha have been established in the Congo and up to 1000 ha in both Colombia and Ghana (Jacobs 1981). Recently, over 2000 ha has been planted in a community forestry program in Sri Lanka (S.J. Midgley, personal communication). Smaller areas have been established in many other countries.

E. tereticornis has been most successful in summer rainfall areas with a moderate to fairly severe dry season. It has proved considerably more drought-resistant than *E. grandis*, but slightly less so than *E. camaldulensis*. The commonly-planted tropical provenances are susceptible to frost. The species survives well in humid tropical savannahs but is troubled by weed competition, particularly the grass *Imperata,* for up to five years after planting, and also by termites.

Optimum annual rainfall appears to be between 800 and 1500 mm, but it has been planted successfully in both lower rainfall (550 mm in Mozambique and Israel) and much higher rainfall (for example, over a range of 1250–3500 mm in Bangladesh).

E. tereticornis will grow on various soils but prefers deep, well-drained, fairly light textured soils. Sensitivity of the species to site quality and cultural practices is high. For example in the Indian state of Tamil Nadu the range of mean annual increment (MAI) in volume overbark at 8 years is from 2.3 to 40 m^3 ha^{-1} yr^{-1}. In India generally, average MAI is about 15 m^3 ha^{-1} yr^{-1} on adequately stocked medium quality sites at age 8–10 years (Chaturvedi 1973). The best provenances on the best sites in Bangladesh yielded over 60 m^3 ha^{-1} yr^{-1} after a rotation of five years, using dense initial plantings (10 000 stems ha^{-1}) (Davidson and Das 1985), but these plots were small and edge effects were not taken into account.

15.3 Provenance variation

Tree form and growth of *E. tereticornis* varies considerably according to provenance (Martin 1978) but until the mid-1960s no range-wide seed collections had been made. Spectacular provenance differences have been obtained in growth rates and stem form in recent trials in Bangladesh, Brazil, Papua New Guinea, and Uganda.

Up to nine provenances were set out in trials at four widely separated sites in Bangladesh during 1978. Results at five years indicated large advantages (140–180 per cent) of the best over the worst provenances in total stemwood volume production. Best provenances were from near Mount Garnet, Helenvale, and Laura between 15° and 19°S in north Queensland. Considerable provenance–site interaction was evident, though this was least for the Mount Garnet source. Consequently it was the Mount Garnet source that was chosen for further introduction to create seed stands in 1983 (Davidson and Das 1985).

In Uganda, after 4.3 years' growth, superior provenances were Bulolo (land race) and Sogeri in Papua New Guinea and Palmer River, Mount Garnet, and Kennedy River in north Queensland. All these provenances reached at least 14 m in height and 10 cm in diameter after 4 years (J. Davidson data).

In early 1975, provenance trials were established in Papua New Guinea at two locations (Dobodura near Popondetta 9°S, and Kunjingini near Maprik), testing the same (except for one) eighteen seed sources of *E. tereticornis* from north Queensland and Papua New Guinea at each site. Early results indicated better performance of Kennedy River and Helenvale sources from Queensland and Mount Obree, Sogeri near Port Moresby, and Toma River sources from Papua New Guinea, at both sites (J. Davidson data).

Eight north Queensland provenances had a clear advantage in growth over two seedlots from Papua New Guinea and one from India in a trial in northern Nigeria at age six years (Otegbeye 1990). Trials in Côte d'Ivoire, Cameroon, and Congo also indicate superiority of the north Queensland sources, particularly Mount Garnet, Cooktown, and Herberton, though no Papua New Guinea sources were included in these trials. In the Congo, production from these provenances is about 30 m^3 ha^{-1} yr^{-1} (J. Davidson data). Similar results have been obtained in Mexico.

A three-year-old provenance trial at Mogi Guaçu, Brazil, testing both northern Australia and Papua New Guinea sources, showed provenances from Laura and Cooktown in north Queensland were better in height and diameter growth than those from Papua New Guinea (Timoni *et al.* 1983).

During 1980, a series of international provenance trials was organized by the IUFRO Working Party on Eucalypt Provenances and CSIRO Division of Forestry (Doran 1983; Pederick 1983). Twenty-one seed sources were made available. Co-operators were found in 15 organizations in 13 countries.

The results of provenance trials in Zimbabwe using the IUFRO seed sources have been reported by Matheson and Mullin (1987). Twenty two *E. tereticornis* provenances, representing the full natural range from Papua New Guinea to southern Victoria, and including two seedlots from Zimbabwe, were tested. Unfortunately, while nearly half the provenances were represented by five or more seed trees, most provenances from the range extremes in Papua New Guinea, southern New South Wales and Victoria were represented by only one or two trees. Trials were established on deep sandy soils at Mtao (altitude 1480 m, rainfall 690 mm) and Chesa (1200 m, 500 mm). The trial at Chesa was not successful: after four years most of the trees were dying, either from drought or disease. At Mtao, the most productive provenances were from northern New South Wales and southern Queensland. These had reached heights of over 6 m at age 42 months. Two provenances from Papua New Guinea performed poorly, with mean heights of 4.4 and 3.2 m, while those from northern Queensland, southern New South Wales and Victoria and Zimbabwe were intermediate. There were highly significant differences between provenances for all attributes studied, including height, diameter, volume, form, and survival. Marked differences were noted between geographically adjacent provenances from north Queensland, as was the case for the *E. grandis* provenances studied by the same authors. A similar pattern was evident in the combined results from the two trial sites at age 30 months, before mortality set in at the Chesa site. Matheson and Mullin concluded that including genetic material from the best of the natural provenances identified in the trials would lead to substantially increased growth in any breeding population in Zimbabwe.

It may be noted here that the climate at Mtao (latitude 19° 20'S) is subtropical, because of its high altitude, so the superior performance of the southern Queensland and northern New South Wales seed sources over those from northern Queensland is not surprising.

A number of north Queensland provenances of *E. tereticornis* performed well and were not significantly different from each other in two trials in India (Chaturvedi *et al.* 1989). The Kennedy River and Lakeland Downs provenances were among the best in both trials after 4–5 years. The trials were planted in 1981 and 1982 at 18°N latitude with 800 mm rainfall and employed 32-tree plots with four replications. Other provenance trials of *E. tereticornis* planted in India in 1981 did not show a distinct advantage for the north Queensland provenances. For example, Banerjee *et al.* (1990) found no geographic pattern of variation among 22 provenances in their performance at five years.

Provenance trials of *E. tereticornis* have been planted in several other countries in recent years. For example, in a trial of 11 species of eucalypts in south China on Hainan Island (latitude 19°N) the Kennedy River provenance was the best of the 15 range-wide provenances of *E. tereticornis* for height at 18 months on an infertile site (Zhou and Bai 1989). It was second among all 79 provenances and about equal to several *E. camaldulensis* provenances in this trial of 18-tree plots and 4 replications. However, another north Queensland provenance from Helenvale, nearer the coast and just south of Cooktown, ranked 54 out of 79. Clearly it will be worthwhile to test more seed collections from the north Queensland region in the continuing effort to find appropriate provenances for particular sites.

15.4 Base populations for selection

Because of the international importance of the northern provenances of *E. tereticornis*, the FAO Panel of Experts on Forest Gene Resources recommended that some form of conservation should be undertaken to ensure their survival and continuing availability of seed, over and above the efforts taken in 1977. Field trips were made by officers of the CSIRO Division of Forestry in 1979–80 with the objective of obtaining broadly-based seed collections (25 trees at each site) of *E. tereticornis* near Helenvale and Mount Garnet, Queensland (Doran 1983).

In 1977, *ex situ* gene conservation stands of Cooktown and Mount Garnet sources were established in the Congo. Mount Garnet conservation stands have been established also in Bangladesh, Fiji, Nigeria and Zambia using seed from the 1979–80 collections.

Following extensive provenance trials, *E. tereticornis* and *E. camaldulensis* were chosen as the favoured species for plantations in the southern lowlands of Nepal (White 1986). The program of improvement for the two species was based primarily on vegetative propagation from a large number of clones (nominally 120 per species with 40 planted yearly on a three-year rotation). The production clones were supported by a larger reference clone bank representing all top-ranking families from the provenance trials. Propagation from between six and 27 clones from each of four *E. tereticornis* provenances commenced in 1984. The provenances were Kennedy River, Palmer River, Morehead and North Lakeland Downs, all from between 16° 07' and 15° 02'S in north Queensland.

16

Eucalyptus urophylla

16.1 Natural forests

Eucalyptus urophylla is one of the two eucalypts which do not occur in Australia; the other is *E. deglupta*. Over most of its range *E. urophylla* is a forest tree usually reaching 25–45 m in height and up to 1 m diameter, with a straight bole for half to two-thirds of the tree height. Exceptional specimens attain 55 m in height and up to 2 m diameter. It is one of the best low-latitude eucalypts for planting in tropical areas (Turnbull and Brooker 1978; Jacobs 1981).

There are extensive stands of *E. urophylla* on the large Indonesian island of Timor above 1000 m and throughout Wetar (Fig. 16.1). More scattered stands are found on the nearby islands of Adonara, Alor, Flores, Lomblen, and Pantar. These islands form the south-eastern end of the Sunda Archipelago. They are isolated to the north and south by the seas of Flores and Timor, so limiting the latitudinal spread of the species to between about 7° 30'S and 10°S. The eastern and western limits are less precise, because it has not been possible to prove the presence or absence of natural occurrences of this species on islands other than the seven listed above. Therefore the natural range is presently set between 127°E and 122°E, a distance of about 500 km, which represents the southernmost and driest part of the long line of islands stretching in a continuous chain for about 1400 km in an easterly direction from South Java (Martin and Cossalter 1975–1976).

The natural range is within the wet/dry tropics climatic region (Fig. 4.2, p. 41) with a dry season varying from about eight months to about two months and rainfall of 600 to 2500 mm (Martin and Cossalter 1976, No. 166, pp. 15–17). In Timor, with the exception of a few locations on the north-eastern coast, *E. urophylla* occurs above 1000 m and up to 2960 m. It occurs over an altitudinal range from about 300 m to about 1100 m in Flores, Adonara, Lomblen, Pantar, and Alor and from 70 m to more

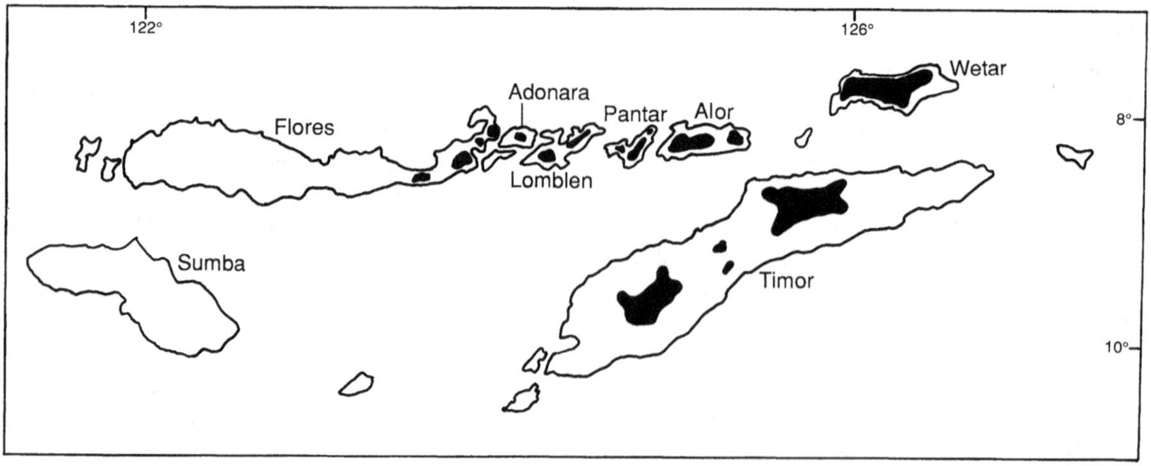

Fig. 16.1. Natural distribution of *E. urophylla* (after Martin and Cossalter 1976, No. 165, p. 9 and Gunn and McDonald 1991).

than 800 m in Wetar. Altitude has a marked effect on mean ambient temperatures: mean monthly maximum temperatures vary from 27–30°C at 400 m to only 17–21°C at 1900 m. Mist and fog are common and rain is abundant (1300–2200 mm with 3–4 dry months) above 1000 m in Timor.

The species is found on mountain slopes and in valleys. Best development occurs between 1000 m and 2000 m altitude in Timor and on favourable sites in the other islands on moist, deep, well-drained soils generally derived from volcanic and metamorphic rocks (Fig. 16.2). *E. urophylla* commonly grows on basalts, schists and slates, but is rarely found on limestone soils. It is frequently the dominant species of advanced secondary mountain forest. It is a notable feature of *E. urophylla* that it is the sole occupant of a very wide range of altitude and microhabitats; such a wide range in Australia would be occupied by a mosaic of several eucalypt species (Pryor and Johnson 1981, p. 524). However, on the lower slopes in Timor *E. urophylla* does form a mosaic distribution pattern with *E. alba* which dominates these drier sites (Martin and Cossalter 1975–1976). The two species are not difficult to distinguish as *E. urophylla* has a distinctive and constant leaf character, a 'drip tip', which is absent in *E. alba*.

Fig. 16.2. Natural stand of *E. urophylla* on Mount Egon, Flores (photo CSIRO).

16.2 Plantations

The Dutch apparently gathered seeds of this eucalypt in the Sunda Isles in 1890 and raised seedlings in Java for planting. No plantations exist now of such an age but individuals of about that age can be seen planted at Bogor and in the streets of old Jakarta, notably in Ancol Avenue near the docks. Those at Bogor are *E. urophylla*, though two specimen trees over 80 years old by a roadside in the Bogor Botanic Gardens were still labelled '*E. alba*' when visited by J. Davidson in 1978 (Fig. 16.3). Those in Jakarta seem to be *E. alba* × *urophylla* hybrids.

The most widespread and largest plantations of *E. urophylla* are in Brazil but early plantings are almost certainly *E.* 'alba', the hybrid crosses of *E. urophylla* with other species in the Rio Claro arboretum (Gurgel and Cavalcanti 1978) and perpetuated through several generations (Section 16.5). The area in Brazil eventually planted with

Fig. 16.3. *E.* 'alba' (*E. urophylla*) in the Bogor Botanic Gardens, one of the two trees which were the likely origin of *E.* 'alba' in Brazil (photo J. Davidson).

'Brazil alba' was probably of the order of 500 000 ha before it was realized that the quality of new plantations was much inferior to earlier plantations because of loss of vigour and irregularity in form due to hybrid breakdown, and also inferior to *E. grandis*.

In some other countries plantations have had as their origin the three plots at Rio Claro (*E. urophylla* labelled as *E. alba*) or F_2 and F_3 generation plantations in Brazil resulting from the same source.

In recent years *E. urophylla* has increased greatly in popularity for plantations in humid and subhumid tropical climates with several months of drought (the wet/dry tropics), for example in parts of Indonesia, Brazil and southern China (Fig. 16.4). However, care is needed in matching this 'new' species to planting sites. For example, in north Queensland, Australia, with an apparently suitable climate at 17°S latitude, the species has failed in several trials after excellent growth in the first year or two. Stem-boring beetles and other insects are the suspected cause of death (D.A. Kleinig, personal communication).

In contrast to its good growth in the wet/dry tropics, *E. urophylla* performed poorly in humid tropical lowlands at Madang, Papua New Guinea, on areas harvested for wood chips. Survival was poor because of apparent inability to overcome intense

Fig. 16.4. Plantation of *E. urophylla* at Dong Men in southern China (photo Xiang Dongyun).

competition from weeds and secondary tree growth under the tending regimes normally employed for species such as *E. deglupta*. A further trial in grasslands at Bulolo, Papua New Guinea, again grew poorly. Similar poor results were obtained in the Solomon Islands. Growth was not much better and survival even less in trials in East Kalimantan, Indonesia. Performance of trees from two high-altitude sources (Queorema and Mt. Tatamailau) was very poor (J. Davidson observations).

However, given that survival could be improved by silvicultural means, seed sources from below 1000 m altitude are still considered to have potential in the wet lowland tropics with no marked dry season, as well as in the wet/dry tropics.

These observations and the measurements in a trial near Edea, Cameroon (I.A.S. Gibson and J. Davidson, unpublished report) confirm that over a rotation of about eight years on a suitable, well-prepared site a mean annual increment of 30 $m^3 ha^{-1} yr^{-1}$ is obtainable following provenance screening. This growth rate is exceeded by selected hybrids of this species with *E. alba*, *E. grandis*, or *E. tereticornis*, followed by their mass propagation by cuttings (Section 22.2.4, p. 232).

16.3 Exploration and genecology

Except for very fragmentary information, the eucalypts of the Sunda Isles were virtually unknown before 1963, when an Australian expedition led by M.R. Jacobs went to Dili on Timor to gather seed and botanical material (Anon. 1964, pp. 47–48). At this time the species was known by the illegitimate name of *E.* 'decaisneana'. Blume based the name *Eucalyptus* 'decaisneana' on a specimen at Leiden labelled 'Eucalyptus obliqua L'Hérit., Timor' (Blume 1849). This specimen actually is of *E. obliqua* for which the incorrect locality was given (see Blake 1953). Therefore, the name *E.* 'decaisneana' could not be attached to the Timor eucalypt.

This invalid name has been perpetuated in literature, for example, Maiden (1903–1933) treated *E.* 'decaisneana' as a synonym for *E. alba* Reinwardt ex Blume and, unfortunately, it has been attached to most botanical specimens and seed collected this century, until the species was formally described by Blake (1977).

A Dutch botanical expedition in 1939 studied the vegetation of West Timor and Wetar and also made brief visits to the islands of Sumbawa, Sumba, Flores, and Sabu. Specimens of the eucalypts collected still exist at the Herbarium in Bogor. All of the sheets of *E. urophylla* were labelled as *E.* 'alba' and all of *E. alba* as *E.* 'platyphylla'. Some intermediate specimens, apparently hybrids between *E. urophylla* and *E. alba*, were incorrectly labelled *E. moluccana*. These labels were still attached when the collection was inspected by J. Davidson in 1978.

Following the 1963 expedition, a succession of Australian teams carried out more extensive exploration of eucalypts in the Sunda Isles. Egon Larsen visited Timor and Flores in 1968 studying over 40 localities where *E. urophylla* and *E. alba* were found. He made the connection between smooth barked *E. urophylla* found locally in the eastern part of Flores with similar trees in three plots in the Rio Claro arboretum in Brazil (Gurgel and Cavalcanti 1978). The small quantity of seed obtained in 1968 was the first reliably source-identified material made available for early trials in Australia and in some other countries including the Congo.

Three years later, in 1971, another team (L.D. Pryor and J.W. Turnbull) went to Timor. This mission was essentially botanical but about 20 more seedlots were collected and later distributed to several countries including Papua New Guinea, Congo, Cameroon, and Malaysia to comprise some of the first formal provenance trials of this species.

Two French expeditions were mounted in the early 1970s. The first, from July to September 1973, covered Timor, Flores, Adonara, and Lomblen, the second (from June to July 1975) travelled to Pantar, Alor, and Wetar. These very thorough collections of seed and botanical material comprised 86 sources of *E. urophylla* and 19 of *E. alba*, totalling 534 parent trees from the seven islands. Details were published in a series of seven well-illustrated papers by Martin and Cossalter (1975–1976). Additional comprehensive seed collections of *E. urophylla* have been made by foresters from Australia, Brazil, France, and Indonesia (Section 16.4).

It is evident some very large trees must have existed in parts of the natural range. On Timor C. Cossalter found on the slopes of Mt Mutis at an altitude of 2100 m some very large living trees and stumps up to 2.5 m in diameter. On cool mountain sites at around 1300 m altitude, on deep soils, tall, straight, slender trees occur in often dense forest stands. In drier areas, on shallow soils of the ridges, the habit is different, trunks are short and thick with many branches forming rounded crowns. The species can grow as a small, gnarled shrub only a few metres or even less than one metre tall at its maximum altitude (Mt Tatamailau, at almost 3000 m) or in peculiar environments: for example, on the crater rim exposed to the fumarole of the Sirung volcano on Pantar Island.

Just as striking is the variation in bark texture. The appearance of the bark differs under the influence of moisture and altitude being mainly smooth at lower altitudes and under drier climatic conditions. Below 1000 m on Alor and Flores trees have smooth trunks with a short stocking of rough bark at the base. However, between 1000 and 2000 m on Timor where conditions are moister and mountain mists occur, the trees are usually covered with rough bark even to the finest branches. A wide range of bark types is found within some stands on other islands.

In a study of variation along an altitudinal transect from 600 m to 2800 m in East Timor, Moura (1977) found that fruit and seed were small at low altitude and much larger at high altitude. On the nearby island of Wetar Martin and Cossalter (1975–1976) and Gunn and McDonald (1991) found equally extreme variation in fruit size in *E. urophylla*, all at low altitude, about 300 m (Fig. 16.5). On a wet site in the north-west of the island the fruits were the smallest found throughout the range of the species, and on a site in the south-west with a much longer dry season the fruit had 25 times larger volume (Martin and Cossalter 1976, No. 166, p. 11). As well as large fruits the *E. urophylla* on the drier site had large leaves, smooth bark on the branches, and poor stem form, while trees on the wetter site had narrower leaves, rough bark, and good form. Very few collections of *E. urophylla* had been made on Wetar before 1990 compared with the thorough collecting for 20 years over the rest of the range, particular in Timor and Flores. The Martin and Cossalter (1975–1976) collections were from only two localities in the west of Wetar and the trials reported by Suseno *et al.* (1985) tested three localities also in the west. It appears the extensive stands in the east of Wetar had never been sampled for provenance trials before the comprehensive collection made by Gunn and McDonald (1991).

When progeny growing in provenance trials in Papua New Guinea were examined, a clearer picture of variation emerged than from the study of botanical material obtained from the parent trees in the wild (Martin and Cossalter 1975–1976). There were trends with increasing altitude of source from broader leaves to narrower leaves, from round to squarish twigs and from large to small lignotubers. Small fruit and small leaves seemed to be associated mainly with rough-barked individuals from a humid climate and large fruit and large leaves with smooth-barked individuals from a drier climate (J. Davidson, unpublished information).

Exploration of *E. urophylla* in its natural habitat shows it is one of the most variable of all the eucalypts. However, the continuous variation in fruit size and other morphological traits has so far inhibited the formal taxonomic separation of distinct forms (for example large fruits and smooth bark, and small fruits and rough bark) into new species or varieties (Fig. 16.5). Two more subjects for further genecological studies are (a) hybrids with *E. alba* as a source of variation and (b) the taxonomic relationship with *E. pellita* and other members of the Series *Transversaria* (Pryor and Johnson 1981).

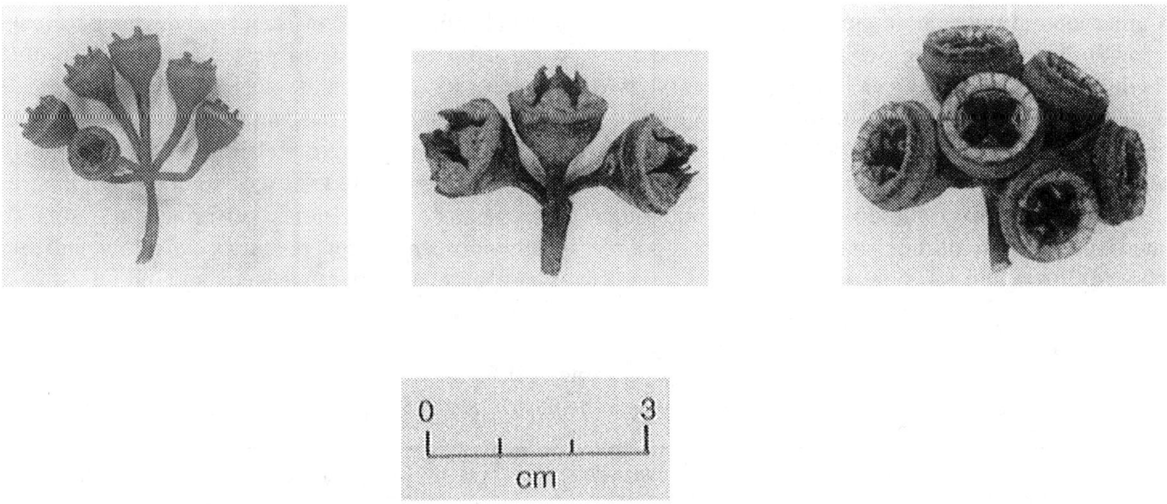

Fig. 16.5. Examples of the enormous range of fruit size and shape in *E. urophylla* from three trees on Wetar (from Gunn and McDonald 1991).

16.4 Provenance variation

Serious attempts were started in the late 1960s to make provenance collections of *E. urophylla* and these have become more extensive with time, the earlier collections being made only from Timor and Flores (Section 16.3).

A trial in Papua New Guinea planted in 1971 at Bulolo (800 m altitude) used some of these early seed collections. By four years of age low-altitude sources from Timor and Flores (less than 600 m) were far better than sources from above 1000 m on Timor, but growth overall was disappointing. The species was unable to suppress vigorous weed and *Imperata* grass competition. However, in Puerto Rico on two wet lowland sites planted in 1971 the species grew quite well (Whitmore 1983). Of the five provenances the best two, which were from 600 m on East Timor, averaged 22 m at eight years.

Other early trials of *E. urophylla* included three plantings in South Africa in 1973 and 1974 of five provenances on three sites (Darrow and Roeder 1983). The dry matter production from *E. urophylla* at eight years was greater than from the taller *E. grandis* in adjacent plantations due to the higher density of *E. urophylla* wood, particularly of the smooth barked form.

In 1973 a trial testing eight provenances in unreplicated plots was established at Mangombe, Cameroon (altitude 30 m, rainfall 2600 mm). The growth obtained from a low-altitude source near Mt Egon, Flores (83 $m^3 ha^{-1} yr^{-1}$ to age 8 years) was spectacular, even conceding there were edge effects in this single plot where stocking was still 1460 stems ha^{-1}. The highest altitude source from Mt Tatamailau (2350 m) failed completely and other medium and high altitude sources performed poorly (I.A.S. Gibson and J. Davidson, unpublished report).

These early trials demonstrated the marked difference in growth between low-and high-altitude sources on some wet tropical sites. However, trials planted in São Paulo, Brazil, in 1968 did not show such a marked altitudinal trend at nine years of age (Pasztor 1978). This may have been because all the seedlots tested (M.R. Jacobs 1963 collections) were from above 1200 m and the faster growing low altitude material was not included.

From extensive collections of *E. urophylla* which were made by the CTFT, France, from 1973 to 1975 (Martin and Cossalter 1975–1976) provenance trials were established in the Cameroon, Congo, Côte d'Ivoire, Gabon, Guyana, Indonesia, Madagascar, and Papua New Guinea among others. Further trials were established from the collections made by a number of Brazilian afforestation companies in 1973 and 1977 (Moura 1977, 1983; Vieira 1983) and by teams from CSIRO Australia and CTFT France in association with the Indonesian Department of Forestry and the Indonesian Forestry Research Institute (Doran and Kleinig 1979; C. Cossalter, personal communication; Gunn and McDonald 1991).

In Congo 101 provenances of *E. urophylla* were tested during the period 1970–1981. Provenances from low altitude on Mount Lewotobi, Mount Egon (both on Flores) and Ulanu Valley (Alor Island) (Table 16.1) have consistently given the best results.

In Brazil trials of 14 provenances of *E. urophylla* were established on 29 sites between 1974 and 1977 (Moura 1983). For the Timor provenances there was a very strong negative correlation between tree height at 3.5 years and altitude of seed source. In all the trials a provenance from Flores performed best, but some other provenances were almost equal to it, and several parts of the natural range were not represented.

Table 16.1
Principal sites of seed collection of E. urophylla *mentioned in Chapter 16 (from CSIRO records).*

	Latitude °S	Longitude °E	Altitude (m)
Mount Egon, Flores	8 38	122 27	400–900
Mount Lewotobi, Flores	8 32	122 48	400–500
Ulanu Valley, Alor	8 18	124 26	600

A distinct trend with altitude of seed source was apparent in a large trial established in Côte d'Ivoire in 1974. The trial contained a total of 142 open-pollinated families representing 52 provenances mostly from Timor (Diabate 1978). At age two years (Diabate 1978) and at six years (Wencelius 1983) very few of the Timor families, even those from low altitude, were among the leaders. Seed sources above 1500 m on Timor performed very poorly. The islands of Alor, Flores, and Lomblen were the best seed sources. The strong trend to slower growth for higher altitude seed sources shown clearly in Fig. 16.6 is typical of the results of *E. urophylla* provenance trials which included higher altitude Timor provenances. The extreme variability between provenances, between families, and within families means there is a great need for improving the species by selection before its potential for plantations can be realized. The tallest *E. urophylla* tree was 31 m at six years, better than *E. deglupta* selected in Côte d'Ivoire (Wencelius 1983).

Further trials of several more provenances were planted in 1977 at Mangombe, Cameroon using samples of most of the French seed collections. One trial of 13 provenances was laid out in randomized blocks of four replications with square plots of 25 trees spaced at 3 × 3 m. Another 31 provenances were planted out in unreplicated plots of 36 trees, five sources being common with the replicated trial. Site preparation was intensive with clearing and windrowing of debris (I.A.S. Gibson and J. Davidson, unpublished report). Provenances from low altitude on Mount Wuko, Mount Lewotobi, Mount Mandiri, and Mount Egon (all on Flores) have performed well. Low altitude provenances from south of Dili (Remexio region) in Timor also have given good growth. However, in a trial of 22 provenances planted in 1982 near Yogyakarta in Java, Indonesia, none of the Timor provenances (all from below 1550 m) was among the best at two years (Suseno *et al.* 1985).

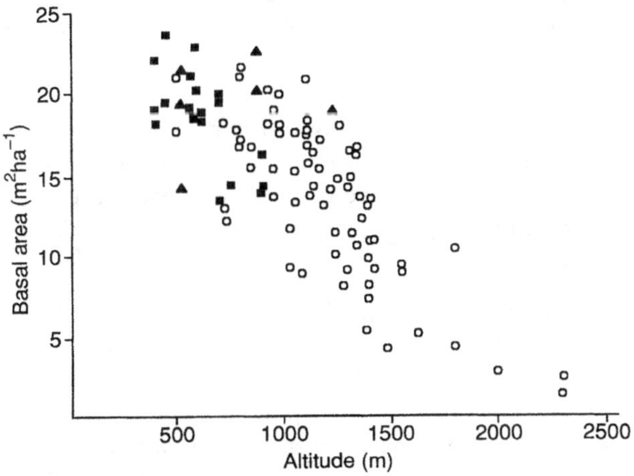

Fig. 16.6. Relation of growth (basal area at 5 years) of individual tree seedlots of *E. urophylla* to altitude of seed source from Timor (circle), Flores (square), and Lomblen (triangle) in a trial at San Pedro, Côte d'Ivoire (after Wencelius 1983).

The Wetar provenance is not represented in most trials of *E. urophylla*. This seems to be due to administrative and transport difficulties for seed collectors to visit the island. It is the southern most island of the large province of Maluku, the provincial capital of which is Ambon, 500 km to the north. Recent trials in Indonesia indicate Wetar is a provenance well worth further collection and testing. For example, in a report on provenance trials by Indorayon in Java and Sumatra the three lowland seed collections from western Wetar were outstanding at age two years among 22 range-wide provenances (Suseno *et al.* 1985; White 1989).

Wood density was examined in a trial of 11 provenances planted in 1982 in Malawi and there were statistically significant differences in whole-tree wood density at five years (Ngulube 1989). The range in density was 564–453 kg m^{-3} and there was a trend to lower density with higher altitude of seed source.

A canker disease was found in *E. urophylla* at Mangombe, near Edea, Cameroon by a World Bank Mission in November 1980. It was caused by *Cryphonectria cubensis* (Bruner) Hodges (formerly known as *Diaporthe* canker), which has caused serious losses in eucalypts in Latin America but which had hitherto not been found on the African continent (I.A.S. Gibson and J. Davidson, unpublished report). Although, in Brazil, *E. urophylla* is much more resistant to *Cryphonectria cubensis* than either *E. grandis* or *E. saligna*, it would appear several provenances are quite susceptible, at least in wet tropical lowlands without a marked dry season, and measures such as selecting resistant provenances and individuals will be required to combat this problem.

16.5 Base populations for selection

Seed was first collected and introduced into Java (where the species is not indigenous) by a Dutch botanist in 1890. In 1919 Navarro de Andrade introduced the species to the Rio Claro arboretum in Brazil under the name *E. alba*, probably using seed from trees growing in the Bogor Botanic Gardens. Indeed, the origin of the *E.* 'alba' in the Rio Claro plots is recorded as Java (Gurgel and Cavalcanti 1978) not Timor or Flores.

The three plots planted in the Rio Claro arboretum in 1919 each consisted of 399 trees spaced 2 × 2 m (Gurgel and Cavalcanti 1978). Seed from these plots, and from plantations derived from them, was used to plant extensive areas in Brazil and other countries between 1940 and 1970, usually under the name 'Brazil alba'. Much of this seed has since proved to be of hybrid origin, because of the ease with which the Rio Claro trees, now definitely identified as *E. urophylla*, hybridize with certain other species of eucalypt growing in the same arboretum and flowering at the same time, particularly *E. tereticornis*, but also *E. grandis*, *E. resinifera*, *E. robusta*, and *E. saligna*. In many cases the performance of the F_1 generation was excellent, but F_2 and subsequent generations have shown poor growth and form. Some hybridization with *E. alba* does take place in the natural range of *E. urophylla* and these hybrids are now showing up in plantations. However, the 'PF_1 hybrid' in Congo is not the product of hybridisation which has taken place in the natural range. This hybrid called sometimes *E. alba* × *urophylla* for simplification, is in fact the F_1 of three *E.* 'platyphylla' trees grown from seed collected on one of the *E.* 'platyphylla' of the Ancol Avenue, Jakarta. The father species of PF_1 is presumed to be a *E. urophylla* hybrid maybe 'Brazil alba'. The place where PF_1 originated is called Loandgili, Pointe Noire (C. Cossalter, personal communication).

In addition to Brazil, introductions have been made to Bali Island (1935), East Java (1937), Bandung (1952), Bogor (1960), Congo (1957 and 1962), Australia (1966), and Papua New Guinea (1971). Plantations arising from these early introductions were among the first base populations for selection though emphasis soon switched to populations grown from the major seed collections of the 1970s. For example, the collections in 1979 by Australian, French, and Indonesian foresters were planned to expand the base populations for future selection (Doran and Kleinig 1979). They aimed to collect from 30 to 50 trees at each of the most promising locations as indicated by early results of provenance trials. Several hectares of base populations are now established in Congo and French Guyana with three of the provenances collected in 1979: Mt Lewotobi, Flores, Mt Egon, Flores, and Ulanu Valley, Alor, totalling respectively 50, 30, and 30 families (C. Cossalter personal communication). In a trial at Dongmen in south China comparing six provenances from Flores and Alor, Pegg and Li (1989) found little difference between provenances in growth rate and concluded 'there appears to be no sound reason for restricting tree improvement work to a particular provenance'. However, the first 28 ha of seed production areas at Dongmen were all of one provenance, from 30 trees at Mount Egon on Flores.

Despite the youth of extensive recent trials, the natural range of *E. urophylla* has been shown to contain great genetic diversity. The natural range can be divided into two main sections (Martin 1983).

1. Southern section: Timor provenances with small fruit and rough bark adapted to a humid tropical climate with a short dry season (2500 mm annual rainfall). For the wet lowland tropics the provenances from low altitude south of Dili (Remexio region) have given good results. Provenances from high altitude, above 1500 m, have usually performed poorly. Inability of all provenances to overcome weed competition is a disadvantage on cleared lowland moist tropical sites.

2. Northern section: Provenances from Flores, Lomblen, Adonara, Pantar, Alor, Wetar, and the north-east coast of Timor mainly with larger fruit and smooth bark (except at the base) adapted to drier conditions (750–1200 mm rainfall with a marked dry season) and low altitude (300–1100 m). This type shows great promise in humid or sub-humid, tropical or sub-tropical areas in Brazil (i.e. in areas with 1100–1500 mm rainfall and a dry season of 1–5 months in the coolest part

of the year). A similar rainfall regime occurs at Loudima in the Republic of Congo where growth of the Mt Lewotobi provenance from Flores was approximately 40 $m^3 ha^{-1} yr^{-1}$ at 5 years.

Among recent trends in the domestication of *E. urophylla* are selection of plus trees within provenance trials, conservation stands, and plantations. Additional selection has been made within individual families where these have been kept separate in provenance and progeny trials. For example, in Indonesia 32 ha of seedling seed orchards have been established since 1980, combining an open-pollinated progeny test and a provenance trial (Arisman 1985). Natural and artificial interspecific hybrids are being increasingly used in programs using mass vegetation propagation of cuttings. Brazil, China, and Congo are leaders in these developments.

17

Eucalyptus viminalis

17.1 Natural forests

Eucalyptus viminalis (ribbon gum, or manna gum) is one of the most geographically widespread eucalypts in south-eastern Australia (Boland *et al.* 1984), and occupies a very wide range of ecological conditions. On well-watered, fertile soils of gullies and hillsides it forms a tall forest, either alone or with other species including *E. amygdalina, E. dives, E. fastigata, E. obliqua, E. radiata,* and *E. regnans.* On drier, poorer sites on the plains and tablelands, on a variety of soil types, it occurs in low, open forests or woodlands, alone or with species such as *E. amygdalina, E. dives, E. obliqua, E. pauciflora,* and *E. radiata.* The geographic range is from southern Tasmania (latitude 43°S), through Victoria, southern South Australia, eastern New South Wales and the Australian Capital Territory, with high altitude stands extending to the Queensland border (29°S) (Fig. 17.1). The altitudinal range is from sea level to 1070 m in Victoria, and to 1370 m further north in New South Wales (Ladiges and Ashton 1974). In Tasmania, the species is widely distributed over the eastern half of the state, from sea level to about 300 m, extending to 500 m in moist gullies (Phillips and Reid 1980). The climate varies considerably over the species' range, with rainfall from about 500 to over 2000 mm per year, and the number of frosts from a few mild ones in the coastal environments to over 60 per year on the high-altitude hillside, valley, and tableland sites. At these latter sites, snow can be expected to lie for several days at a time in most winters (Costin 1954; Boland *et al.* 1984). As in the case of *E. obliqua,* this wide ecological and geographic range is associated with major phenotypic variation and population differentiation.

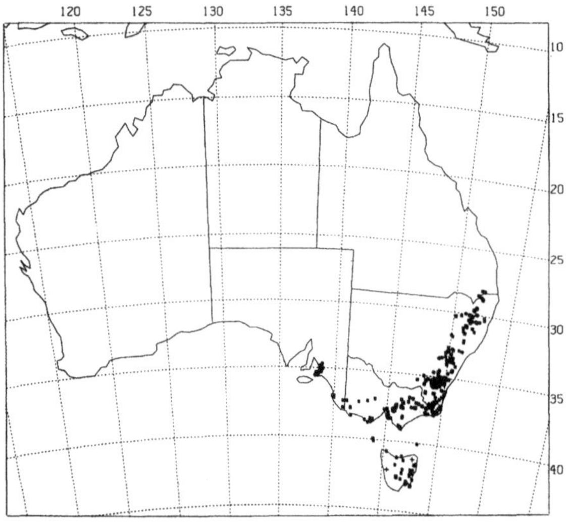

Fig. 17.1. Natural distribution of *E. viminalis.*

E. viminalis belongs to the subgenus *Symphyomyrtus.* In the common form occurring in moist gullies and on hill slopes with good moisture availability, it is a tall forest tree, having a clean white trunk with a stocking of dark brown bark at its base. Strands of decorticating bark often hang from the forks of the branches (hence the common name, ribbon gum). In north-eastern Tasmania, trees 90 m in height have been recorded (Fig. 17.2), making it the second tallest hardwood in the world, after *E. regnans* (Boland *et al.* 1984). There are normally three flower buds per inflorescence, and the juvenile seedling has opposite, sessile, lanceolate green leaves.

In other environments, there is a wide variation in tree shape (Fig. 17.3) and other phenotypic characteristics. Taxonomic authorities differ in their treatment of varieties and subspecies. *E. viminalis* subsp. *pryoriana* (Pryor and Johnson 1971) occurs on the southern coast of Victoria and is identified by the rough bark covering most of the trunk and sometimes the lower limbs. It is treated as a separate species, *E. pryoriana* L. Johnson, by Brooker and Kleinig (1983) and Chippendale and Wolf (1981). Brooker and

Fig. 17.2. Giant *E. viminalis* 92 m tall near Fingal in north-eastern Tasmania (photo D.A. Kleinig).

Fig. 17.3. Variation in shape and size of trees of *E. viminalis* from coastal heathland (right) to fertile inland sites (left) (after Ladiges and Ashton 1974).

Kleinig also identify *E. viminalis* subsp. *cygnetensis* Boomsma, occurring in South Australia and south-western Victoria, on the basis of seven-flowered umbels and rough bark extending up the trunk. Hybridization between *E. viminalis* and many other species has been recorded (Griffin *et al.* 1988), and hybrids with *E. rubida* and *E. dalrympleana* are very often found at the boundaries of the species' natural distributions.

Central Victorian populations of *E. viminalis* were studied by Ladiges and Ashton (1974), who found a trend of decreasing adult tree height (Fig. 17.3), increasing height of the rough bark stocking of the lower trunk, and decreasing leaf size, with decrease in rainfall of the sample site. Some trees at the lower-rainfall sites had inflorescences with more than three buds per umbel.

The relationship between *E. viminalis* and *E. dalrympleana* in Tasmania, where the two species are hard to distinguish, was examined by Phillips and Reid (1980). They examined adult trees and raised seedling progenies for 12 populations along a 120 km transect from sea level to the Central Plateau. There was continuous, clinal variation in juvenile leaf shape from the long, thin, lanceolate leaves typical of coastal *E. viminalis* to the broad, cordate leaves characteristic of *E. dalrympleana* on the Central Plateau. Adult leaf, bud, and fruit characteristics varied from population to population, but with no consistent trend along the transect. They concluded that in Tasmania clinal intermediates between the two species were perhaps the most common types. This was in contrast to the situation on the mainland, where intermediates occur in a narrow hybrid zone, about 50 m wide, between the areas occupied by the two species (Mendonza 1968). The name 'vim-dal' was suggested by Barber (1956) to identify the intermediate Tasmanian types.

17.2 Plantations

E. viminalis is an important plantation species in temperate regions of a number of continents, because it combines reasonably rapid growth, good form and acceptable wood characteristics with some degree of frost resistance (Hunt and Zobel 1978). In trials comparing the early growth of 36 eucalypt species near Mount Gambier (37°49'S) in South Australia it proved one of the fastest-growing species, with only *E. globulus* and *E. nitens* showing more rapid growth (Cotterill *et al.* 1985). Its main use in overseas plantations is as a short-rotation fibre crop for paper pulp and hardboard production (Hillis and Brown 1984). We estimate the total area of plantations of this species to be of the order of 200 000 ha.

There were extensive plantations of *E. viminalis* in the Eastern Transvaal Highland of South Africa, but these were devastated by an insect pest accidentally introduced from Australia, the eucalyptus snout beetle (*Gonipterus scutellatus*) in the 1920s. *E. viminalis* was the species of eucalypt most heavily attacked (Darrow 1984; Richardson and Meakins 1986). This led to it being abandoned as a commercial species in South Africa despite its good growth rate and frost resistance in this environment. Introduction of the mymarid egg parasites of the beetle was successful in limiting damage to *E. viminalis* to an acceptable level except in severe climates where the parasites do not survive. Thus, plantings in the cold climate of Lesotho in the 1970s also failed because of eucalyptus snout beetle attack. However, there was considerable variation in susceptibility to insect attack (Richardson and Meakins 1986).

Large areas of *E. viminalis* were planted in the Black Sea region of the USSR, but were devastated by a harsh winter in 1949–50, and the species is no longer grown there (Linnard 1969).

E. viminalis is the most important commercial eucalypt in frost-prone areas in southern Brazil (Machado de Fonseca *et al.* 1979) where more than 50 000 hectares of plantations have been established (A.R. Higa, personal communication). *E. viminalis* is the favoured commercial species in the province of Buenos Aires, Argentina, because of its good growth rate, adequate frost resistance, high quality of the wood for fibreboard, and good form for the production of poles (Mendonza and Alliani 1984). It is also commercially significant in some Mediterranean countries such as Italy and Spain, and in India. It has performed well in trials in many other countries (Hillis and Brown 1984; Jacobs 1981).

17.3 Genecology

E. viminalis varies greatly in its appearance in natural forests, for example, in tree height and form (Fig. 17.3). Accordingly, detailed genecological studies involving field and laboratory experiments have been carried out on a number of Victorian populations from habitats with differences in moisture availability, degree of waterlogging, and soil properties (Ladiges 1974a, b; Ladiges and

Ashton 1974, 1977; Ladiges and Kelso 1977). They found in reciprocal transplant trials that seedlings generally performed best in their own environments, showing adaptations to the particular environmental conditions involved. A population from a seasonally waterlogged solodic soil showed the best survival and height growth on this soil type, and also in pot trials simulating waterlogging. Superior drought resistance of populations from low rainfall sites, evident in greater survival and lower damage levels in the field trials, was confirmed in laboratory experiments. Seedlings of the low rainfall populations had greater root/shoot ratios, and could survive greater leaf water loss without wilting. There were differences between adjacent populations on soils derived from granite and basalt in a low rainfall area (580 mm per year), with the population on granite more resistant to drought, coinciding with apparently greater drought stress on this soil type. There was also a genetic component to the between-population differences in juvenile leaf shape and the development of lignotubers, the higher rainfall populations showing the least lignotuber development.

Seedlings of a Victorian population from a calcareous soil grew significantly faster than populations from nearby non-calcareous soils, when raised in glasshouse trials using a calcareous soil (pH 7.5). On non-calcareous soils (pH less than 7) there were no significant differences between the populations (Ladiges and Ashton 1977).

Ladiges and Ashton (1974, 1977) concluded that ecological races (edaphic ecotypes) of *E. viminalis* existed, and that they had arisen from the natural selection of plants in particular habitats. They suggested that the general similarity of the tall forest populations in central Victoria might indicate that selection pressures in the moist habitats occupied by this type were similar, while there was a greater diversity of habitats in low rainfall areas leading to more widely differentiated populations in these areas.

17.4 Provenance variation

17.4.1 Growth rate

Twenty-two Australian provenances of *E. viminalis,* covering much of the species' natural range, and 19 open-pollinated families of selected trees of two Argentinian land races of the species, were tested in a well-replicated provenance trial at Ramallo, in the north of the Argentinian province of Buenos Aires (Mendonza 1974). Mean tree heights of the 41 seedlots at 3.5 years ranged from 6.5 to 9.5 metres. Although there were statistically significant differences, there was no overall geographic pattern evident in the growth of the Australian provenances, several of which were poorly sampled as they were represented by only one or two seed trees. However, the three provenances from the south-east of South Australia (subsp. *cygnetensis*) ranked 18, 19, 22 among the 22 Australian provenances (Table 17.1). The local land race was somewhat slower growing than the provenances from natural stands in Australia. The average height of the 19 Argentinian families was 7.85 m compared with 8.16 m for the 22 Australian provenances. Also the 10 top ranking seedlots included only one of the Argentinian families (Mendonza 1974).

Considerable differences in growth between Australian provenances were also demonstrated in a number of trials established from 1971 to 1974 in southern Brazil (Machado de Fonseca *et al.* 1979). Individual trials at different sites tested up to 10 provenances, but most provenances were represented by seed collected from fewer than five trees, and several by only one. Victoria, Tasmania, and South Australia were represented by no more than 2, 3, and 4 seed trees respectively in any individual trial. Provenances from New South Wales consistently survived better and outperformed those from South Australia, although significant levels of difference were not reported. Volume yields of the best provenances reached 30–40 m^3 ha^{-1} yr^{-1}. One Brazilian land race from Canela in the state of Rio Grande do Sul, was included in a trial at Telemaco Borba in Parana state, and performed nearly as well as the best Australian provenance, with a volume yield of 36 m^3 ha^{-1} yr^{-1}. The local land race was considered by the authors to be unsuitable for use in a breeding programme as it was highly variable in its growth rate, form, and frost resistance, and was suspected to have a narrow genetic base. The authors concluded that certain Australian provenances identified in the tests were

Table 17.1
Variation among 22 Australian provenances of E. viminalis *in height growth at age three years in a trial near Ramallo in Argentina, compared with the progeny of 19 selected Argentinian trees (from Mendonza 1974, with amendments from CSIRO seed collection records).*

Origin	State	Latitude °S	Longitude °E	Altitude (m)	Height (m)
Cann River	Vic	37 45	149 15	15	9.5
Orange	NSW	33 20	149 03	1000	9.4
Blundell's Flat	ACT	35 20	148 50	800	9.0
Tumut	NSW	35 20	148 25	250	8.9
Warburton	Vic	37 45	145 42	90	8.9
Uriarra	ACT	35 14	148 53	730	8.7
Yarram	Vic	38 32	146 42	150	8.7
Bombala	NSW	37 08	149 11	850	8.7
Kangaroo Island	SA	35 40	137 38	110	8.6
Tallaganda	NSW	35 21	149 13	660	8.2
Armidale	NSW	30 40	151 25	700	8.2
Kingston	Tas	42 59	147 25	240	8.1
Mylor	SA	35 04	139 00	340	8.1
Rylstone	NSW	32 55	150 20	1100	7.9
Argentina land race	–	–	–	–	7.8
Cabbage Tree	Vic	37 47	148 40	10	7.7
Bago	NSW	35 36	148 10	850	7.6
Penola	SA	37 24	140 50	70	7.5
Glencoe	SA	37 50	140 50	65	7.4
Mt Crawford	SA	34 42	138 58	450	6.7
Brindabella Ranges	ACT	35 27	148 50	1100	6.5
Mt Gambier	SA	37 50	140 50	50	6.5

ACT Australian Capital Territory, NSW New South Wales, SA South Australia, Tas Tasmania, Vic Victoria

considered worthwhile, but further seed imports and tests would be required because of the small number of trees sampled for most of the provenances that were examined.

It may be concluded that the value of the provenance trials reported above has been reduced because of inadequate sampling of the provenances tested, many being represented by five or fewer families, and also from inadequate sampling of the geographical range of the species. In particular, Tasmanian and Victorian populations appear to have been inadequately represented in the trials.

17.4.2 Frost tolerance

Between-provenance differences in frost resistance have been demonstrated by a number of workers. In a study of three populations from southern New South Wales, Paton (1972) found that seedlings of the population from intermediate altitude (300 m) were more damaged by artificial frost than the seedlings from provenances at sea level and 1100 m elevation. As part of their studies on Tasmanian populations, Phillips and Reid (1980) exposed glasshouse-raised seedlings from the two extremes and the middle of their Derwent valley transect to natural frosts. The seedlings from the coastal population typical of *E. viminalis* suffered most damage (mean leaf damage score of 63 per cent), and the high altitude population typical of *E. dalrympleana* suffered the least (16 per cent), while the intermediate population averaged 23 per cent damage.

A provenance trial of 25 Australian seedlots covering the full latitudinal range of *E. viminalis* from northern New South Wales to Tasmania was carried out at Jessievale State Forest in the Eastern Transvaal highveldt of South Africa (Darrow 1984). Height growth, frost resistance, and survival varied significantly between provenances after 2 years, with mean height of provenances varying from 2.6 to 4.6 m, and survival from 38 to 82 per cent. Some seedlots from southern New South Wales and the Australian Capital Territory combined good frost resistance with excellent to fair growth, while Tasmanian provenances had generally poor frost resistance. However, many of the provenances tested were represented by only 1 or 2 parent trees in the seedlot, the 5 Tasmanian provenances being represented by a total of only 11 parent trees. This inadequate sampling of provenances limited the conclusions that could be drawn as to the merits of the various seedlots.

The *E. viminalis* seedlots tested by Darrow were also included in provenance trials at three sites in Natal, South Africa by Nixon (1984). In the Natal trials, severe winter frosts occurred when the plants were about 0.5 m high. The provenance rankings were similar to those obtained by Darrow, the seedlots from southern New South Wales and the Australian Capital Territory receiving least frost damage, and those from Tasmania being most damaged.

Provenance trials at Parana and Irati in the Brazilian state of Parana are reported by Restrego and Stohr (1983). Twelve Australian provenances of *E. viminalis* were tested, eight of these being from New South Wales and the Australian Capital Territory. Two local land races of the species were also included. The trial sites were subject to winter frosts, which killed many of the plants in their first year, so that survival at 12 months only averaged 20 per cent. These preliminary results indicated that the local land races and one South Australian provenance appeared to survive best and grow fastest.

A testing program near Bainbridge, Georgia, in south-eastern U.S.A., evaluated 98 seedlots (some representing only one family) for growth and cold-hardiness over the period 1972–1980 (Hunt and Zobel 1978; Jahromi 1982). Substantial differences in growth rate and cold-hardiness were noted, although it is not clear whether differences were statistically significant. In these trials difference in growth rate between seedlots is confounded with difference in frost damage and the ability to regrow after damage, as most trees in the trials received severe damage during winter freezes. Several seed sources showed good resilience (ability to recover from frost damage to foliage) with little growth loss resulting after up to 80 per cent canopy damage. *E. viminalis* had the best overall growth performance of 20 eucalypt species tested in these trials. At least one tree in each of several *E. viminalis* seedlots tested combined above-average frost resistance and good growth rate (Jahromi 1982). Grafts from outstanding individuals in these trials were planted in a clonal seed orchard as part of a breeding program.

It is unfortunate that the field trials for growth and frost resistance of provenances reported above, which represent a major commitment of effort, did not succeed in clearly identifying the best *E. viminalis* provenances for the regions concerned. The main reasons appear to be that the provenances tested were inadequately sampled and did not fully cover the geographic range of the species, and that in some cases tests were not properly replicated. It seems likely that more extensive seed collections in the natural environment, including collections from stands close to the edge of frost hollows over the latitudinal range of the species from Tasmania to northern New South Wales, may locate frost-resistant and fast-growing provenances, families and individuals which are needed for plantation and breeding programs in the areas where the species is being tested.

17.4.3 Insect attack

There are large intraspecific differences in the damage caused by the eucalyptus snout beetle (*Gonipterus scutellatus*) in Lesotho (Richardson and Meakins 1986). For 11 *E. viminalis* provenances tested, mean damage at age 68 months varied from 53 per cent to nearly 100 per cent (Fig. 17.4). Tasmanian provenances had significantly lower mean damage scores, and variability was greater in these provenances. Some individual trees from Tasmania were almost undamaged, whereas the lowest individual damage scores for trees from mainland provenances exceeded 70 per cent (damage below about 50 per cent had little effect on height growth). A similar trend of better resistance in Tasmanian provenances was found in the closely related *E. dalrympleana* (Fig. 17.5).

The availability of *E. viminalis* resistant to the insects would be of great commercial significance for production of firewood and mine timber in Lesotho, as the species appears very well suited to the high altitude sites there, relative to other insect-resistant

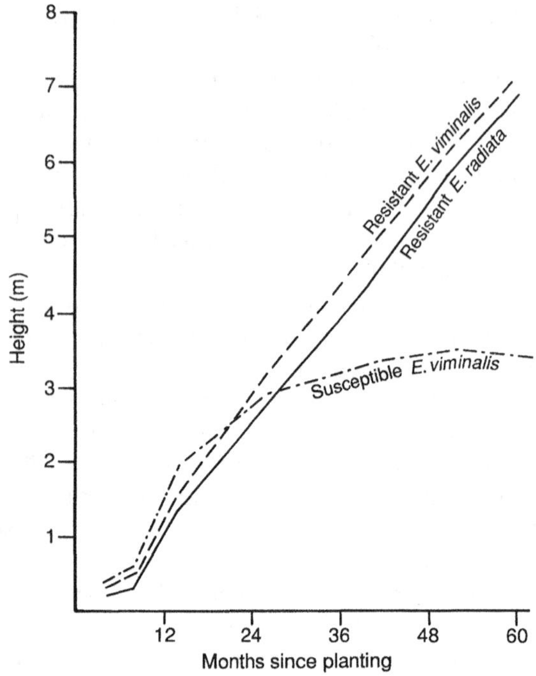

Fig. 17.4. The effect of attack by snout beetle on the height growth of susceptible and resistant provenances of *E. viminalis*, and a resistant provenance of *E. radiata* (after Richardson and Meakins 1986).

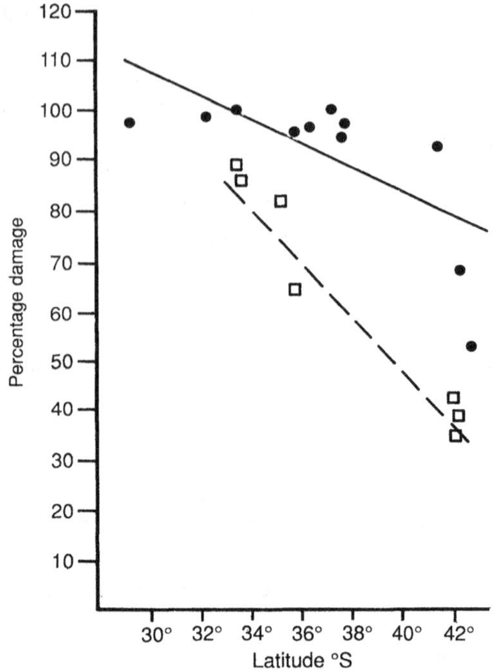

Fig. 17.5. Comparison of snout beetle damage on *E. viminalis* (circles) and *E. dalrympleana* (squares) with latitude of seed source (after Richardson and Meakins 1986).

eucalypts which are slower growing. Further investigations, including a selection program to take advantage of the genetic variation in resistance, have been recommended (Richardson and Meakins 1986).

17.5 Base populations for selection

Given the extensive range of natural habitats of *E. viminalis*, and the demonstrated genetic differences in growth rate, form, resistance to insect attack, frost and drought tolerance, and adaptation to different soil types, careful selection of seed sources for specific planting environments will obviously pay off.

In Brazil, there is an IBDF registered seed stand of 370 trees at Trombini, near Canela in São Paulo state, which produces much of the country's commercial seed. The origin of the stand is not known, but it may well have a limited genetic base (Fonseca *et al.* 1979).

Extensive seed collections of *E. viminalis* were made by the CSIRO Division of Forestry and EMBRAPA (Brazil) in 1984, concentrating on potential seed sources in New South Wales which showed early promise in preliminary field trials in Brazil. Collections were made from 25 trees at each location. However, there has been no complete range-wide collection. Currently available seedlots mainly represent the tall forest form of the species that grows in well-watered, relatively fertile and not particularly frost-prone environments. While trees from this habitat type may be expected to produce the best growth rates under favourable environmental conditions, it would seem to be worth acquiring seedlots from other habitats occupied by the species for breeding programs aimed at combining good growth with increased tolerance of frost, drought, adverse soil conditions, insect attack, and so on. Resampling of many areas is also needed to provide seedlots with adequate numbers of parent trees.

It may be noted that good performance with regard to all the various attributes reviewed above under the heading 'provenance variation' is unlikely to be found in the one provenance. For example, the studies in South Africa and Lesotho suggest that resistance to eucalyptus snout beetle is greatest in Tasmanian provenances while frost resistance may be greatest from high-altitude provenances from southern New South Wales and the Australian Capital Territory. Breeding strategies for high-altitude plantations in southern Africa might therefore require inter-provenance hybridization to 'repackage' the two types of resistance, derived from different provenances, into improved progeny.

Overall, the species is well conserved in its natural environment, with many stands included in national parks, but some of the ecotypes found in particular habitat types such as those identified by Ladiges and co-workers may not be adequately protected.

18

Breeding strategies and breeding plans

18.1 Concepts of breeding strategies and breeding plans

The tree breeder's work program to achieve a supply of genetically improved planting stock consists of many activities, most of them seasonal: collecting seed, running a nursery, making grafts and cuttings, planting and measuring progeny tests, statistical analysis, interpreting results, preparing budgets and plans for future years, and so on.

The breeder's efficiency will be increased by putting all the current activities into order, providing a framework of ideas for the future, and writing a plan for several years ahead. A breeding strategy is an overview or philosophy of the management of genetic improvement of a tree species used in man-made forests. A well-planned strategy makes one aware that current activities can have a great influence on future opportunities for genetic gain, and can help avoid the cost and embarrassment of having to scrap an existing program and start again.

For example, the procedure used by *Pinus radiata* breeders in the 1950s, when a number of breeding projects started in Australia and New Zealand, was simply to select about 30 outstanding trees (plus trees), make many grafts of each tree, establish them in seed orchards, and in most cases plant progeny tests. Such clonal seed orchards were the first steps in almost all the world's tree breeding projects. This approach originated with the pioneering work of Scandinavian foresters (Larsen 1934, 1956). The use of orchard seed certainly resulted in improved *P. radiata* plantations (for example Wright and Eldridge 1985). However, the opportunity for further improvement in the next generation of selection which started in the 1970s was limited by the small number of unrelated trees available for selection in the progeny tests and orchards. A fresh start was made in the 1970s with several hundred newly selected plus trees and a new, long-term breeding strategy (Burdon *et al.* 1978; Cotterill 1984). Breeding programs with poplars (Mohrdiek 1983) and with the southern pines in south-east USA have a similar history.

A breeding strategy is the overall concept of how to go about breeding. Its essential element is population improvement by a combination of a particular type of selection and a particular type of mating, starting with a well-adapted and broad genetic base. One example could be selection within and among families in progeny tests using index selection, and mating of selected trees by controlled pollination in a diallel mating design. Another example could be mass selection in special plantations in which families are not identified, and mating by open pollination.

As an ongoing and recurrent process, a breeding strategy accumulates benefits over successive generations of the cycle of testing, selection, and mating (Fig 18.1). The testing process aims to select indi-

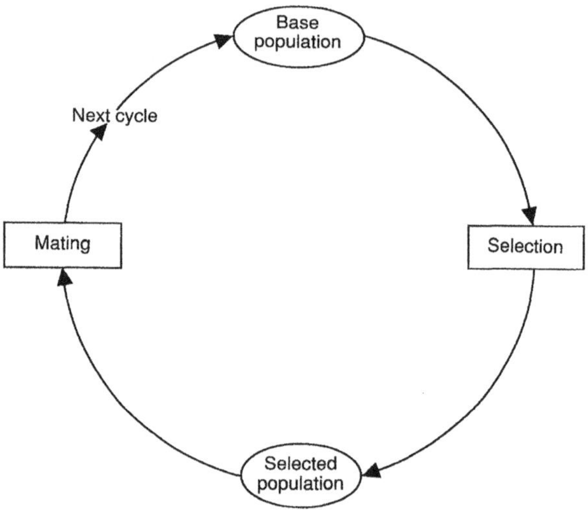

Fig. 18.1. A complete cycle of selection and mating (after Namkoong *et al.* 1988).

viduals in which there is the largest number of favourable genes. Mating of these individuals allows the favourable genes to recombine so that new and better individuals arise, carrying even greater numbers of favourable genes. Each generation the benefits are captured for use in plantations by mass propagating the best currently available genetic material through seed or cuttings.

Every effective breeding strategy involves the maintenance of a hierarchy of three major types of population (Fig. 18.2; Libby 1973) which can continue to meet the demand for genetically improved planting stock for a fourth population, the wood-producing plantation. These four populations are as follows.

1. ***Base*** or gene resource ***population***: the natural forests of eucalypts and some plantations (millions of trees), in which selection can be carried out; these broadly based genetic reserves will continue to be the source of a wide range of genetic variation to meet future needs.
2. ***Breeding population***: the selected trees and their progeny (perhaps 500 families of 200 trees each) in a series of progeny tests and clonal archives in which the breeding cycle of selection and mating (Fig. 18.3; White 1987) will be repeated over many generations. This is the tree breeder's main area of work.
3. ***Propagation population***: the intensely selected trees (commonly fewer than 100 trees) in seed orchards or cuttings multiplication areas where the combinations of genes selected in the breeding population are mass produced as genetically improved planting stock.

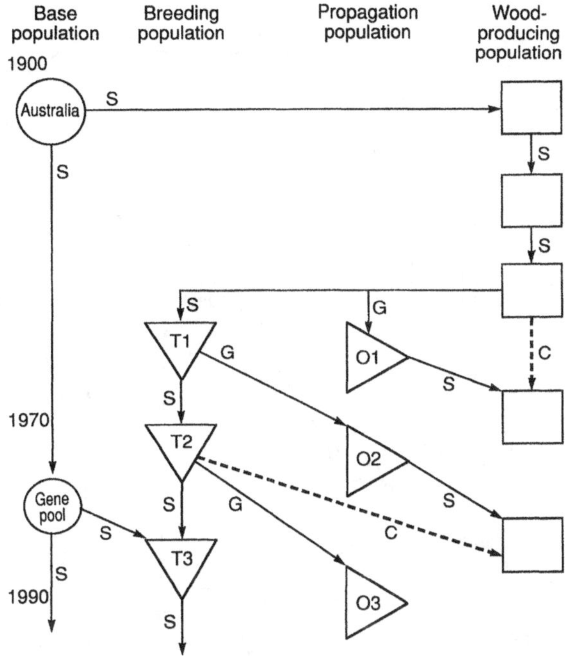

Fig. 18.2. Flow chart for domestication and breeding of eucalypts. S seed, C cuttings, G grafts, T progeny test, O seed orchard. A hypothetical time scale is indicated in the left margin (after Eldridge 1984, based on Libby 1973).

4. ***Wood-producing population***: in future the large wood-producing plantations will not themselves be the source of improved planting stock; this will come from the propagation population derived from a well-managed breeding population.

Several terms are in current use to describe each of these four populations (Table 18.1).

Table 18.1

Terms used to describe the four populations considered in breeding strategies.

	Libby (1973)	Eldridge (1984)	Shelbourne *et al.* (1986)	White (1987)
1.	original variability	base	gene resource	external, infusion
2.	selection	breeding	breeding	base, selected, breeding
3.	packaging	propagation	seed production	production
4.	production forests and plantations	wood-producing	wood-producing	operational plantations

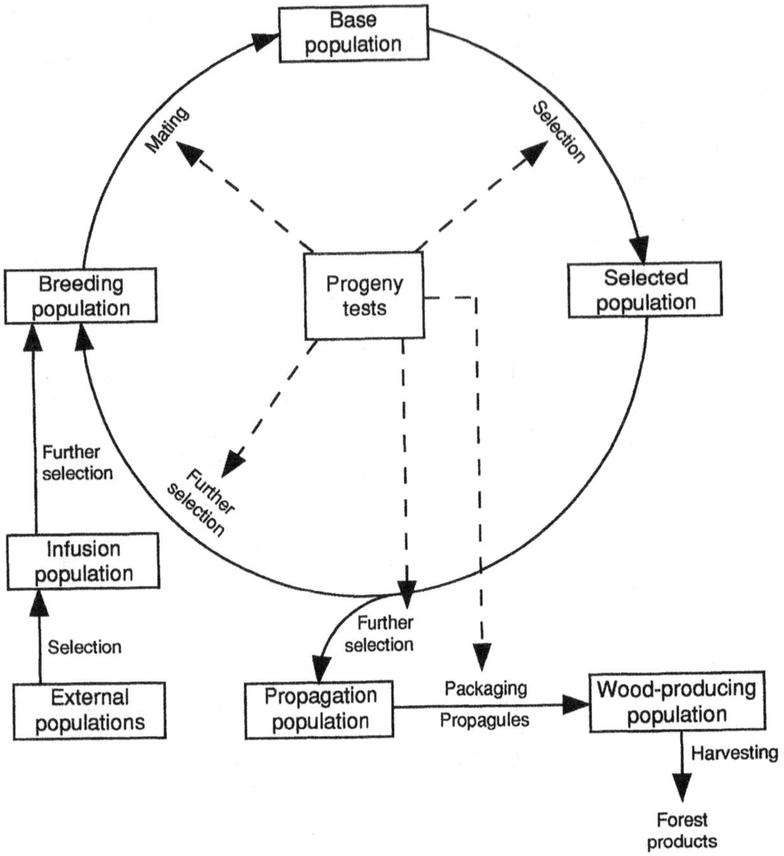

Fig. 18.3. The major components and activities of the breeding cycle as represented by White (1987). The 'infusion' and 'external' populations are the 'base' population of Fig. 18.2. The 'selected', 'breeding', and 'base' populations are the 'breeding' population of Fig. 18.2. Progeny tests provide information (dashed lines) used in several stages of the breeding cycle (after White 1987).

Like a forester's working plan for management of an area of forest the breeding plan puts the strategy to work, and specifies objectives and what has to be done each month for several years ahead. In some cases, particularly in Florida (Franklin and Meskimen 1984) and Zimbabwe (Barnes 1984b), populations 1, 2, and 3 have been physically combined into one plantation which is converted from a provenance/ progeny test into a seed orchard. However, awareness of the concept of the three distinct populations is essential in planning the operations of genetic improvement.

There are three types of people involved in tree breeding (Slee 1985) and in preparation of breeding plans:

(1) managers who set objectives and control funding;

(2) tree breeders who carry out the work;

(3) forest geneticists who provide new ideas.

Sometimes two of these functions, usually the last two, are combined in one person, but rarely all three.

Few tree-breeding programs in the world had their breeding strategy and breeding plans written down in 1980. There is now a widespread recognition that without written, critically reviewed plans, expensive programs are too dependent on the continued employment of enthusiastic individuals. If such a person leaves the job (as manager, breeder, or geneticist) the program suffers. Furthermore, a well-prepared plan is essential for adequate long-term budgeting. A carefully chosen breeding strategy is likely to be stable over many years, even decades. A breeding plan is more dynamic and is likely to be reviewed frequently and perhaps revised substantially at two- to five-year intervals.

One of the major prerequisites to choosing a long-term breeding strategy and writing an efficient breeding plan is a set of good estimates of the genetic parameters of the eucalypt population which is to be improved (Section 18.3).

18.2 Eucalypt breeding strategies

Since eucalypts are largely outcrossing (Section 19.3, Table 19.2, p. 194) they are still highly heterogeneous and variable in plantations even after the several generations of selection under domestication to which some species have been subject. It is the function of well-planned, long-term eucalypt breeding strategies to manage broadly-based and constantly improving breeding populations in which inbreeding is minimized. Thus outcrossing and heterozygosity are promoted in the propagation population and increasingly vigorous growth is sought for the wood-producing plantation population.

Unlike the breeding strategies for some of the major annual grain crops, such as wheat, rice, barley, and oats (Simmonds 1979) it is not feasible to have pure uniform inbred lines as the product of eucalypt breeding. Uniformity in plantations, if required, can be achieved by the use of clones in those species which can be mass propagated vegetatively (Chapter 22). However, for future genetic gain it is essential to promote and maintain variability in the breeding population in spite of the tendency to reduced variability which is the inevitable consequence of intense selection.

In many agricultural crop plants all lines are closely related, having ancestors in common and the wild progenitors, if not extinct, are vastly inferior in utility and productivity. In contrast, the eucalypts and other forest trees are valuable as wood crop plants even in the wild state, and their original genetic resources are largely intact. The immense natural variability of the eucalypt base populations constitute 'a sack of uncut diamonds' (Evans 1976) still to be exploited by exploration, evaluation, selection, and breeding, including interspecific hybridization.

In eucalypt breeding, unlike breeding of many agricultural crop plants, there is at present little incentive to employ sources of artificial variation such as polyploids and induced mutation. Genetic engineering offers exciting prospects of being able to add or remove specific genes in genetically advanced clones to confer resistance to certain diseases, insects, or herbicides. Other prospects include making interspecific hybrids between species which cannot be crossed by current methods. All these procedures require efficient tissue culture and clonal propagation techniques which have been pursued intensively for only a few eucalypt species.

Genetic engineering, the transfer of genes between organisms, should be considered as a supplement to an advanced program of eucalypt population improvement, not a replacement. At the time of writing no practical application of genetic engineering had been achieved in eucalypt breeding. To help decide whether or not to invest in genetic engineering, it would be appropriate to make an economic analysis of the relative efficiency of genetic engineering and more conventional breeding procedures for each trait to be improved (Herdt 1988).

There are numerous alternative breeding strategies for population improvement of eucalypts. In relation to the concept of the four populations described in Section 18.1 (Fig. 18.2) the breeding strategy mainly concerns efficient management of the breeding population. It is derived from an appropriate base population which continues to offer infusions of genes of individuals unrelated to those in the breeding population.

Sometimes the management of the breeding population may be adjusted for the needs of the propagation population. For example, a whole unit of the breeding population may be converted into a seed orchard (Franklin and Meskimen 1984; Barnes 1987). Alternatively, for clones of rooted cuttings, a special line or nucleus of the most advanced members of the breeding population may be created to provide the small number of outstanding crosses needed for mass vegetative propagation (Cotterill *et al.* 1989).

18.2.1 Developing a breeding strategy

How is a strategy developed? Developing a breeding strategy for a eucalypt planting project, large or small, requires consideration of a number of questions, many of which are addressed in other parts of this book. Here are ten important questions, among many.

1. Is there any need for a breeding program? Eucalypt growers often notice that some trees in a plantation are much better than others, and there may be varieties not yet tested. It is this recognition of variation, together with dissatis-

faction with the growth rate and quality of present plantations, which leads to investment in tree breeding.

2. What are the characteristics of the planting sites and what products are required (fuelwood, poles, pulpwood, sawlogs, shelter from wind, etc.)? (Sections 1.4, 20.1).
3. Do the available species, provenances, varieties, or clones match well with the intended range of site conditions and with the community's expectations? (Chapters 4 and 5).
4. What traits are to be selected for? How are these traits to be assessed? Are they easy to measure? What are the selection criteria? (Selection, even intense selection, will only be successful for a maximum of 3 or 4 independent traits and little improvement could be expected in any trait when selecting for each of 10 or 20 traits, unless they are positively correlated) (Sections 20.1, 20.2).
5. What is the economic importance (economic weight) of the traits being considered? (Section 20.2).
6. What evidence is there that these traits have enough genetic variation and high enough heritability for selection to be effective? (Section 18.3.2).
7. How much money is available to invest in breeding and for how long is the investment likely to be sustained?
8. Is there a long-term breeding objective over several decades, or a short term objective, or both?
9. What are the alternative breeding strategies and are they biologically and silviculturally practical?
10. For each strategy, what is the expected genetic gain per decade in the chosen traits, and what are the financial costs and benefits?

18.2.2 Examples of early eucalypt breeding strategies

In retrospect, the first steps in breeding eucalypts before about 1970 do not appear to have always included a careful consideration of all relevant questions but were more often taking advantage of opportunities which existed. Then gradually there was an improvement in the quality of planning and execution of the breeding programs. Looked at another way, the level of understanding of the whole area of tree-breeding strategy and technology has improved greatly since about 1970.

Here are nine examples of eucalypt breeding strategies commenced in the 1960s and 1970s.

1. *E. grandis* seed orchard established 1961 in São Paulo state, Brazil, with open-pollinated seedlings, was based on selection in a partly hybrid population. Heavy thinning on performance in the orchard was intended to produce seed which would result in stands of almost pure *E. grandis* with greater uniformity and reduced segregation (Pasztor 1962). This orchard's plan for selection within and among open-pollinated families was the model for many seedling seed orchards of pure species of eucalypts (Eldridge 1975b).
2. *E. regnans* seed orchard was established in 1970 by APM Forests Pty Ltd in Victoria, Australia, with open-pollinated seedlings of trees selected in natural forests nearby (Eldridge 1971; Richmond 1971). This is a very successful orchard (Section 21.2) and although it was based originally on 120 selected trees, the genetic base for future generations of selection has been expanded in recent years (Cameron *et al.* 1989).
3. *E. grandis* breeding started in Florida in 1961 by planting open-pollinated seed of the best trees selected locally (Franklin 1986). Pedigreed seed from the best trees in the best families in the progeny test was used to establish the next generation, together with seed from selected trees in plantations in other countries and from natural forests in Australia. However, there was no systematic range-wide provenance testing. Over four generations of increasingly orderly and expanding activity great success was achieved (without great expense) both in genetic gain (Meskimen 1983) and in commercial seed production from the heavily thinned progeny tests converted to seed orchards. Unfortunately, an extreme freeze in 1982 killed the trees and brought the plantation project to an end.
4. *E. grandis* grafted seed orchards were started in 1965 in South Africa (Section 21.3) based on the Scandinavian model widely used at that time. Selection of the plus-trees in plantations em-

phasized reduced end-splitting of logs. Difficulties with graft incompatibility were overcome and yield of seed was very large. Subsequent generations of population improvement of *E. grandis* in South Africa have employed a much wider genetic base of selections from plantations and newly imported provenances (Van Wyk 1985a).

5. *E. globulus* breeding started in Portugal in 1966 when Billerud selected plus trees in plantations of the Portuguese land race for size, form, and cellulose content (Dillner *et al.* 1971). Grafted seed orchards were established with 25 clones, together with progeny tests for later culling of the orchards. There were problems with poor seed yield from the orchards, particularly those furthest inland. A new approach to breeding which commenced in 1985 with provenance trials, controlled pollination, hybrids, and vegetative propagation, now employs nucleus breeding as the breeding strategy (Cotterill *et al.* 1989).

6. *E. deglupta* grafted seed orchards were first made in Papua New Guinea in 1973 using scions from 28–50 of the 150 plus trees selected for growth, form, and wood quality in plantations in New Britain (Davidson 1978b, 1983b). Seedling seed orchards were also established. Both types of orchard produced abundant seed in the second year. Progeny tests were established with seed from the 150 plus trees. Fast growth and early flowering in the progeny tests, leading to a turnover in breeding generations of only two years, would have assured rapid genetic gain by simple methods. However, breeding was suspended when the proposed plantation projects did not eventuate.

7. Hybrids of *E. gunnii* with *E. dalrympleana* and other species have been greatly improved in frost resistance and growth rate by AFOCEL in the south of France (Potts and Potts 1986). Frost hardy, fast growing seedlings were selected in the nursery and propagated by rooted cuttings, followed by testing the clones in field trials. Provenance testing and population improvement of the parental species were not part of the initial activities but were added later.

8. Hybrids which appeared fortuitously in eucalypt plantations near Pointe Noire, Congo, were the basis for development by CTFT of the successful technique for mass propagation of eucalypts by rooting cuttings of basal coppice after felling selected trees. Starting in 1972, the first hybrids used were *E. tereticornis* with *E. grandis*, and *E. alba* with a number of species including *E. grandis* and *E. urophylla* (Delwaulle 1985a). The hybrids are much better adapted to this equatorial environment than any pure species (Martin 1989). Since 1978 further clonal selection has been based on carefully planned hybrid combinations by controlled pollination of parent trees selected in provenance trials.

9. Hybrids of *E. grandis* and *E. urophylla* are common in plantations of the Brazilian landrace of *E. grandis*. Outstanding hybrid individuals were propagated as rooted cuttings from basal coppice when the plus-trees were felled. The clones were greatly superior in yield and quality to the land race (Brandão *et al.* 1984). This well-known clonal selection programme of Aracruz Florestal started in 1975. In the next decade the program developed into provenance testing and population improvement of pure *E. grandis* and pure *E. urophylla* in sublines to provide select parents for hybridization (Campinhos and Ikemori 1989).

None of these nine examples of pioneer eucalypt breeding programs of the 1960s and 1970s started with a broad genetic base population of 200 or more selected, unrelated trees of well-adapted provenances. And none started with a written, long-term strategy allowing for continued efficient genetic improvement over several generations of selection and mating. The broader base and long-term strategy came later in these programs. Eucalypt breeders starting new programs in the late 1980s learned from the experience of their predecessors. They not only insisted on an appropriate base population but, by publishing their breeding strategies, they have invited comment and amendment from interested colleagues around the world, for example *E. robusta* in Florida, USA (Dvorak *et al.* 1981), *E. urophylla, E. camaldulensis, E. grandis,* and *E. tereticornis* in China (Nikles 1987), *E. globulus* in China (Raymond 1988a), *E. globulus* in Portugal (Cotterill *et al.* 1989), and *E. globulus, E. nitens,* and *E. regnans* in Victoria, Australia (Cameron *et al.* 1989).

Often a published 'strategy' includes some details of the breeding plan, the means of implementing the strategy. In this section we concentrate on the relatively brief statements of breeding strategy and leave for a later section (Section 18.5) an account of how a breeding plan might be developed from the breeding strategy.

Eucalypt breeding strategies can range from simple and cheap to complex and expensive. Increasingly profound understanding of the genetics of eucalypts does not necessarily lead to a complex strategy. Indeed, Cotterill (1986b) strongly advocates the KIS principle ('Keep It Simple') when choosing a strategy, although the behind-the-scenes investigations and comparison of alternative strategies may have involved the most advanced quantitative genetics.

18.2.3 Examples of advanced eucalypt breeding strategies

As eucalypt breeding has become more advanced, estimates of genetic parameters of the eucalypt populations to be improved are being used to make the necessary calculations of genetic gain and to develop selection indices. Choosing between alternative strategies on the basis of estimated genetic gain and cost requires a combination of skills in statistics, genetics, biology, silviculture, and economics (Cotterill 1986a; Shelbourne 1991), which we will not go into in detail here.

Here are five examples of well-planned breeding strategies introduced in the 1980s for breeding eucalypts. Note that the great expense of controlled pollination is avoided, or at least minimized in each of these examples.

1. Mass selection with open pollination in a single population without keeping family identity (Shelbourne 1969, 1991):

(a) collect open-pollinated seed from about 300 unrelated trees in natural forests of suitable provenances with little or no selection, or in suitable plantations with a selection intensity of perhaps 1 in 1000;

(b) establish a plantation of 5–10 ha (or more than one for security against loss) with a single seedlot composed of a mixture of equal amounts of seed from all trees;

(c) thin early and heavily to convert to a seed production area and promote seed production on the retained trees;

(d) select the best 50–100 trees in the seed production area, collect open-pollinated seed and establish the next generation;

(e) repeat for 2 or 3 generations.

This strategy has been recommended frequently for eucalypts (for example Boland 1981) to provide a reliable and cheap source of somewhat improved seed while more complex and expensive strategies are developed.

In the first generation genetic gain in growth would be substantial because the neighbourhood inbreeding of the natural forests would be broken (Section 19.6.2). Generally genetic gain will be less than for the more complex strategies but could be substantial for traits of high heritability (Barnes 1984b, p. 626). The cost per unit gain per year may be minimal as there are no measurements, no computing, and few records. In later generations genetic gain would be small unless intense selection is applied, more than 1 in 10. With intense selection this is a fairly short-term strategy (about 2 or 3 generations) because inbreeding effects are likely to reduce genetic gain to zero within a small number of generations unless new unrelated material is infused. However, the second generation of such a large, mass selected, single population could be used as the basis for a more complex strategy, such as example 2 or 3 below.

2. Recurrent selection for general combining ability with open-pollination in a single population keeping family identity (Franklin and Meskimen 1984; Franklin 1986):

(a) collect open-pollinated seed from about 300 trees as in example 1, above;

(b) establish a progeny test of about 10 ha with small plots (1–5 trees), well replicated, and about 30–50 seedlings per family;

(c) measure and cull the first generation progeny test as early as possible (at age 3 years for *E. grandis*);

(d) collect open-pollinated seed from the best trees in all but the worst families (at age 4 years for *E. grandis*) to establish the second generation;

(e) cull the first generation heavily (at age 5 years for *E. grandis*) for future seed production by removing families identified as having low general combining ability at the first measurement of the second generation (after one growing season for *E. grandis*);

(f) each new generation is composed of the best trees of the previous generation, plus a small infusion of new material from any other promising source.

This strategy involves a single plantation per generation which serves sequentially as a progeny test of trees selected in the previous generation, as a basis for selection and breeding for the next generation, and finally as a commercial seedling seed orchard (Franklin 1986). In other words, it combines the base, breeding, and propagation populations. It requires a high level of skill and efficiency in establishment and assessment of tests, and in data analysis for selection on both individual and family merit. Large genetic gain in growth rate and frost tolerance was achieved over 20 years in four generations of selection of *E. grandis* in southern Florida without excessive cost (Meskimen 1983) (Section 21.2).

3. Multiple population breeding (Barnes 1984b). There are advantages in subdividing a breeding population to adapt to future selection goals and to avoid inbreeding in future plantations. If the several sub-populations are selected for different criteria the procedure is called multiple population breeding. But if the same selection criteria are applied to the several sub-populations they are called sublines (Burdon and Namkoong 1983). In practice the important thing is subdividing the breeding population, not the distinction between multiple populations and sublines:

(a) collect open-pollinated seed from about 300 trees as in example 1, above;

(b) allocate seedlots to sub-populations; if the 300 families were all from one provenance they could be arbitrarily divided into groups or sub-populations; more often each group would be based on a suitable provenance or land race;

(c) for each sub-population establish one or more plantings of open-pollinated families in progeny tests with well-replicated small plots, and up to 50 seedlings per family, with 25–50 families per sub-population, each planting in a subregion of the breeding region;

(d) each plantation is a sub-population which can be managed for commercial seed production after heavy thinning to form a 'breeding seedling orchard' (Barnes and Mullin 1989);

(e) to make the next generation collect open-pollinated seed from the best tree in each family;

(f) in later generations, as inbreeding increases within each sub-population, inbreeding effects in the wood-producing population can be minimized by establishing a seedling seed orchard with open-pollinated seed from a small number of selected trees from each of the several sub-populations.

Mating is by open pollination and selection is by combined index selection with family identity retained. This is another example of recurrent selection for general combining ability as in example 2, above (Namkoong *et al.* 1988). The special advantage of multiple population breeding strategy is that pedigree control can be at the level of sub-populations instead of individual trees, and consequently there is the capacity to revert to a much simpler procedure if funds and key personnel are withdrawn. Each of the sub-populations could be thinned and re-established for the next generation without retaining family identity or making assessments or data analysis and with only a small reduction in genetic gain (Barnes 1984b, p. 626). The management of the breeding population in this strategy is almost the same as 'open-pollinated mating with combined index selection' in sublines described by Cotterill (1986a). There is a difference in the propagation population in that each of Barnes' multiple populations becomes a seed orchard, while the selected trees from Cotterill's sublines are crossed for the propagation population, either by (a) controlled pollination of selections of different sublines for foundation stock for mass vegetative propagation, or (b) by grafting to make a seed orchard containing one or a few clones from each subline.

4. Nucleus breeding for the production of outstanding families for mass vegetative propagation after population improvement in a pure species (Cotterill *et al.* 1989):

(a) establish a main breeding population with about 300 parents using open-pollinated mating and combined index selection, as in example 2, above;

(b) from the best individuals in the best families establish a small nucleus breeding population of about 40 outstanding parents;

(c) make controlled pollinations in the nucleus and obtain greater genetic gain than in the main breeding population which employs open pollination;

(d) raise the control-pollinated families in progeny tests and select the best individuals in the best families by combined index selection for mass vegetative propagation;

(e) the tendency of the nucleus towards inbreeding is reduced by infusion of open-pollinated seed of a few of the best individuals from the main breeding population.

Nucleus breeding concentrates the expense of controlled pollination on a small number of the very best individuals to obtain maximum genetic gain per generation. Full pedigree of both parents is maintained in the nucleus. The improvement of the main breeding population continues at a somewhat slower rate by open pollination. Maintaining the main breeding population by controlled pollination seems to be prohibitively expensive and impractical for eucalypts. Nucleus breeding has been successful with sheep breeding since the early 1970s (Jackson and Turner 1972), and seems to have good prospects for application to eucalypt breeding.

5. Interspecific hybrid breeding (Van Wyk *et al.* 1989; Campinhos and Ikemori 1989):

(a) establish a breeding population of each parent species with families identified as in example 2, above;

(b) repeat selection and open-pollinated mating for a number of generations to improve each parental population;

(c) select outstanding individuals of each species and make interspecific crosses by controlled pollination;

(d) raise the F_1 hybrid families, select the best individuals in the best families and mass propagate as rooted cuttings, or by tissue culture;

(e) further selection of the best clones may be carried out in clone trials.

If additive gene effects predominate (general combining ability) in the interspecific hybrid population, this procedure provides the opportunity to make a new variety combining traits which do not occur in a single species, for example fast growth, drought resistance, and wood suitable for pulping in *E. grandis* × *camaldulensis* (Van Wyk *et al.* 1989). If non-additive gene effects predominate (specific combining ability), the more complex procedures of reciprocal recurrent selection would be required (Allard 1960; Matheson 1990; Vigneron 1991). In either case, difficulties in mass vegetative propagation, by rooted cuttings or by micropropagation, limit the use of certain otherwise attractive combinations, such as *E. nitens* × *globulus*. Also, there may be no advantage in growth rate in interspecific hybrids where pure species are well adapted; hybrid vigour is environmentally dependent (Martin 1989, p.85) and is generally confined to sites which are marginal for the pure species concerned. Therefore, one should not always expect to repeat the spectacular gains in wood production achieved with cuttings from hybrids at Aracruz and Pointe Noire as the technology, though it may appear to be relatively simple, is not necessarily transferrable to other environments and species.

18.3 Genetic parameters

In manipulating quantitative characters the plant breeder deals with the nature, magnitude, and interrelationships of genetic and non-heritable variation (Allard 1960). This variation is expressed by estimates of genetic parameters, each of which is 'a numerical quantity which specifies a population in respect to some characteristic' (Allard 1960, p. 470).

The main activity in improvement of eucalypt breeding populations is recurrent selection for general combining ability (see Figs 18.1, 18.2, 18.3). This form of selection seeks to increase the frequency of favourable genes in populations and so to improve traits of economic value. Thus the genetics of eucalypt breeding is predominantly quantitative genetics. This is because eucalypt breeders have selected only for quantitative traits for which variation is continuous, such as growth rate and stem straightness. No selection has been reported for any important, simply inherited, qualitative traits in the improvement of eucalypt breeding populations. High yield of certain leaf oils may become the first qualitative trait selected in eucalypts (Doran and Brophy 1991; Barton *et al.* 1991).

In this section we give an outline of estimates for eucalypts of the following genetic parameters: additive and non-additive genetic variance, heritability, genetic and phenotypic correlation, juvenile–mature correlation, genotype–environment interaction, inbreeding coefficient, and inbreeding depression. To obtain a realistic appraisal of rate and magnitude of improvement by selection and to compare possible alternative strategies, it is essential to have good unbiased estimates of genetic parameters from well-designed field trials. Finally, we outline how these parameters are used in predicting the effect of selection.

For more detail the reader is referred to a number of books which will help provide a basic understanding of tree breeding and forest genetics. For plant breeding in general we suggest Allard (1960), Brewbaker (1960), or Simmonds (1979), particularly the sections on breeding outcrossing plants (the sections on inbreeding and chromosomes are less relevant). For forest tree breeding Wright (1976) and Zobel and Talbert (1984) are practical texts and Namkoong *et al.* (1988) is more advanced. For a deeper understanding of genetic principles underlying selection and breeding Falconer (1989), Namkoong (1979), Mayo (1987), Hallauer and Miranda (1981), and Becker (1985) are suitable, provided the reader has some training in statistics. Clear accounts of the statistics and genetics involved in selecting trees efficiently in progeny tests are presented by White and Hodge (1989) and Cotterill and Dean (1990).

18.3.1 Additive and non-additive genetic variance

The genetic components of continuous variation are additive, dominance, and epistatic variation. Gene effects are said to be 'additive' when many genes each of small effect all influence the same trait: the more genes there are for that trait in an individual the bigger the effect. However, the several additive genes for one trait do not all have equal effects: some have much larger effects than others. A trait with this type of inheritance is described as polygenic or quantitative, in contrast to single gene or qualitative traits.

'Dominance' is the masking effect of one allele by another allele at the same gene locus. A simply inherited dominant trait, such as evergreen-ness in poplar, is very useful to the tree breeder as it is so easily incorporated in a population (Pryor and Willing 1983). The progeny of all crosses with a tree homozygous for this dominant trait are evergreen. No trait with this convenient type of inheritance has yet been found to be of economic importance in eucalypt breeding. Dominance is also important as part of non-additive effects in quantitative inheritance.

'Epistasis' is due to interaction between genes at different loci. Epistatic effects have not yet been estimated for eucalypts, but epistasis usually has been found to be small or negligible in studies of other plants and is usually considered with dominance to be part of 'non-additive' variance.

Additive genetic variance can be estimated from analysis of open-pollinated progeny in suitable field trials using the 'general combining ability' (GCA) of many individuals whose progeny are under test, that is, the average performance of each individual parent in crosses with several other trees. (GCA can be estimated more precisely with sets of several control-pollinated progenies in appropriate mating designs.) A very good cross can be predicted confidently by mating any two parents which have high general combining ability, as shown by their progeny performance (Tables 18.2, 18.3; Van Wyk 1990). Outstanding performance of a particular cross or an outstanding individual in a plantation may be due to non-additive gene effects. Such high 'specific combining ability' (SCA) is a matter of chance and is difficult to plan for or predict. Furthermore, the superiority is not carried on to the next generation because the favourable interaction between genes breaks down as random assortment and recombination take place during sexual reproduction.

Even if non-additive variance exceeds additive variance, as it did for volume in the study of *E. grandis* by Van Wyk (1976, 1990), selection for GCA will still be effective as it operates on whatever additive variance is present. For example, the performance of nearly all the 71 families in the progeny test (Table 18.2) was close to the volume expected from the average performance of each parent over several

Table 18.2

Mean tree volume growth (m^3 per tree) for 71 full-sib families of an incomplete diallel (20 × 20) of E. grandis *at 8.7 years. The ranking of the best five families is in brackets. The parents are ranked by mean volume of their progeny, female on the left side and male along the top. Note that families with high volume are mainly at the top left of the table and those with low volume are mainly at the bottom right (from van Wyk (1985a); more detail on the same data is in van Wyk (1990)).*

	4	15	22	6	58	17	101	19	10	45	37	35	39	66	44	38	60	36	64	Mean
4		0.77(2)	–	0.58	0.68	–	–	–	–	–	–	–	0.47	–	–	–	–	–		0.623
6	0.73(5)	–	–	–	–	–	–	0.44	–	–	–	–	–	–	–	–	–	–		0.599
50	0.79(1)	–	0.68	–	–	–	0.45	0.71	0.50	–	–	–	–	–	0.43	–	–	–		0.597
37	0.74(4)	–	–	–	–	–	–	–	–	–	–	–	–	–	0.38	–	–	–		0.557
15	–	–	–	0.76(3)	–	–	–	–	–	–	–	–	0.71	0.46	0.34	–	0.48	–		0.552
58	0.65	–	–	–	–	0.52	–	–	–	–	–	–	0.44	–	–	–	–	–		0.539
35	0.42	0.61	–	–	–	–	–	–	–	–	0.58	–	–	–	–	–	–	–		0.535
45	–	0.64	0.53	–	0.58	–	0.57	–	–	–	–	–	–	–	0.44	–	–	–	0.34	0.515
38	–	–	–	–	–	–	0.51	–	–	–	–	–	–	–	–	–	–	–		0.512
101	0.52	–	–	–	–	0.51	–	0.51	–	0.50	0.49	–	–	–	–	0.47	–	–		0.501
66	–	–	–	–	–	–	–	0.49	–	–	–	–	0.46	–	–	–	–	–		0.474
22	–	–	–	–	0.49	–	–	–	–	–	0.48	–	–	0.44	–	–	0.43	–		0.459
36	–	–	–	0.46	–	–	–	0.43	–	–	0.52	–	0.46	–	–	0.37	–	–		0.446
17	–	–	–	–	0.46	–	–	–	–	–	–	–	–	–	0.53	–	–	–	0.34	0.440
60	–	–	–	–	–	–	–	0.44	–	–	–	–	–	–	–	–	–	–		0.439
44	0.54	0.32	0.39	0.49	–	–	–	–	–	–	0.43	0.49	–	–	–	–	0.34	–		0.428
64	–	–	–	0.44	0.44	–	–	–	–	–	–	–	0.35	–	–	–	0.43	–		0.414
10	–	–	–	0.35	0.45	–	0.52	–	–	–	–	–	–	0.36	–	–	0.28	0.39	0.23	0.369
Mean	0.627	0.583	0.535	0.516	0.515	0.515	0.515	0.506	0.503	0.501	0.499	0.490	0.480	0.425	0.425	0.423	0.390	0.389	0.302	

Table 18.3
The best five of the 71 full-sib families in Table 18.2 showing the high rank of each of their parents for general combining ability

Progeny rank	Parent rank		Parent number
	♀	♂	♀ ♂
1	3	1	50 × 4
2	1	2	4 × 15
3	5	4	15 × 6
4	4	1	37 × 4
5	2	1	6 × 4

crosses. Consider the best cross 50 × 4 which had an observed value of 0.79. The female parent 50 had an average volume over six crosses of 0.597, which is 0.107 greater than the overall mean of 0.490. The male parent 4 had an average volume over seven crosses of 0.627, 0.137 greater than the overall mean. The expected value for 50 × 4 is thus

$$0.490 + 0.107 + 0.137 = 0.734$$

close to the observed value of 0.79. Similarly for average family 10 × 37 and the worst family 10 × 64 the observed and expected values were very close (0.49 and 0.510, 0.23 and 0.181).

The widest deviations from the expected values, such as 15 × 39 and 35 × 4, are considered to be examples of specific combining ability resulting from non-additive gene action. (Deviation is either above or below the expected value, although the breeder is usually interested in the cross which is above the expected value.) However, the breeder's objective of making several superior crosses without excessive cost would be easily met by mating best with best. Choice of the best could be done in a short-term open-pollinated progeny test to rank the 20 parents in Table 18.2 for their general combining ability.

For population improvement the commonly used procedure of recurrent selection for general combining ability consists of selection for additive effects only. No advantage can be taken of dominance or epistasis. This will be of little importance if, as is commonly supposed, large amounts of additive genetic variance are available in almost all traits in eucalypts and other forest trees, as well as non-additive variance. This is because domestication and breeding are at such an early stage in tree crops for wood production. In contrast, genetic conservation measures have come so late in most 'old' agricultural crop plants that little additive genetic variance remains and breeding is mainly with non-additive genetic variance.

When a plus tree in a eucalypt plantation is felled and mass propagated by cuttings from basal coppice shoots, immediate advantage is being taken of both the non-additive and the additive gene effects which led to that individual being outstanding. In general non-additive genetic variation is accessible to tree breeders only through vegetative propagation (Zobel and Talbert 1984, p. 312). An alternative, controlled pollination on a large scale to repeat proven superior crosses, is not feasible for the eucalypts at present.

18.3.2 Heritability

Heritability is a measure of how strongly a trait is influenced by genetics and how much by environment (Hanson 1963). It is the statistical expression for the relative contributions of genotype and environment to the phenotype. When heritability is high, gain from selection will be high. Heritability in the narrow sense (h^2) is the ratio of the additive genetic variance (V_A) to the total phenotypic variance (V_P):

$$h^2 = V_A / V_P.$$

Phenotypic variance is the sum of additive variance, non-additive variance (V_{NA}), and environmental variance (V_E) (Falconer 1989, p. 125):

$$V_P = V_A + V_{NA} + V_E$$

To obtain a reliable estimate of h^2 it is essential to have a large number of families and to minimize V_E by efficient statistical design of the progeny test from which the estimates are made.

An estimate of heritability only has meaning when several qualifications are stated. These include whether heritability is:

(1) used in the broad or narrow sense;
(2) calculated from variance components or parent–offspring regression;
(3) based on individual values or family means;
(4) assessed in one or several environments;
(5) assessed when trees are very young or near rotation age;

(6) estimated with confidence limits depending on the number of parents, offspring, and replications;

(7) for a particular population.

The extent of inbreeding among parents and in the progeny being tested are major sources of error by inflating estimates of heritability in eucalypts (Griffin and Cotterill 1988).

It should be emphasized that heritability is only a derivation from the analysis of variance (or from parent-offspring regression) and it is the significance of the genetic differences between and within families which is of primary interest. If the families do not differ significantly, an estimate of heritability is not justified.

With all these sources of error it may be asked whether heritability is worth estimating? The answer is certainly yes, provided the sources of error are understood and controlled, because heritability is so useful for index selection and for estimating genetic gain from alternative breeding strategies.

Estimates of heritability (h_i^2 narrow sense, individual heritability) for a number of eucalypt species are shown in Table 18.4. These estimates are mainly from open-pollinated progeny, some from trials which were large and well designed and others smaller and less efficient. Estimates of individual heritability based on open-pollinated progeny of trees in natural forests in which the original report used a coefficient of relationship among open-pollinated progeny of $\frac{1}{4}$ (assuming a half-sib relationship) have been reduced in Table 18.4 using a coefficient of 1/2.5 assuming 30 per cent selfing (Griffin and Cotterill 1988). These estimates would still be inflated and imprecise because of neighbourhood inbreeding when open-pollinated seed was collected in natural stands (Squillace 1974). For estimates of h_i^2 based on open-pollinated seed collected in plantations in which neighbouring trees are unrelated, the coefficient of relationship is probably closer to $\frac{1}{4}$. The most precise estimates of h_i^2 are from progeny tests of the control-pollinated progeny of unrelated trees.

Much remains to be done before reliable estimates of heritability are available for the major eucalypt plantation species. However, Cotterill and Dean (1990) note that estimates of genetic gain from index selection do not require very precise estimates of genetic parameters and it is enough to know, for example, that heritability is low, about 0.1–0.2, or high, about 0.3–0.5.

18.3.3 Genetic and phenotypic correlations (trait–trait correlation)

If taller trees are usually also straighter then these two desirable traits are positively correlated and selection for one will lead to improvement in the other. For example, King and Wilcox (1988) reported that selection of *E. nitens* only for diameter would produce a substantial correlated response in tree form because the genetic correlation of the two traits was high, 0.65. However if there is a negative correlation such that taller trees usually have wood of lower density it would be very difficult to improve a population in both traits by recurrent selection. It may be possible to find rare individuals which are correlation breakers and mass propagate these tall dense individuals by cuttings from basal coppice. For example, in a study of genetic variation in relation to penetration of wood preservatives of *E. grandis*, Van Wyk (1985b) found unfavourable genetic correlations between the penetration and retention of preservatives and the generally desirable traits of volume production and wood density. However, he was able to select several individuals which were correlation breakers suitable for mass propagation by cuttings.

Genetic correlation is an expression of correlated response to selection: if one trait is selected, what response is there in another trait? It is based on estimates of genetic covariance between traits from a progeny test. Phenotypic correlation, which includes genetic and environmental components, is simply calculated from the corresponding measurements of two traits, say the height and straightness for each of several trees. Genetic correlation is much more important to the breeder as it is the correlation of breeding values (Falconer 1989, p. 314). (Breeding value is the value of an individual judged by the mean value of its progeny.) Sometimes there is a large difference between genetic and phenotypic correlation coeffi-

Table 18.4

Published estimates of narrow sense heritability (h_i^2) from control-pollinated (c–p) and open-pollinated (o–p) progeny tests of six eucalypt species. Also shown are standard error (s.e.) of the estimate of heritability, age when the trial was measured, and number of families in the trial.

Trait	Species		$h_i^2 \pm$ s.e.	Age (yr)	Families	Source
Volume per tree	*E. globulus*	o–p	0.19 ± 0.07	6	45	(1)
	E. grandis	c–p	0.10	1	51	(2)
		c–p	0.53 ± 0.21	8	71	(3)
		o–p	0.31	5	529	(4)
	E. regnans	c–p	0.18 ± 0.17	4	21	(5)
		o–p	0.45 ± 0.28	4	13	(5)
Height	*E. globulus*	o–p	0.12 ± 0.05	6	45	(1)
		o–p	0.29 ± 0.09	4	36	(11)
	E. grandis	c–p	0.11	1	51	(2)
		c–p	0.45 ± 0.18	8	18	(10)
		o–p	0.39	5	529	(4)
	E. nitens	o–p	0.07	2	46	(6)
	E. regnans	o–p	0.13	4	16	(7)
		o–p	0.23	6	14	(7)
		c–p	0.18 ± 0.16	4	21	(5)
		o–p	0.43 ± 0.27	4	13	(5)
	E. tereticornis	o–p	0.16	4	15	(8)
Diameter	*E. globulus*	o–p	0.24 ± 0.07	6	45	(1)
	E. grandis	c–p	0.30 ± 0.16	8	18	(10)
		o–p	0.39	5	529	(4)
	E. obliqua	o–p	0.57	13	216	(9)
	E. regnans	o–p	0.10	10	64	(7)
		c–p	0.19 ± 0.17	4	21	(5)
		o–p	0.46 ± 0.28	4	13	(5)
	E. tereticornis	o–p	0.11	4	15	(8)
Sectional area	*E. globulus*	o–p	0.13 ± 0.05	18	36	(11)
Crown form	*E. globulus*	o–p	0.29 ± 0.08	6	45	(1)
	E. grandis	c–p	0.33 ± 0.11	8	71	(3)
	E. regnans	o–p	0.15	7	64	(7)
Stem form	*E. globulus*	o–p	0.22 ± 0.07	6	45	(1)
	E. grandis	c–p	0.46 ± 0.18	8	71	(3)
	E. nitens	o–p	0.07	3	38	(6)
Wood density	*E. grandis*	c–p	0.45 ± 0.18	8	18	(10)
Fibre length	*E. grandis*	c–p	0.54 ± 0.19	8	18	(10)

(1) Volker *et al.* (1990), (2) Van Wyk (1976), (3) Van Wyk (1990), (4) Reddy and Rockwood (1989), not adjusted as for (7), (5) Griffin and Cotterill (1988), (6) Purnell (1986), changed from family heritability (h_f^2) to individual heritability (h_i^2)using the formula of Wright (1976, p. 249), (7) Eldridge (1972), estimates reduced by using 1/2.5 as the coefficient of relationship instead of $\frac{1}{4}$, as recommended by Griffin and Cotterill (1988), (8) Kedharnath and Vakshasya (1978), reduced as in (7), (9) Matheson *et al.* (1986), (10) Malan (1988), (11) Borralho *et al.* (1992).

Table 18.5
Estimates of genetic correlations (upper triangle) and phenotypic correlations (lower triangle) for 45 open-pollinated families of E. globulus *at 6 years (from Volker et al. 1990).*

	Height	Diameter	Volume	Stem form	Branch size
Height		0.55	0.70	0.43	–0.38
Diameter	0.66		0.98	0.07	–0.45
Volume	0.75	0.95		0.13	–0.28
Stem form	0.09	–0.03	0.0		0.76
Branch size	–0.04	–0.38	–0.33	0.21	

cients, and even a different sign. Table 18.5 shows estimates of phenotypic and genetic correlations from a progeny test of *E. globulus* (Volker *et al.* 1990). Estimates of genetic correlation are seldom very precise because they are notoriously subject to sampling errors (Falconer 1989).

18.3.4 Juvenile–mature correlation (age–age correlation)

Breeders rarely have the luxury of waiting until rotation age before selecting the best provenances or the best trees both for mass propagation and for the next generation of the breeding population. Half rotation age is a common age for final assessment of eucalypts, and many attempts, often unsuccessful, have been made to select at a much earlier age, even in the nursery or in a controlled environment facility. Juvenile–mature correlation is expressed as genetic or phenotypic correlation coefficients between the pairs of data sets measured at different ages. The earlier the desired age of assessment and selection the more the eucalypt breeder needs the help of efficient physiological techniques. A better correlation between seedling growth rate and wood yield at rotation age may be obtained by including seed size, appropriate nutrient and water regimes, non-destructive estimates of leaf area or of plant volume and hence weight, and the use of relative growth rate. The cost of such physiological procedures with seedlings or small cuttings should be balanced against the cost of waiting another year or two until field trials of a fast-growing species have grown large enough to produce a reasonable correlation with yield at rotation age.

Juvenile–mature correlation, also known as age–age correlation, is only a form of trait–trait genetic correlation: tree height at one or two years could be one trait and the all-important volume at final harvest another trait. It is an active field of research and promises great benefits in two important aspects of eucalypt breeding: firstly for reducing the number of seedlots in long-term species and provenance trials (Chapter 5), and secondly for early identification of the best parents for the next generation of recurrent selection in an efficiently managed breeding population. To shorten generation interval it will also be necessary to induce early flowering.

18.3.5 Genotype–environment interaction

When the same set of genotypes is planted in trials on distinctly different sites it is sometimes found that the ranking of the genotypes changes greatly. A clone or family or provenance which is the best on a moist, fertile site may be among the worst on a dry, infertile site. The statistical expression for this change in ranking is called genotype–environment interaction (GEI) and is seen in provenance trials as provenance–site interaction. It is estimated by analysing data from several trials of the same clones or seedlots and by estimating the correlation between growth on one site and growth on another site (Matheson and Cotterill 1990).

An extreme case is the excellent growth of northern provenances of *E. camaldulensis* on tropical summer rainfall sites and their poor growth on mediterranean winter rainfall sites, and the converse with southern provenances. Clearly the northern provenances are well adapted to the tropical environment and the southern provenances are not (Section 6.4.1). Some species with a smaller geographic range and less variable natural environment have shown little provenance–site interaction, for example *E. pilularis* (Burgess 1975), while others such as *E. nitens* have substantial interaction, in that case with respect to disease resistance (Section 12.4.4). There appears to be no published information as yet showing significant GEI for eucalypt clones or seedling families, though some might be expected.

When evaluating the practical importance of GEI one can estimate the additional costs and benefits of dividing the breeding activities into regions within which GEI is minimal (Barnes 1984a; Matheson and Raymond 1984). Also one can calculate how much of the potential gain would be lost by ignoring the interaction for certain sites. It can be expected that some of the genotypes will contribute more to the interaction than others. These interactive genotypes could be rejected leaving a set of genotypes which are relatively stable over the major planting sites.

18.3.6 Inbreeding and inbreeding depression

When mating takes place between relatives the offspring are inbred to an extent depending on the degree of relatedness of the parents. The degree of relatedness is called the coefficient of inbreeding (F). Inbreeding depression (A) is the proportion by which inbred trees grow more slowly than trees resulting from outcrossing unrelated individuals.

If a breeding program starts with a base population of too few parent trees and intensive selection is carried out for several generations, a time soon comes when the genetic gain from selection is not enough to balance the inbreeding depression from mating close relatives. A numerical example will illustrate the problem, and its solution (Table 18.6). Assume there is a 5 per cent decrease in growth rate for each 0.1 increase in F, i.e. $A = -0.05\ (F/1.0)$, as the eucalypts appear to have a similar rate of inbreeding depression to other forest trees (Van Wyk 1976; Sedgley and Griffin 1989). The increase in F per generation for mass selection and polycross matings is $\frac{1}{2N}$, approximately, where N is the number of female parents starting each generation (Falconer 1989, p. 64). Table 18.6 gives two examples of inbreeding depression,

Table 18.6
Inbreeding depression (percentage decrease in growth rate, A) in each generation of mass selection with control-pollinated polycross mating and a constant number of parents (N) for each generation. The N parents are each used as male and female. Increase in the coefficient of inbreeding F is $\frac{1}{2N}$*, and* $A = -0.05\ (F/0.1)$.

Generation	$N = 50$		$N = 4$	
	F	A	F	A
1	0	0	0	0
2	0.01	–0.5	0.12	–6
3	0.02	–1.0	0.25	–12.5
4	0.03	–1.5	0.37	–19
10	0.10	–5.0		

with the decrease in the value of a trait as a percentage of its original outcrossed value.

Clearly inbreeding will cause little trouble when there are many unrelated parents (or only distantly related) say 50 each generation, because genetic gain in almost any trait can be expected to be much more than a loss of 0.5 per cent per generation due to this level of inbreeding. However, with a very small number of intensively selected parents, say 4, it may be difficult for genetic gain from selection for a single trait to match a loss of 6 per cent per generation due to inbreeding. Further research is required before being able to say whether the assumptions used here are of the correct magnitude for particular eucalypt species, and thus choose with more confidence whether 20 or 50 parents would be enough each generation to avoid excessive inbreeding in each subline of a large, long-term breeding population.

Inbreeding can usually be avoided altogether in trees intended for mass production in plantations because inbreeding effects are eliminated by a single generation of outcrossing. If the breeding population is divided into separate sublines (Section 18.3.3, example 3) each subline may eventually show signs of inbreeding depression, but not the plantations derived from crossing the sublines. This is because the progeny of a cross between two unrelated but inbred parents will be completely outcrossed and grow much better than the parents.

18.3.7 Use of genetic parameters in selection

The simplest expression for genetic gain through mass selection is half the product of the intensity of selection (i) and two genetic parameters, the phenotypic standard deviation of the population (σ_P) and an appropriate estimate of narrow sense heritability (h^2):

$$\text{gain} = \tfrac{1}{2}\, i\, \sigma_P\, h^2$$

The factor $\times \frac{1}{2}$ is included because only the female parents are selected. Intensity of selection (i) is expressed in units of standard deviation and is obtained from tables, for example from Falconer (1989, p. 354):

No. trees selected / no. trees scanned	1/4	1/10	1/20	1/50	1/100
i	1.27	1.76	2.06	2.42	2.66

The amount of variation in a population can sometimes be guessed close enough to make a rough calculation of gain. For example, when variation in several traits (volume, straightness, etc.) is expressed as coefficient of variation, i.e. standard deviation of means divided by the general mean for the trait, it is often about 10–15 per cent (Eldridge 1972). Similarly, narrow sense heritability is often about 0.1–0.2 for growth traits and about 0.4 for wood density (Table 18.4).

Consider the genetic improvement of a population by the separate selection of two traits, wood density and tree diameter. Density is expensive to measure and therefore a selection intensity of 1 in 4 (i = 1.27) would be appropriate. Variation in density is not extreme; for a mean density of 400 kg m^{-3} the standard deviation may be 40 kg m^{-3}, i.e. the coefficient of variation is 10 per cent and σ_P = 40. Heritability is usually high (about 0.40). Genetic gain per generation for density would be

$$\text{gain} = \tfrac{1}{2} \times 1.27 \times 40 \times 0.40 = 10.16 \text{ kg m}^{-3}$$

$$\text{per cent gain} = (10.16/400) \times 100 = 2.5 \text{ per cent}$$

Diameter is cheap and easy to measure and a selection intensity of 1 in 100 (i = 2.66) would be appropriate. There is considerable variation between trees in diameter, say 20 ± 4 cm (σ_P = 4). Heritability of diameter is low (about 0.10). Genetic gain per generation for diameter would be

$$\text{gain} = \tfrac{1}{2} \times 2.66 \times 4 \times 0.10 = 0.532 \text{ cm}$$

$$\text{per cent gain} = (0.532/20) \times 100 = 2.7 \text{ per cent}$$

Such rough calculations can give the tree breeder an idea of how much gain to expect per generation for a single polygenic trait by simple mass selection with open pollination. With higher values for i, h^2, and σ_P a gain of 5–10 per cent in the population could be expected. More elaborate and expensive breeding procedures of controlled pollination (selection of both parents), progeny testing, and index selection would be required to exceed 20 per cent gain per generation in a single trait.

Gain per generation is reduced if two adversely correlated traits such as diameter and density (Malan 1988) are both selected at the same time. The number of traits, even independent traits, being selected at one time must be small otherwise each trait can only be selected at very low intensity. Gain may also be reduced by genotype–environment interaction, as discussed in Section 18.3.5.

Genetic gains of much greater magnitude, 50–100 per cent in wood volume, have been obtained by wise choice of species or provenance and by mass vegetative propagation of selected outstanding individuals. However, both these spectacular increases in plantation production are once-only events. Having obtained a well-adapted variety far more productive than the inferior variety used previously, the next rounds of selection will produce more modest gains. The small gains by population improvement will be repeatable and accumulate each generation provided the breeding strategy is well thought out, avoids inbreeding, and is based on good estimates of the genetic parameters of the species. The more genetic information is available the more confidently the breeder can choose on a rational basis between several alternative breeding strategies using such methods for estimating genetic gain as those set out by Cotterill (1986a) and Shelbourne (1991).

18.4 Breeding plans

Having decided, with expert advice if necessary, which breeding strategy will provide the greatest genetic gain per decade at an acceptable cost for a particular breeding program, the eucalypt breeder can now write a detailed breeding plan of month-by-month operations for several years ahead.

One of the first published breeding plans for any forest tree was written for the North Carolina State University–Industry pine tree improvement co-operative by Talbert (1979). Another was written for the Southern Tree Breeding Association Inc. (STBA), a *Pinus radiata* breeding co-operative established in 1983 (Cotterill 1984). A more detailed plan for radiata pine breeding in New Zealand was produced by Shelbourne *et al.* (1986). Although Cotterill's basic objective of rapid turnover of generations employing a large initial base population, combined index selection, and efficient mating remained the same, the breeding plan was substantially revised and expanded in 1987 and 1989. This is to be expected as the breeder takes the opportunity to apply new ideas.

The subjects included in the STBA breeding plan would be appropriate for any eucalypt breeding plans. The largest section, apart from a catalogue of genetic material, is the detailed timetable for the next five years, presented as a list and a wall chart.
The subjects are as follows:

- Executive summary
- Objectives — a formal statement from the constitution of the cooperative, requiring rapid, continuing genetic improvement.
- Periodic review of plan — an annual review matching objectives to achievements, and a more searching review every five years or so.
- Historical background — including a catalogue of genetic material in progeny tests and clone banks.
- Seed requirements — to set the scale of operations.
- Seed orchards already established.
- Plans for mass propagation by cuttings.
- Outline of the long-term breeding strategy.
- Sublines — what they are composed of and how they will be managed.
- Gene pools — the broad base of unselected material.
- Monitoring genetic gain — yield trials.
- Timetable for five years — details of the monthly activities such as pollination and seed collection, and progeny test establishment, maintenance, assessment, checking data, analysis, selection, and reporting.
- Personnel and facilities required.
- Research needs.

A forward budget could well be part of the five-year breeding plan, although STBA takes the budget as a separate function.

The breeding plan is best written by the breeder who will be responsible for its implementation. If the breeder has insufficient expertise in quantitative genetics an outside consultant may be required to help choose the appropriate breeding strategy (the selection procedure and mating design). The breeding plan requires detailed local knowledge of the biology and silviculture and of the organizational structures and people in the plantation industry which the plan will serve. A joint writing effort by an outside consultant and the local breeder would be appropriate in some cases.

18.5 Realized genetic gain

The justification for breeding eucalypts is the genetic gain actually realized in improved plantations and the financial return on the investment in breeding. Most tree-breeding projects have proceeded without adequate yield trials to show just how much gain has been realized in growth rate, stem and branch form, wood quality, and resistance to diseases, pests, or frost, and how well predicted gain corresponds with realized gain.

For an unequivocal estimate of realized gain it is necessary to compare the commercial-scale product of breeding (seed from an orchard in full production, or cuttings from an operational nursery) with a well-defined and current commercial alternative (the previous generation of selection, a land race, or unselected seed from the wild). Though a yield trial may contain only two or three seedlots it must be carefully designed.

The requirements include: (a) a statistical design adequate to detect small differences; (b) uniform nursery conditions; (c) appropriate site preparation, spacing, and tending; (d) large plots; and (e) age of the trials at times of assessment sufficient to relate results to rotation age with reasonable confidence (Eldridge 1982).

Estimates of realized gain from progeny tests or from parts of extensive plantations are less precise than yield trials for cost–benefit analysis of breeding programs and for checking whether predicted gain matches realized gain.

The most notable estimates of realized genetic gain in eucalypt breeding have been for cuttings of *E. grandis* × *urophylla* in Brazil, cuttings of various hybrids in Congo (Section 22.1), and a *E. grandis* seed orchard in Florida. Of these, the only published report on realized gain based on a specially designed yield trial is of *E. grandis* in south-west Florida, USA (Meskimen 1983).

The Florida trial was planted in 1979 with four replications of 49-tree plots comparing 3 seedlots designated Ancestral (4 unselected imported seedlots), Commercial (the 34 best families of 529 in a progeny test, equivalent to seed from a seedling seed orchard), Premier (the 6 best families, all in the fourth generation of selection).

The genetic gain of the Premier seedlot over Ancestral was 66 per cent in height and 159 per cent in volume per hectare, at 2.5 years. Rotation age is 8 years. At the same age the progeny test with 529 families had predicted a smaller gain: 26 per cent in height and 126 per cent in volume per hectare. This may have been partly due to the very young age of both trials and to the yield trial being planted two years later at a wider spacing and on a poorer site than the progeny test (mean heights at 2.5 years 4.8 m and 7.1 m respectively).

A freeze of –8.5°C at 1.5 years left Premier trees with about 25 per cent of their leaves functional while Ancestral trees had all the crown and most of the branches and bole damaged. Regrettably the Florida yield trial was severely damaged in January 1982 by a catastrophic freeze of –11.5°C which killed the trees to ground level, though they sprouted again from the base (Meskimen *et al.* 1987).

For most eucalypt breeding programs the precise monitoring of realized genetic gain through specially designed yield trials is a subject still to be addressed.

19

Reproductive biology of eucalypts

A good understanding of the biology of sexual reproduction of eucalypts is necessary for the breeder to plan seed production efficiently. Because of the variation among species, it is necessary when making controlled pollinations to have detailed information on the flowering biology of the particular species of interest. It is also necessary for the eucalypt breeder collecting seed to be well aware of the extent and effects of inbreeding and hybridization. Those readers seeking an understanding of the processes involved in sexual reproduction of tree crops generally are referred to the book by Sedgley and Griffin (1989). These authors have presented details of breeding systems and of flower, fruit, and seed production in both horticultural and silvicultural crops. They examined the ecological context in which these attributes have developed and in many cases used eucalypts as examples.

From the point of view of domestication and breeding of eucalypts for wood production the distinctive characteristics of the sexual reproduction of the eucalypts are:

(1) flowers are hermaphrodite (male and female in the one flower);

(2) pollination is by animals (mainly insects and birds) not by the wind;

(3) outcrossing is favoured by various mechanisms including protandry, which means that the stigma is not receptive until some days after the pollen has started shedding from the anthers;

(4) the breeding system is predominantly outcrossing but there is also a considerable degree of self fertility as for many other tree-crop species;

(5) fruits are dry woody capsules and seeds are very small in the species used for fast-growing plantations (100 000 to 600 000 kg^{-1});

(6) seed production varies enormously under the influence of such non-genetic factors as spacing, site and seasonal conditions, effectiveness of pollen vectors, and predation of the developing seed crop by insects;

(7) species differ greatly in many aspects of flowering and seed production so that detailed information for one species does not necessarily apply to others, even those which are closely related.

In this chapter we review a large amount of information on all these subjects accumulated in recent years.

19.1 Flower morphology

A distinctive feature of eucalypt flowers, when open, is their lack of clearly recognizable sepals and petals. The showy part of an open eucalypt flower is the mass of stamens, the petals and sepals having been shed already (Fig. 19.1). The fused petals and sepals form the cap or operculum which provides a cover for the flower bud (Figs. 19.2(e), (f), Fig. 19.5). Indeed the name *Eucalyptus* comes from the Greek words 'eu' meaning 'well' and 'calyptos' meaning 'covered' referring to the operculum. Apart from having an operculum, flowers of the eucalypts are similar to those of several other genera in the large family *Myrtaceae* to which the genus *Eucalyptus* belongs. The eucalypt inflorescence may bear several flowers (commonly 1, 3, 7, or 11), arising in the axils of leaves. Much larger numbers occur in terminal

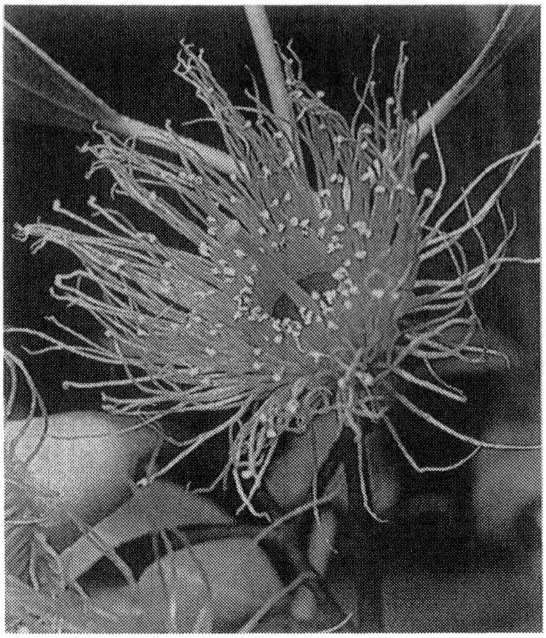

Fig. 19.1. The showy part of an open eucalypt flower is the mass of stamens (*E. leucoxylon*) (photo A.G. Edward).

inflorescences of some species (Figs. 7.4, p. 76, Fig. 20.1, p. 205) (Pryor 1976; Brooker and Kleinig 1983, 1990). When the developing inflorescence first becomes visible, the individual flower buds cannot be seen as they are enclosed by involucral bracts (Fig. 19.2(a), (b)).

The flower bud, which may be stalked or sessile, is in the form of a hollow receptacle surmounted by the lid-like operculum, which, in different species, can take shapes like a cone or a hemisphere (Jacobs 1981, pp. 589–609). The operculum is either a single-layered structure (in subgenus *Monocalyptus*) or double-layered (*Symphyomyrtus* and *Corymbia*). Detailed studies by Drinnan and Ladiges (1991) have shown that the operculum of eucalypts may be derived in more than one way, and not always from the same structures. In the most common case, where there are two layers, the inner layer is evidently made up of four fused petals and the outer of four fused sepals. The outer operculum completely covers the inner one and may fall away several days or months before flowering. The inner operculum covers the pistil and stamens until the flower opens.

Following the opening of the flower, the eucalypt shows the essential features of a normal bisexual flower. Inside the ring of tissue to which the operculum was attached is a band (the staminal ring) bearing the stamens (the male reproductive structures, the androecium) which unfold after the operculum falls away (Fig. 19.3(a), (b), Fig. 19.5). Further inwards from the staminal ring is the hollow cup-like receptacle. Buried in the base of the receptacle is the ovary, surmounted by the style and stigma (collectively the female reproductive structures, the gynoecium). The style and stigma extend beyond the rim of the receptacle (Fig. 19.3(b)).

The retention of the operculum until just before pollen shed (anthesis) and the separation of the anthers from the stigma inside the flower mean the stigma is well isolated from pollen until the operculum is removed. Emasculation of flowers can be achieved just before anthesis by removal of the operculum and staminal ring (see Section 20.5.1).

In most species each stamen consists of a filament supporting an anther, but in some species, such as *E. sideroxylon* and others in section *Adnataria* of subgenus *Symphyomyrtus*, some of the outer filaments are sterile. Filaments commonly are of different lengths in one flower, the inner ones being the shortest (Moncur and Boland 1989).

Differences in anther shape, as well as in the place where the filament is attached to the anther, occur in *Eucalyptus* and this feature has been used in making up keys to the species (Bentham 1867; Blakely 1965). The generalized anther structure is a pair of anther cells placed close together with a narrow connective, and dehiscing by longitudinal slits (Fig. 19.3(e)). A specialized type of anther which dehisces by terminal pores or slits is a feature of some species, including *E. melliodora* and *E. sideroxylon*. In the eucalypt bud before the operculum is shed, the stamens may be erect, as in *E. camaldulensis* (Fig. 19.5) and *E. tereticornis*, wholly inflexed as in *E. citriodora*, *E. grandis* (Fig. 19.2(e)), *E. deglupta* and *E. urophylla*, or variously twisted and inflexed. In those species with inflexed stamens in the bud the style is at some time during anthesis surrounded by a ring of many stamens and the stigma may come into contact with some of the unopened anthers as they brush past. At this stage the stigma is still unexpanded and not receptive, although the pollen is mature and able to germinate if placed on a receptive stigma (Pryor 1976; Hodgson 1976a;

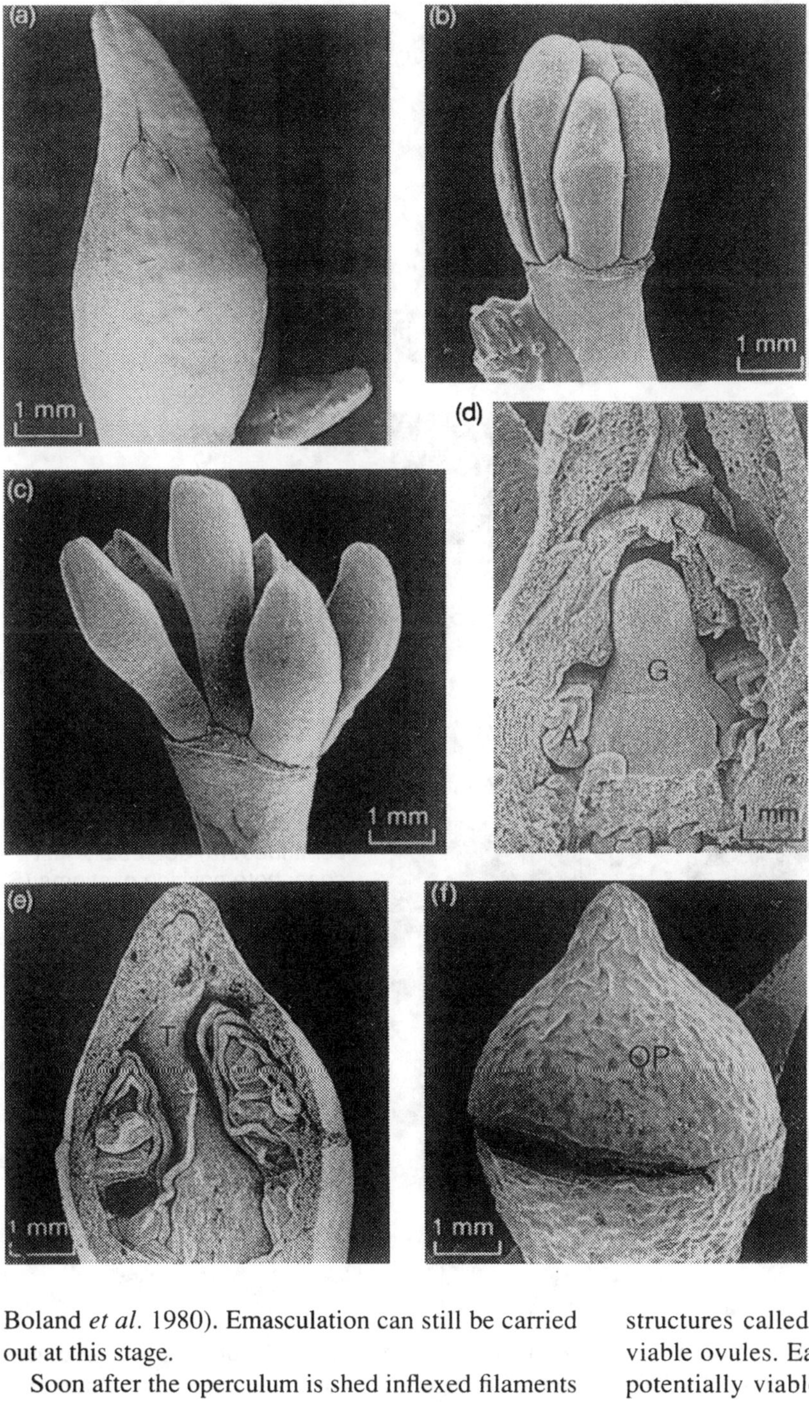

Fig. 19.2. Sequence of floral development in *E. grandis.* (scanning electron micrographs by M.W. Moncur): (a) inflorescence (umbel) in leaf axil, at the stage when the bud is first visible; (b) involucral bracts just shed, flower buds still close together; (c) flower buds spreading out; (d) section of bud as in (c) showing gynoecium (G) and androecium (A) at an early stage of development; (e) section of bud just before operculum shed; note the reflexed stamens and the bend in style (T); (f) operculum (OP) splitting prior to shedding.

Boland *et al.* 1980). Emasculation can still be carried out at this stage.

Soon after the operculum is shed inflexed filaments unfold radially and the anthers dehisce exposing pollen (Fig. 19.3(e), (f)) usually as irregular aggregates with a granular appearance.

The ovary consists of the chambers which enclose three types of structure (Boland *et al.* 1980a): sterile structures called ovulodes, viable ovules, and non-viable ovules. Each chamber usually contains 4 or 5 potentially viable ovules, depending on the species (for example *E. camaldulensis* 4, *E. deglupta* 3–5, *E. grandis* 5, *E. regnans* 5–6, *E. urophylla* 3 or 4, but 8 in *E. woodwardii*). A central vertical column bears the placentae, one per chamber, on which the ovular structures are borne in longitudinal rows.

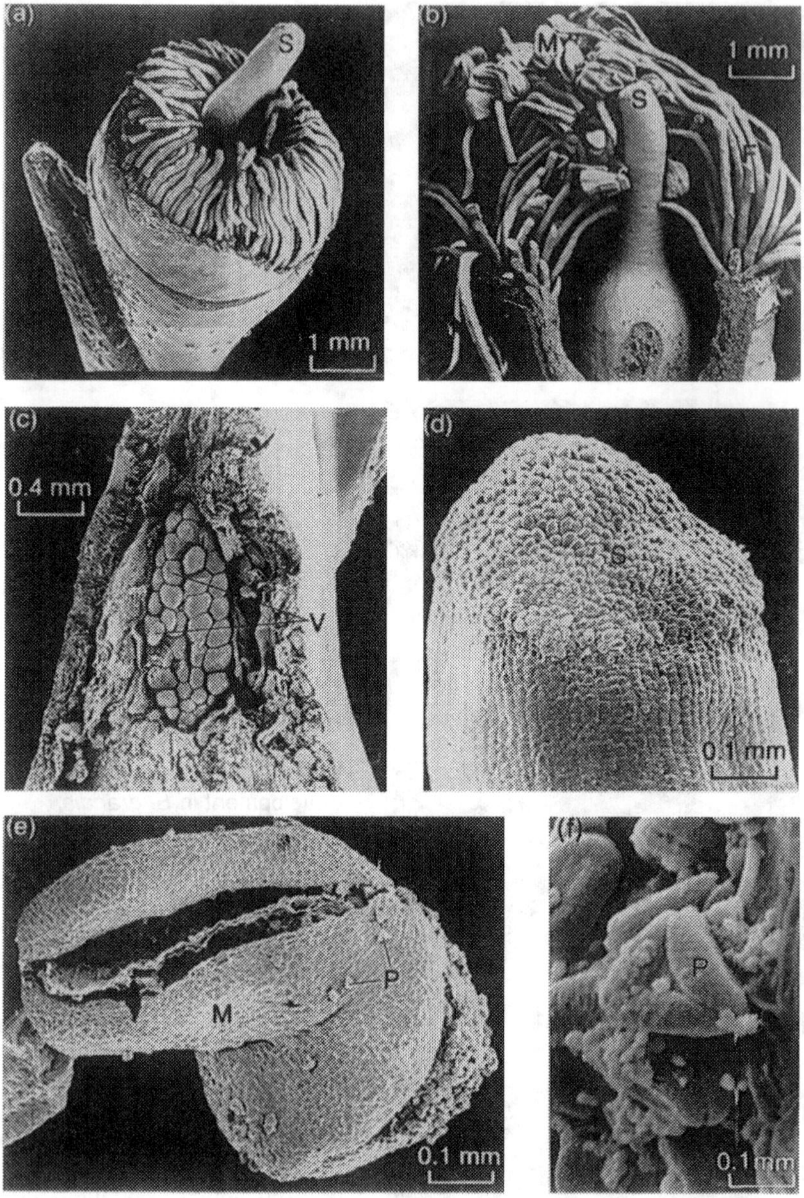

Fig. 19.3. Sequence of floral development in *E. grandis*, continued: (a) operculum just shed; note position of stigma (S) and anthers still at base of receptacle; (b) filaments (F) have straightened out; note position of anthers (M) and stigma (S); (c) ovary dissected to show ovules (V); (d) stigma (S) not receptive; (e) anthers (M) at stage (b) releasing pollen (P); (f) pollen grains (P); the small particles on the pollen grains are artefacts of the SEM process.

The mature fruit of the eucalypts is a woody capsule with triangular valves at the top (Fig. 16.5, p. 149, Fig.19.4(f)). There is a wide range of size and shape among the species (Jacobs 1981, pp. 589–609). It is important for the eucalypt breeder to know something of the mechanism of seed shedding of the particular species to ensure efficient and complete extraction of seed (see Section 19.4). The non-fertilized ovules and the ovulodes develop into chaff and are shed first while much of the viable seed forms deep in the mature ovary and is shed from the capsule last. Sometimes the lowermost seeds are pressed tightly into the base of the capsule and may not be shed naturally.

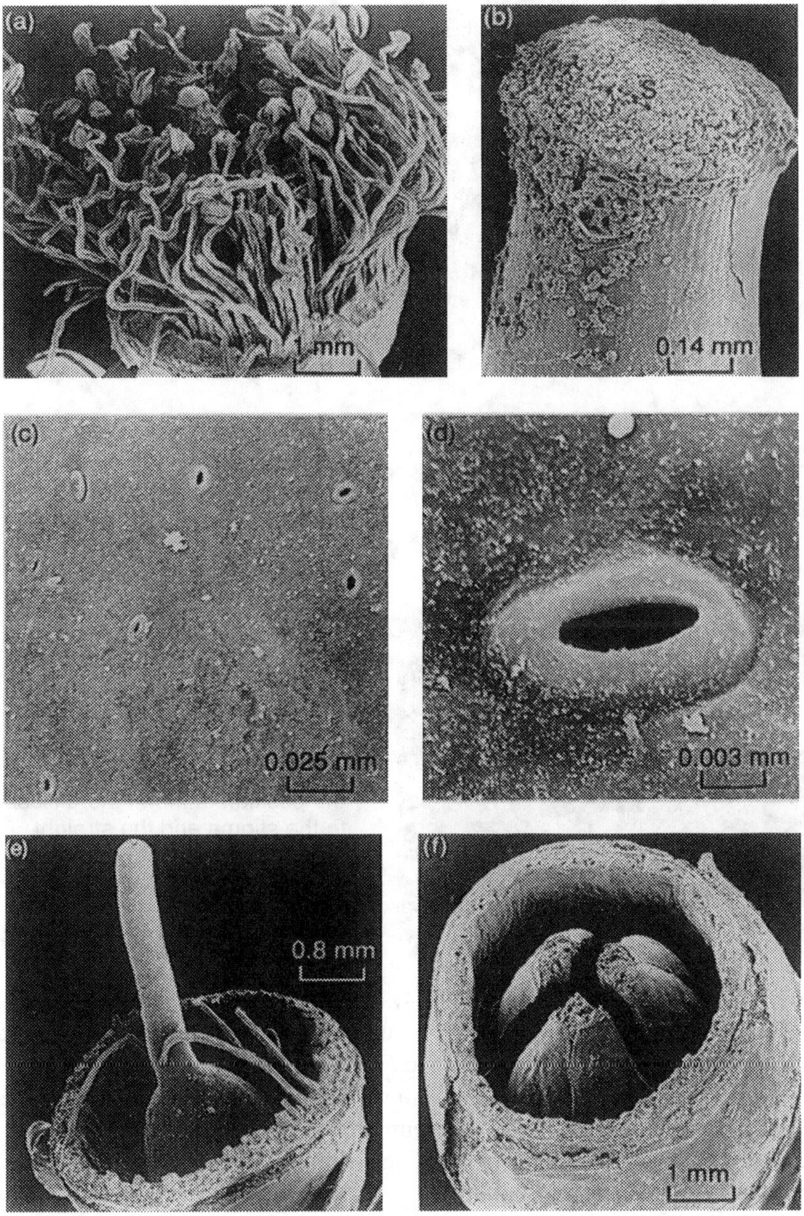

Fig. 19.4. Sequence of floral development in *E. grandis*, continued: (a) Filaments arch out, some collapsing; pollen has been released; (b) Stigma (S) now receptive, note secretion; (c) Nectar pores on inside of receptacle; (d) Single nectar pore, note likeness to stomata; (e) All filaments have abscissed; (f) Mature capsule, valves open.

19.2 Flowering phenology

19.2.1 Age of the tree at first flowering

Age of the tree at first flowering is of importance in tree breeding in that it determines the speed of generation turnover and therefore the rapidity of genetic gain. Pederick (1976a) noted that cultivated trees of *E. leucoxylon* will flower as early as age two years. In *E. grandis* first general flowering in a plantation occurs at two to three years. Some species from cooler climates, such as *E. diversicolor*, *E. globulus*, *E. nitens*, and *E. regnans* do not flower heavily, even at wide spacing, until about 7–10 years. *E. dunnii* is perhaps the slowest to start flowering with only a few trees in flower at 10 years (Graca 1987).

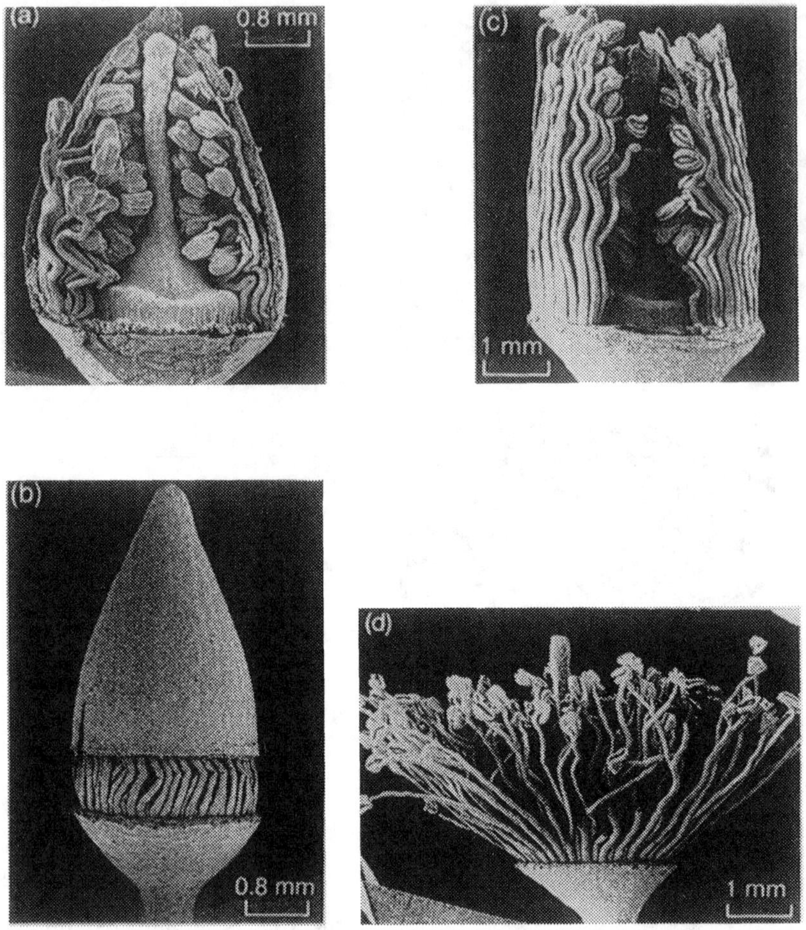

Fig. 19.5. Sequence of floral development in *E. camaldulensis* (scanning electron micrographs by S. Ngamkhajornwiwat and M.W. Moncur): (a) Section of flower bud just before operculum shed showing that the anthers are not confined to the base of the receptacle as they are in *E. grandis*, Fig. 19.2(e), 19.3(a); (b) Operculum shedding; (c) Operculum just shed; note the position of the anthers close to the stigma and the straight filaments; (d) Filaments have spread out; note position of stigma.

While the age of the tree at first flowering is variable between species it is also very variable within species. Occasionally two-month-old seedlings of *E. deglupta*, *E. grandis*, *E. tereticornis* and *E. urophylla* bearing flowers have been found in nurseries in Brazil and elsewhere. Such plants developed normally and flowered each following year. Descendants of these plants were just as precocious, suggesting that early flowering is strongly inherited (J. Davidson observations).

19.2.2 Season of flowering

Within any population of eucalypts the main flowering periods for individual trees may be quite out of phase. Such situations have been observed for *E. deglupta* (Eldridge 1976; Davidson 1983b), *E. regnans* (Eldridge 1970; Ashton 1975a), and *E. pilularis* (Florence 1964) in studies of season and duration of flowering in natural stands. Griffin (1980) observed that on average *E. regnans* trees flowered for 43 days and the order in which individual trees flowered was consistent from one year to the next. The first-flowering and last-flowering trees in a stand were reproductively isolated because their flowering times did not overlap.

The published lists of season of flowering (Blakely 1965; Boland *et al.* 1980) provide only a rough guide as flowering in natural forests has been shown to vary widely with site, with provenance, from tree to tree within a species, and to some extent from year to year (Burgess and Griffin 1991). Trees at a site with cooler temperature, for example at high altitude, will often flower later than those at warmer sites at lower altitudes. Season of flowering in natural forests is not a good guide for exotic plantings. In the Rio Claro arboretum in Brazil (Gurgel and Cavalcanti 1978) some *E. tereticornis* trees can be found in flower in every month of the year, leading to hybridization with several related species in neighbouring plots of the arboretum, an outcome which would be abnormal in the native habitat. In *E. deglupta*, which normally has two main flushes of flowering each year, some trees bear a few flowers all year round. It was noted that this lowland tropical species flowered more copiously if planted in highland areas, a feature which was used to advantage by locating seed orchards above 1000 m altitude in Papua New Guinea (information from J. Davidson). Eucalypts growing as exotics often have a more extensive flowering season than in natural forests.

Provenance variation in flowering season has been observed in trials of several species. For example, in a provenance trial of *E. globulus* in Tasmania, Volker *et al.* (1992) found that trees originating in the Bass Strait provenance region flowered so much later than trees from the south-east region of Tasmania that the two groups were to a large extent reproductively isolated (Section 10.4.1). Similarly, in a provenance trial of *E. camaldulensis* in Zimbabwe, Mullin and Pswarayi (1990) found distinct differences between groups of provenances in flowering season and duration.

19.2.3 Chronology of flower development

Phenology and related anatomy for *E. grandis* in South Africa was studied in detail by Hodgson (1975, 1976a). Hodgson recognized three stages in flower bud development for *E. grandis*. The first was the 'umbel bud' stage, when the inflorescence was enclosed in its involucre of bracts. The second was when the involucral bracts had been shed, revealing individual buds with a two-layered operculum and the third was the 'flower bud' stage when only the inner operculum was present. Each of these stages lasted about six weeks to give a total of about four months from appearance of the inflorescence bud to anthesis. Depending on the particular species and subgeneric group the time from the first appearance of the complete inflorescence bud to anthesis may be from four weeks to two years or more. For example, in *E. regnans* this period lasts 27–29 months (Ashton 1975a).

Shedding of the operculum at anthesis is preceded in many species by a change in its colour. This change is from green to yellowish in *E. grandis* and *E. regnans*, green to reddish-yellow in *E. nitens*, green to cream in *E. deglupta* and *E. camaldulensis*, green or pale brown to cream in *E. urophylla*, and green to pale brown or cream in *E. citriodora* for example. This colour change indicates bud maturity and is an important indicator for the timing of emasculation.

For controlled pollinations (Section 20.5) it is important to understand exactly what stage flower development has reached when deciding whether to put pollination bags on or take them off and when to apply pollen. Moncur and Boland (1989) recognized 13 stages in flower development of *E. melliodora* from splitting of the operculum to shedding of the stigma.

At anthesis an increase in the length of the style completes the detachment of the operculum at the top of the receptacle where an abscission layer had formed previously. Then as the stamens begin to unfold or extend they function as levers in lifting or tilting the detached operculum. In several species such as *E. citriodora, E. deglupta, E. grandis* (Fig. 19.3(a)), and *E. urophylla* the stamens are still relatively tightly inflexed when the operculum is finally shed. A few hours later in these species the anthers pass very close to the stigma to be completely unfolded after about 24 hours. In *E. deglupta* anthesis of individual flowers in an inflorescence is spread over some 36 hours, on a branch over 1–2 weeks, and on the whole tree over as much as two months (J. Davidson observations). In some clones of *E. grandis* there is a tendency for style elongation to be small or absent resulting in the inner operculum remaining on the bud for indefinite periods. There are also clones exhibiting early operculum abscission where both the inner and outer operculum are cast off together at an early stage (Hodgson 1975, 1976a).

For *E. regnans* Eldridge (1970) noted anthesis may spread over one week within a single inflorescence and that flowers in the more shaded parts of the crown opened 2–3 weeks later than in the more exposed areas. In a later study Griffin (1980) reported the spread of anthesis for *E. regnans* to average 6 days in the upper crown and 14 days within the side branches.

A stigma with a blunt tip is characteristic of a large number of species in the subgenera *Symphyomyrtus* and *Monocalyptus* surveyed by Boland and Sedgley

(1986). However, when the stigma first appears after the operculum falls it is somewhat tapered in outline and dark green in *E. camaldulensis, E. deglupta, E. grandis* and *E. regnans*. Within the next 3–4 days, it becomes more rounded in shape and a lighter colour and in many species these features develop still further, until by about the sixth day the stigma may be very swollen, shiny white, and sticky in appearance (Fig. 19.4(b)). This condition lasts for a few days, then the stigma develops a brown speckled or plain dark brown appearance. The style normally withers or breaks off at the base after 1–3 weeks. The most receptive stage for germination of pollen on the stigma seems to be just before it reaches the shiny and sticky stage (see Section 19.3.4).

It is important to be aware that the chronology of eucalypt flowering does not depend only on species or individuals. Environmental conditions such as temperature and availability of moisture and nutrients have large effects on flowering.

19.3 Pollination and fertilization

19.3.1 Pollen production and germination

When the anthers dehisce and the pollen is exposed it is found to be (in nearly all species) in irregular granular aggregates caused by the sticky nature of the individual grains. The sparseness and sticky properties of eucalypt pollen are in marked contrast to conifer pollen which is dry and blows freely in the wind as yellow clouds.

Pollen shed from individual anthers is mostly complete in *E. deglupta, E.grandis*, and *E. regnans* two days after anthesis. Although microscopic examination reveals small amounts of pollen on the anthers up to the fourth day, these pollen grains are usually dead. In fact, it is wise to collect pollen at anthesis, or even just before, when one can be sure the pollen is viable.

Germination of pollen on the sticky and receptive stigma (Fig. 19.4(b)) is generally rapid and, when mating is compatible, the pollen tube grows down through the style to reach an ovule and effect fertilization (Sedgley and Smith 1989). Hodgson (1976a) found that pollen could remain on the stigma of *E. grandis* flowers for 2–3 days before germinating. However, Griffin and Hand (1979) suggest that the pollen of *E. regnans* does not have the same capacity because pollen deposited in the first days after anthesis will die before the stigmatic surfaces become receptive 12–14 days later.

19.3.2 Pollen vectors

It appears that pollination in all eucalypt species is effected by animal vectors, particularly insects. A number of species with very large flowers (more than about 2 cm diameter) are considered to be predominantly pollinated by birds (Griffin 1982). Among these the pendulous yellow flowers of *E. stoatei* have characteristics which appear to favour exclusive bird pollination. Large insects cannot enter the flowers as the stamens form a dense dome over the floral cup and access is possible only to the probing beaks of honeyeaters (Hopper and Moran 1981). *E. stoatei* and other species with very large colourful flowers are small, slow-growing, multi-stemmed species which are not used in plantations. They usually have a small number of flowers, sometimes only 100 per tree.

Griffin and Ohmart (1986) considered that 'the flowers of perhaps half the species in the genus produce sufficient food reward to attract birds'. There are many observations that small-flowered species are visited by birds (for example Churchill and Christensen 1970; Ashton 1975a; Bond and Brown 1979; Hopper 1980). Ford *et al.* (1979) estimated that at least 200 species could, in part, be pollinated by birds with honeyeaters and brush-tongued parrots (Fig. 19.6) among the most likely to effect pollination (Barber and Jackson 1957; Christensen 1971). The significant difference between birds and insects is that birds are potentially more effective in outcrossing because they are capable of a wider range of movement and are still active in rather inclement weather conditions. The major plantation species have small flowers (with the exception of *E. globulus*), large numbers of flowers on each tree (sometimes over 1 million), white filaments, and no apparent adaptation to particular vectors. The absence of specialization is reflected in the great variety of visiting insects including *Coleoptera, Diptera, Hemiptera, Hymenoptera, Lepidoptera,* and others (Pryor and Boden 1962; Ashton 1975a; Hodgson 1976a; Ireland and Griffin 1984; Griffin and Ohmart 1986).

Fig. 19.6. Eucalypt flowers can be pollinated by birds, such as the varied lorikeet, *Psitteuteles versicolor* (photo Graeme Chapman).

Eucalypt flowers produce nectar which is not only an attractant and reward to pollen vectors but also an important source of honey in many countries. The European honey-bee (*Apis mellifera*) is regarded as an effective substitute for natural pollinators in some natural forests as well as in seed orchards and plantations (Krug and Alves 1949; Eldridge 1963; Harris 1966; May 1969; Loneragan 1979; Sedgley and Griffin 1989).

Dominant pollen vectors may also vary in time and space. For example, Ashton (1975a) observed flocks of honeyeaters visiting *E. regnans* but in a more isolated stand some years later, Griffin (1980) noticed a lack of birds over several flowering seasons. Nonetheless Griffin found adequate seed-set did occur, presumably through the activity of insects (Griffin and Cotterill 1988).

Much research remains to be done before it is established which insects, birds, and mammals actually pollinate a particular eucalypt in natural forests or plantations and which are the most effective pollinators. It is not enough to record the visitors to flowers. It is essential to develop techniques to show how much viable pollen each species carries to receptive stigmas and over what distance (Griffin 1982). In a study of pollen vectors in a remnant of a natural stand of *E. regnans* Griffin and Ohmart (1986) developed an index of pollen deposition potential (PDP = number of insects caught × pollen load per insect × frequency of stigma contact). They judged that the larger flies (tachinids, calliphorids, syrphids) were the most important pollen vectors and that bees, wasps, and several other kinds of insects also play an important role (Table 19.1). They found that both the diversity and activity of insects was greater under calm sunny conditions.

Table 19.1
Pollen deposition potential (PDP) of the 10 most effective functional groups of insects visiting flowers of Eucalyptus regnans, *from Griffin and Ohmart (1986).*

Functional group	Pollen load (grams/insect × 10^{-2})	PDP	% Total PDP all groups
Large tachinid flies	840	252	22.9
Medium tachinid flies	35	174	15.8
Medium syrphid flies	79	161	14.7
Large calliphorid flies	140	126	11.4
Medium wasps	64	109	9.9
Medium muscoid flies	8	61	5.5
Large wasps	148	44	4.0
Large syrphid flies	125	38	3.4
Medium bees	30	37	3.3
Ichneumonid wasps	40	24	2.1

PDP = No. insects caught × pollen load per insect × stigma contact factor

Davidson (1983b) is of the opinion that the green tree ant could have a pollinating role in tropical *E. deglupta* as it collects eucalypt nectar, unlike most other ants (Janzen 1977). The adult green tree ants, though not hairy, become covered in a sticky film, while foraging among inflorescences in search of nectar and excretions from aphids and scale insects. Their larvae are also sticky and are carried long distances by the adults to provide silk-like secretions to bind green leaves together to form nests. Colonies range for several hundred metres along aerial pathways and may extend over more than 100 trees where the canopies touch or interlace, so they could be responsible for wide movement of pollen.

There is no experimental evidence of wind pollination in eucalypts. Pryor (1976, p. 26) suggested that 'the lack of adherence of the pollen in clumps' in *E. tereticornis* and *E. blakelyi* may lead to some degree of wind pollination though the presence of nectaries and retention of other normal flower structures suggests these species are also primarily pollinated by animal vectors.

19.3.3 Distance of pollen dispersal

The size of an interbreeding population will be affected by the distance that pollen is transported, which is itself dependent on the vector involved. In natural forests, pollen movement has been estimated by assessing the extent of interspecific crossing since F_1 hybrid seedlings can be distinguished from non-hybrid parental type seedlings. *E. fastigata* and *E. robertsonii* often occur naturally in adjacent stands. Although their flowering periods usually differ, occasional overlap occurs with a few individuals of both species flowering concurrently. Pryor (1951) collected seed from trees of the two species that had flowered at the same time. An estimate of the pollen dispersal was obtained by examining the numbers of interspecific hybrids (easily identifiable from seedling morphology) in the progenies of pairs of trees separated by varying distances. Similar studies were made of *E. regnans* and *E. macrorhyncha* by Ashton and Sandiford (1988) (see Section 19.6.3).

Isozyme techniques have been employed by W.T. Adams for the same purpose in a 13-year-old seedling seed orchard of *E. regnans* in eastern Victoria, Australia. The mean effective pollen dispersion distance was estimated to be 42 m (Sedgley and Griffin 1989, p. 237). Pacheco *et al.* (1986) used an isotope of phosphorus (^{32}P) to more accurately follow the movement of pollen. They labelled two trees with the isotope in the centre of a seed orchard of *E. saligna* in Brazil and placed four beehives nearby. They were able to detect the appearance of this ^{32}P-labelled pollen up to 300 m distant, but mainly within 100 m.

19.3.4 Receptivity

There have been two main bases of assessment of receptivity: germination of the pollen on the stigma and seed yield. The ovary contains many ovules but relatively few ordinarily become seed through open pollination. For example, Griffin *et al.* (1987) showed that in *E. regnans* only 9 per cent of ovules became seeds, 1.48 full seeds from 16.7 fertile ovules, on average. The considerable potential for controlled pollination to increase reproductive efficiency was demonstrated by Hodgson (1976a). He obtained 34 seeds per capsule after controlled pollination of *E. grandis* and only 8 seeds per capsule from open pollination.

Subjective observations by Hodgson (1976a) on swelling and stickiness of the stigma as a guide to receptivity seemed to point out that this indicator appeared rather late to be of help in judging the best time for artificial pollination (Fig. 19.4(b)) unless one was proficient in detecting the very beginning of these processes around day 4 after anthesis. A more objective strategy to establish the optimum time of receptivity would be to use time in days from anthesis arrived at by experimental means rather than rely only on subjective indicators of the receptive period for artificial pollination.

Accordingly, Hodgson conducted experiments on stigma receptivity of *E. grandis* based on number of seeds set per capsule after artificial pollination at anthesis and on each of 10 subsequent days. Seed set rose sharply from anthesis to peak around days 4 to 6 and fell away equally sharply. In complementary experiments Hodgson (1976a) studied stigma receptivity on the basis of numbers of pollen grains germinated on stigmas artifically pollinated at anthesis and on each of four days thereafter. Stigma receptivity appeared to rise rapidly after day 2 and was at a maximum on day 4.

In similar experiments with *E. regnans* Griffin and Hand (1979) showed the optimum time for artificial pollination of *E. regnans* in terms of seed yield per flower pollinated was 10–14 days after anthesis. As in *E. grandis* the time lag between maximum receptivity and the change in shape and development of stickiness of the stigma was also apparent in *E. regnans*. Unless skill is developed in recognizing the very first changes in stigma morphology indicating receptivity, it will have only limited practicability in the field to judge the time of artificial pollination of *E. regnans*, except to show when it is too late to apply pollen. These observations were confirmed in a detailed anatomical study by Sedgley *et al.* (1989) on pollen tube growth, ovule penetration and seed development in *E. regnans*.

The development of receptivity of the stigma of *E. woodwardii* also has been examined experimentally as part of a comprehensive study of sexual reproduction of this species which included pollen tube growth and ovule and seed development (Sedgley and Smith 1989; Sedgley 1989). They found that both stigma secretion and ability to support pollen germination reached a peak at 7 days after anthesis. However, they found no change in style length after anthesis, in contrast with *E. grandis* and *E. regnans*.

Polunina (1959) found stigmata of three other species of *Symphyomyrtus, E. cinerea, E. macarthurii*, and *E. occidentalis*, became sticky 2–3 days after anthesis. There is some evidence of temperature dependence in the difference in rate of development of receptivity of stigmas. Hodgson (1976a) noted a difference in temperature between high- and low-altitude planting sites was sufficent to induce a difference of about two days in development of receptivity in *E. grandis* (longer at 1300 m than at 760 m).

19.3.5 Barriers to selfing

In *E. grandis*, when self and cross pollens were applied in mixture by Hodgson (1976a), fertilization occurred in favour of the cross pollen. Hodgson observed no incompatibility on the stigma, but there may have been inhibition of the pollen tubes of the self pollen in the style. Chaperon (1978) is of the opinion that, for some tropical eucalypts at least, the degree of selfing may be reduced by competition between pollen tubes in the style. However, Sedgley and Smith (1989) have shown by detailed anatomical studies of *E. woodwardii* flowers that pollen tube growth of self and outcross pollen is approximately the same. The difference came at the stage of penetration of ovules by pollen tubes when fewer ovules were penetrated following self-pollination than cross-pollination.

Preferential outcrossing in *E. regnans* was studied by Griffin *et al.* (1987) using isozyme techniques and anatomically by Sedgley *et al.* (1989). When an equal mix of self and outcross pollen was applied to receptive stigmas an average 81 per cent of seeds were found to be from outcrossing. There was some evidence to support the conclusion that inbred embryos had lower viability and competitive ability in the developing fruit. In other words, inbreeding depression is expressed at this very early stage as well as in germinated seedlings (Section 19.6.1).

A form of pollen sterility was demonstrated in *E. pulverulenta* by Pryor (1976). The pollen produced was defective in that the grains when shed had no protoplasmic contents. The condition is determined by inheritance and was found at low frequency in the natural population studied. Similarly, some individuals of *E. grandis* were found to be male sterile. No released protein could be detected by the immunofluorescence technique and pollen grains did not germinate on sucrose agar (Pryor 1976). In effect the flowers on a single tree which produces abnormal and sterile pollen are biologically unisexual and the female component functions normally with viable pollen from other plants and will set seed in the usual way. Conversely, in rare variants where female structures are reduced (Carr and Carr 1972) the trees are functionally males.

19.4 Fruit maturation and seed set

19.4.1 The post-pollination phase

From fertilization the time taken for the ovule to develop into a viable seed varies from species to species, but the process is normally complete by the time the fruit has become a brown, woody capsule and sometimes while the capsule is still partly green. To decide when the seed is mature enough to collect, cut through a fruit with sharp secateurs and see whether the seeds at the base of the chambers have attained the colour associated with maturity (brown or black, depending on the species).

The capsule opens by valves which begin to form before maturity of the ovary. Fruits can be opened by drying as soon as the valves are fully formed. However, at this early stage the brown 'seed' extracted may be only chaff, comprising the ovulodes and unfertilized ovules, while the fertilized ovules deeper within the ovary are still pale and immature.

19.4.2 Time between fruit maturation and seed shed

Even under similar conditions the time between fruit maturation and seed shed varies considerably from species to species. Species with thin walled fruits such as the tropical *E. deglupta* and *E. papuana* begin to shed seed within a few days of maturation. In contrast the mature seed crop will remain on *E. obliqua, E. regnans,* and *E. delegatensis* trees for two years or more (Pederick 1960; Cunningham 1960; Moran and Brown 1980).

Species which shed their seed within a few weeks pose many more problems in harvesting from seed orchards than those which hold their seed for a year or more, because the former require more frequent monitoring of the developing crop and more intensive effort to collect seed quickly before it falls to the ground. For *E. torelliana* grown in Brazil, daily observations are required to judge collection times because capsules of this species suddenly open their valves and quickly shed all seed. At the same location, by contrast, *E. citriodora* retains its seed long enough to allow ample opportunity for collection of mature capsules for drying. The times that mature seed crops of different species remain on the trees in Australia are given in Appendix 2 of Boland *et al.* (1980a).

19.4.3 Size and frequency of seed crops

Another important factor in planning and management of seed orchards and seed production areas is the variation in the size and frequency of seed crops. *E. grandis* in South Africa produces heavy seed crops every year. *E. deglupta* sets seed twice each year with an abundant crop normally followed by a less abundant one. Species such as *E. camaldulensis, E. grandis* and *E. saligna* usually bear heavy seed crops in Australia only every 2–3 years (Boland *et al.* 1980a). In *E. regnans* this period is every 2–4 years (Ashton 1956; Cunningham 1960). In exotic climates, patterns may be completely different.

19.4.4 Seed shed

From the examination of several species of different subgeneric groups, Cremer (1965a) concluded that seed shed depends on separation of seed from the placenta, the widening of the chambers, and the opening of the valves on drying of the fruit. These changes take place as a result of the action of rings of hygroscopic tissue in the chamber wall of at least some species. For example, the valves of *E. grandis* and *E. deglupta* can be made to close again immediately after seed shed by placing them in water. Natural seed shed is hastened by hot dry weather, drought, or fire (Cremer 1965a; Boland *et al.* 1980a). Bateman (1961) observed that trees on moist sites in northern Australia carry their seed longer than adjacent ones on dry sites.

In no species are the seed ejected (Boland *et al.* 1980a). Not all the seed are shed from the fruit; in many species some fruit fall with some seed still enclosed. The amount held in this way differs with species and was found to be up to 24 per cent for *E. regnans* (Cunningham 1960) and 43 per cent for *E. delegatensis* (Grose 1961). *E. deglupta* normally totally sheds its seed.

19.5 Problems with flowers, fruit, and seed

Flower buds are easily lost from failure to grow, insect and disease attack, or by mechanical damage from wind, frost, or drought. This happens at any period from the initiation of flower primordia onwards. These losses may not be of concern in natural stands or for bulk seed collection in plantations but loom larger in importance in a eucalypt breeding program where flowering material may be scarce already, and expensive to establish or produce in seed orchards.

Insects attack flower buds and cause them to fall off. In a young forest of *E. regnans* Ashton (1975a) found that up to 40 per cent of buds had fallen by the beginning of their second summer and in some years 25 per cent of remaining buds were galled from insect attack. It is commonly observed that only a small proportion of the massive production of buds in eucalypts eventually produces fruits. In *E. gomphocephala* a bud-weevil attacks immature flower buds causing them to fall, usually from November to April (Anon. 1966).

Insects also attack the seed within the fruit. Some seed-infesting insects lay their eggs in the flowers; the eggs hatch in the developing fruit where the larvae feed on the fertilized ovules (Boland *et al.* 1980a). *Megastigmus* spp. are common seed destroyers in several species of eucalypts (Grose 1957; Drake 1974). Seeds destroyed in this way usually appear as hollow enlarged shells (Boland *et al.* 1980a). Losses of up to 15–20 per cent of usable seed have been recorded (Pederick 1960). The larval stage of a beetle is known to eat seed within *E. diversicolor* fruit in Western Australia (White 1971).

Birds and other animals may damage seed crops by stripping the flowers or destroying the fruit in their search for seed. Ashton (1975a) observed heavy localized attacks on fruit by large parrots. He assessed the losses in one native forest where he found a quarter of the fruit set had been decapitated and bisected by birds. In Papua New Guinea lorikeets do considerable localized damage to inflorescences of *E. deglupta* when foraging for nectar and pollen. Boland *et al.* (1980a) illustrate damage to *E. olsenii* caused by birds.

Damage to seed and fruit of eucalypts by fungi appears to be less common than damage by insects. Drake (1974) reported that some fungi may infect fruit on the tree and damage their contents. However, fires even of low intensity can bring about almost total premature seed release in some species (Cremer 1965a).

19.6 Inbreeding and hybridization

19.6.1 Inbreeding in natural populations

Pryor (1976) described the breeding system of the eucalypts as one of 'preferential outcrossing which is reinforced by a gene-controlled incompatibility mechanism which impedes or prevents selfing'. There is evidence that inbreeding is a common occurrence both through selfing and through the mating of neighbouring close relatives in natural stands.

Early qualitative work on assessing selfing in trees concentrated on searching for deviant types of seedlings which might be used as genetic markers. The suitability of these abnormal types was often limited because of poor definition and low numbers even in some experimental material known to be selfed. There was also the tendency for abnormalities to develop in one season but not in others and some abnormalities were expressed later in life rather than in very young seedlings.

Deviant seedlings of several types are found in eucalypts (Venkatesh and Sharma 1974; Hodgson 1976b; Boland 1977). Using albino seedlings Krug and Alves (1949) estimated natural outcrossing was 77 per cent in Brazilian plantations of *E.* 'alba' (probably *E. urophylla*). These were easily classified and occurred at an early age, two desirable attributes for a genetic marker. Using 'curled' seedlings as markers in *E. regnans*, Eldridge (1970) estimated 72 per cent outcrossing in the lower crown of one tree.

Several eucalypt species have been proved to be self-fertile by using controlled pollination. Of 16 *E. regnans* trees tested 15 were self-fertile to some extent (Eldridge 1970). Hodgson (1976b) found 44 of the 45 *E. grandis* clones tested to be self-fertile to some extent and average outcrossing to be about 70 per cent (range 90 to 67 per cent). Florence (1964) found fertilization of one tree of *E. pilularis*, which varied from others in timing and intensity of flowering, resulted almost exclusively from self-pollination. While acknowledging the capacity for selfing, Pryor (1957) maintained 'there must be rather a high amount of intra-specific out-crossing, whatever the precise mechanism, to maintain the genetic diversity which *Eucalyptus* species usually show'. Pryor also listed nine species (*E. bicostata, E. blakelyi, E. caesia, E. cinerea, E. macrorhyncha, E. maidenii, E. polyanthemos, E. pulverulenta* and *E. rubida*) which were fully self-fertile, only one tree of one species, *E. bicostata*, being reported as self-incompatible. Other eucalypts shown experimentally to be self-fertile include *E. globulus* (P.W. Volker personal communication), *E. gunnii* (Potts *et al.* 1987), and *E. nitens* (Tibbits 1988, 1989). Cass Smith (1970) reported that *E. ficifolia* (subgenus *Corymbia*) is entirely cross-pollinated in nature. Preliminary evidence for *E. deglupta* indicates it may be self-incompatible (Davidson 1978b, 1983b).

Quantitative estimates of outcrossing rate using allozyme variants (Brown 1978, 1979) have been obtained for several eucalypt species (Table 19.2) and outcrossing has been shown to predominate, averaging about 75 per cent in seed from natural populations. A caution is appropriate because the experimental material consists of viable seed and young germinating seedlings. In several of the studies summarized in Table 19.2 it was found that the outcrossing rate was lower in seed than in seedlings. This indicates a greater proportion of selfs just after fertilization and fewer selfs later in the vegetative life of the population because inbreeding depression could be expected to lead to a continual loss of selfed individuals through competition and other lethal and perhaps unknown effects (Barber 1958; Franklin 1971). Field observations of flowering and pollination vectors in eucalypts (for example Ashton 1975a; Griffin 1980; Hopper and Moran 1981) suggest self-pollination rates and thus self-fertilization rates might be found to be higher if determined just after fertilization. While the allozyme studies listed in Table 19.2 indicate the average outcrossing rate within eucalypt populations is about 75 per cent, there is evidence of variation among certain individuals from obligate selfing caused by cleistogamy (Venkatesh *et al.* 1973) to obligate outcrossing caused by self-incompatibility (Pryor 1976), male sterility (Davis 1969), or female sterility (Carr *et al.* 1971; Carr and Carr 1972).

Studies of between-population variability in the mating system of eucalypts have been made in three species. Four populations of *E. obliqua* varied significantly in outcrossing rate (Brown *et al.* 1975) while four populations of *E. delegatensis* did not (Moran and Griffin 1983). Three samples separated by several kilometres, but from a continuous stand of *E. pauciflora* were analysed for their mating system

Table 19.2

Estimates of outcrossing rate (t) of populations of eucalypts using allozyme variants for 11 species in natural populations and two in plantations.

Species	No. of populations	*t*	Source
Natural populations			
E. caesia	2	0.62	Moran and Hopper (1983 and unpubl.)
E. delegatensis	4	0.77	Moran and Brown (1980)
E. grandis	21	0.84	J.C. Bell (unpubl.)
E. kitsoniana	2	0.77	Fripp (1982)
E. obliqua	4	0.76	Brown *et al.* (1975)
E. pauciflora	3	0.70	Phillips and Brown (1977)
E. pulverulenta	3	0.70	Peters *et al.* (1990)
E. regnans	1	0.74	Moran *et al.* (1989)
E. rhodantha	1	0.59	Sampson *et al.* (1989)
E. saligna	1	0.77	J.C. Bell (unpubl.)
E. stellulata	1	0.77	G.F. Moran (unpubl.)
E. stoatei	1	0.82	Hopper and Moran (1981)
Mean		0.74	
Plantations			
E. citriodora	1	0.85	Yeh *et al.*(1983)
E. regnans	1	0.91	Moran *et al.* (1989)
Mean		0.88	

parameters by Phillips and Brown (1977) using seed and seedling allozyme markers. Outcrossing rates were similar for the seed markers but for the seedling markers the two lower-altitude populations had significantly lower rates of selfing (Moran and Griffin 1983).

Outcrossing estimates have been obtained in three sequential seed crops in a population of *E. delegatensis* (Moran and Brown 1980). The combined estimate was 77 per cent outcrossing for the population but the estimates for individual crops varied significantly from 66 per cent in the youngest crop to 85 per cent in the oldest, suggesting a more rapid decrease in viability of selfed seed than outcrossed seed during the years of storage in capsules on the tree.

Detailed studies of the decrease in growth due to inbreeding (inbreeding depression) have been made by comparing the growth of selfed, out-crossed, and open-pollinated progenies of *E. grandis* (Hodgson 1976a, b; Van Wyk 1983), *E. gunnii* (Potts *et al.* 1987), *E. nitens* (Tibbits 1988), and *E. regnans* (Eldridge and Griffin 1983; Griffin and Cotterill 1988). All these studies confirmed that self-pollination produces fewer seed with reduced germination and slower growth of seedlings and young trees compared with outcrossing. Open-pollination was intermediate. Although inbreeding depression may be only a few per cent in a young, open-grown stand (Section 18.3.6), it should be remembered that in a dense stand any inbred young tree that drops behind in growth compared with its outcrossed and faster-growing neighbours will soon be overtopped, suppressed, and eventually eliminated. This is a general phenomenon in forest trees (Sedgley and Griffin 1989).

19.6.2 Neighbourhood inbreeding

The neighbourhood inbreeding referred to in Sections 1.4.2, 3.4, and elsewhere can be understood from the example of a stand of large old trees regenerated naturally after wildfire. Well-stocked stands of *E. regnans* around 200 years old may have an almost continuous canopy with only about 40 'dominant' trees per hectare (Ashton 1975b, 1976). These trees can be 70–80 m tall with crowns 10–20 m in diameter (Fig. 19.7). Since the distance over which seed is dispersed is about the same as tree height (Cremer 1977), the seed from one large tree can spread over 1–2 ha. And as spacing and seed crop vary among trees in a stand, it is not difficult to imagine the seed from one very large crown being the source of a distinct group of offspring, as noted by Richmond (1971). He was searching for plus trees in a 30-year old stand of *E. regnans* which had regenerated after the disastrous fires of 1939. 'It was noted during the search programme that trees of like form occurred in definable groups, covering 1 to 5 acres. These groups were thought to belong to one family whose characteristics were usually sufficiently obvious to enable definition of borders where two groups merged. In some cases the original, fire-killed mother tree was still standing above its 30-year-old progeny. Where this occurred it was noticed that the physical characteristics of the progeny bore a strong resemblance to the standing mother tree' (Richmond 1971).

Members of such a group are closely related having the same mother and a reasonably high chance of having the same father. In the next generation one can imagine the mating of two full-sibs which became dominant within such a group. There appear to be ample opportunities for increasingly close inbreeding within a neighbourhood over several generations. However, selection at all stages of the life cycle (fertilization, seed development, seedling and tree growth) favours the heterozygous products of outcrossing which helps to maintain the vigour of the population and avoid the tendency of the species to evolve a more truly selfing breeding system (Brown *et al.* 1975).

Consider a plantation made with seed collected from many well-separated trees in a natural forest. When seed is collected from a tree in such a plantation the neighbours with which it has mated are very unlikely to be relatives (Fig. 19.7). Now neighbourhood inbreeding (mating with close relatives) has been broken. A higher degree of outcrossing and heterozygosity prevails than in the natural forest due to outcrossing to unrelated individuals. Selfing may also be reduced due to the smaller and closer crowns of plantation trees compared with an old natural forest.

The increase in wood production attainable simply by using seed from plantations of wide genetic base instead of from natural forests is confidently expected to be of the order of ten per cent (Sedgley and Griffin

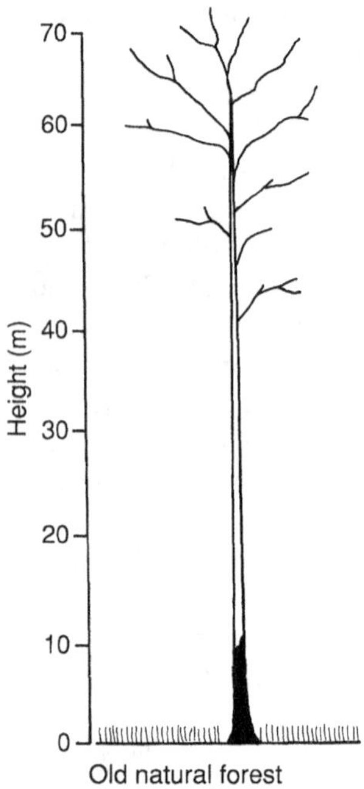

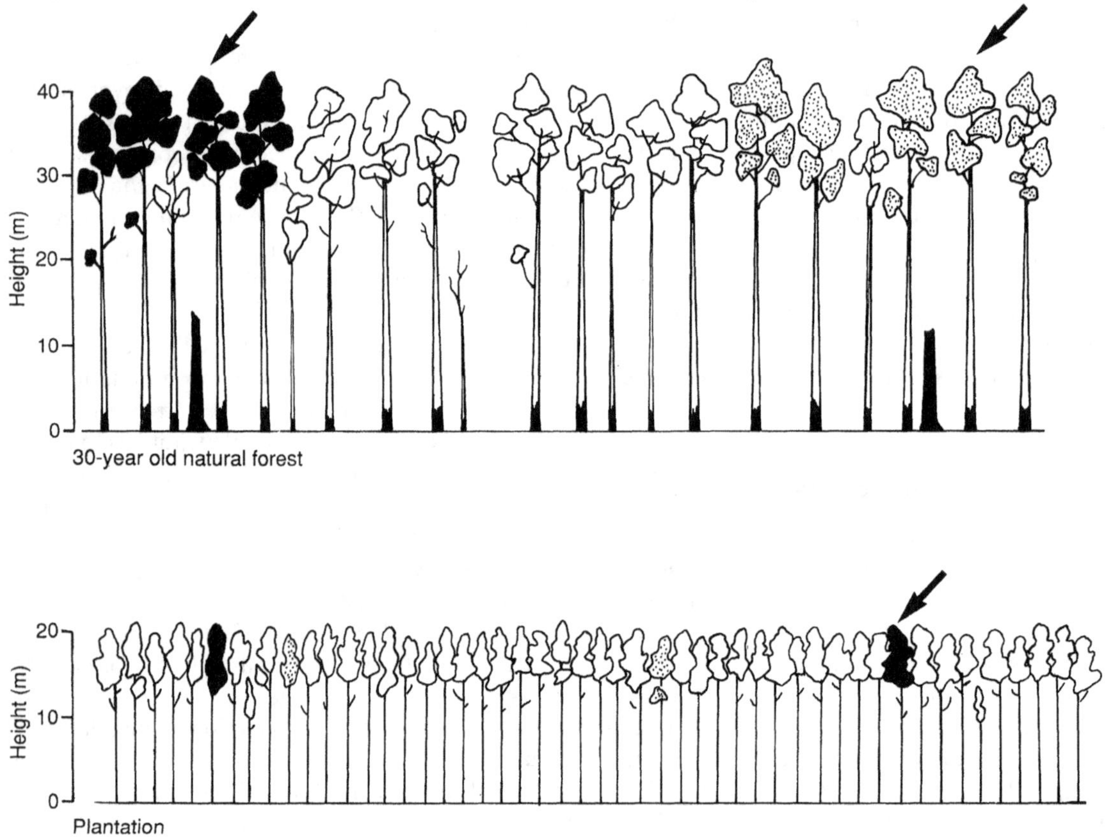

Fig. 19.7. Neighbourhood inbreeding in a natural stand and increased outcrossing in a plantation. After an old natural stand is killed by fire the seedlings growing under a large tree may be mainly from that tree. At about 30–40 years a group of neighbouring trees may be from the same mother tree, now a dead stump (full sibs, half sibs, selfs). Since 'family groups' exist in the stand (one shaded black, another shaded grey), seed collected from a tree in such a group (arrow) is likely to have been pollinated by neighbouring close relatives. In a plantation the relatives are widely dispersed among unrelated trees and seed collected from any tree (arrow) is most unlikely to have been pollinated by a relative (drawn by S.M. House based on spacing and height data for *E. regnans* (Ashton 1975b, 1976)).

1989, p. 218). This prediction is based on observations of inbreeding depression in field trials with selfed, outcrossed, and open-pollinated progeny of *E. regnans* (Griffin and Cotterill 1988; Griffin 1990).

19.6.3 Natural hybrids

Eucalypt breeders will come into contact with hybrids in the course of their work in seed collections in natural forests and plantations. It is advisable to know something of their occurrence and origin. Knowledge of the ecology of natural hybrids also helps to explain some of the potential disadvantages of producing hybrid eucalypts from seed, and the preference for mass production of hybrids by cuttings (Chapter 22).

Patterns of interspecific hybridization in eucalypts have been comprehensively reviewed by Griffin *et al.* (1988). Their survey of recorded natural and artificial hybrids supports the observation by Pryor and Johnson (1971) that attempts to make hybrids between species in different subgenera have not been successful and that within subgenera frequency of natural hybridization generally reflects the hierarchy of taxonomic affinities. There is no confirmed case of success in manipulated crosses between subgenera except a report from Congo (Martin 1987) of a successful cross of *E. alba* (subgenus *Symphyomyrtus*) and *E. deglupta* (subgenus *Telocalyptus*, excised from *Symphyomyrtus* by Johnson (1976)). There has been some debate over the separation of this species from Symphyomyrtus (Ladiges and Humphries 1983). As expected, geographic proximity of species pairs was shown to be necessary for natural hybridization, with all but five of the 520 recorded natural hybrids being between pairs that co-occur within one or more grid squares of 10 × 10 minutes (approx. 23 × 19 km) (Griffin *et al.* 1988).

Across the genus natural hybridization is relatively restricted. A total of 289 out of 528 species reviewed by Griffin *et al.* (1988) were recorded as hybridizing. However, only 15 per cent of the combinations possible between geographically adjacent species pairs in the same series have been recorded and, of such records, 37 per cent were based only on a single herbarium specimen, which is not conclusive proof that hybridization has occurred. Nonetheless, there is ample evidence that natural hybrids do occur. A number of detailed studies have identified suspected F_1 hybrid adult trees, raised progeny from them, and examined their seedling morphological and chemical traits. Such studies have shown that the hybrids are intermediate in comparison with the parent species, and produce highly variable segregating progeny, with the majority of seedlings intermediate in their traits between those of the parent species and some tending towards either parent.

Natural hybrids may occur as single individuals near the junction of two species, often in a local microsite that provides an intermediate environment which may favour the survival of hybrids. An example of this pattern of hybridization between *E. obliqua* and *E. pulchella* in Tasmania was analysed by Potts and Reid (1983). Where there is an abrupt habitat change with no intermediate site, numbers are probably kept small because there is no 'hybrid habitat' (Anderson 1949) where the hybrids are favoured. An example given by Pryor (1950, 1976) is *E. pauciflora* and *E. dives*, with the former confined to granite parent material near the species boundary and the latter restricted to small ridges of shale, slate or phyllite. The geological boundary is sharp, with equally abrupt soil changes. Interspecific hybrids are rare except where clearing or grazing alters the ecology of the site, creating a 'hybrid habitat'.

Sometimes the hybrids may be separated from one of the parent species by a considerable distance. For example, *E. macrorhyncha* × *regnans* hybrids were found in *E. macrorhyncha* forest up to 5 km from the nearest *E. regnans* stands in the Cathedral Range, Victoria (Ashton and Sandiford 1988). Nectar-feeding birds (honeyeaters and lorikeets) were considered the likely pollinating agents in this case. Ashton and Sandiford suggested that such hybridization, on a large scale into a small stand of one of the parent species, could be one mechanism leading to the development of new eucalypt species through subsequent natural selection over several generations favouring and stabilizing an intermediate form.

In some cases there may be a 'hybrid swarm' at the junction of the parent species' ranges, where many hybrid individuals and a range of phenotypes suggest backcrossing of hybrids to the parent species. The genetics and population dynamics of a hybrid swarm in Tasmania between *E. amygdalina* and *E. risdonii* have been examined by Potts and Reid (1985c). They found evidence of asymetric gene flow, with *E. risdonii* invading the *E. amygdalina* stand by both pollen and seed migration. Hybrid individuals were vegetatively vigorous but reproductively less fit than the pure species in this situation. Similarly, Drake (1980, 1981a, 1981b) found that *E. crebra* × *melanophloia* hybrids were reproductively less fit than the pure species. However, *E. crebra* × *populnea* hybrids studied by Drake were reproductively successful and well adapted to the ecotone between these two species.

Certain species pairs exhibit several types of intergradation. For example, *E. regnans* and *E. obliqua* often occupy adjacent habitats in Victoria and Tasmania. There may be a sharp boundary between the two species, with occasional clearly identifiable hybrids, or a narrow zone of trees intermediate in their bark and fruit characteristics suggestive of a hybrid swarm, or a gradual change over several kilometres from *E. regnans* to *E. obliqua*, associated with a gradient of decreasing rainfall (Ashton 1958, 1981; Ashton and Sandiford 1988). Some extensive intermediate populations in south-east Tasmania and south-west Victoria appear to be relatively stable genetically, producing progeny only slightly more variable in bark characteristics than those from 'pure' *E. regnans* and *E. obliqua* stands. In progeny trials some of the intermediate populations grew as rapidly as the faster-growing species, *E. regnans* (Griffin and Eldridge 1980).

Species pairs from the same subgenus that occur naturally in the same or adjacent stands usually have some separation in flowering time, which reduces the frequency of hybridization (Boland *et al.* 1980a, Appendix 2; Potts and Reid 1983). In the case of *E. obliqua* and *E. regnans*, Ashton (1981) noted that hybrids were more common on the *E. obliqua* side of the boundary, consistent with their flowering biology. *E. obliqua* flowered earlier than *E. regnans*, and given the protandry of eucalypt flowering, late-flowering *E. obliqua* parents would be the likely maternal progenitors of the hybrids.

20

Selection and breeding

20.1 Selection criteria

Most of the eucalypt plantations of the world are managed as short rotation coppice crops (5 to 15 years) to provide large quantities of roundwood at a low cost. Wood from young eucalypt plantations is used for firewood, charcoal, poles and posts, mine timbers, pulpwood for paper industries, and for reconstituted wood. Only a very small proportion of the world's eucalypt plantations are allowed to grow to the greater age and large diameter required for sawn timber.

As a broad generalization, the principal selection criteria for plantation eucalypts are health, growth rate, stem straightness, and wood density, in that order of priority. Selection for straighter stems provides better poles and some advantage in stacking and handling logs for firewood or pulpwood. Increased dry wood per unit area is a common goal, by seeking increased wood density as well as increased volume.

20.1.1 Pulpwood

The principal requirements for eucalypt pulpwood, apart from abundance and low price, are a basic density less than 600 kg m^{-3} and a minimum of extractives (Higgins 1984). The length of eucalypt fibres of mature wood is about 1 mm (Hillis 1972), compared with about 3 mm in most conifers. However, in low-density eucalypts the fibre walls are thin enough to flatten and bond well to their neighbours during the formation of a sheet of paper. Eucalypt pulp yield in terms of the number of tonnes of dry wood required to produce one tonne of dry pulp, an important factor in the economics of pulp mills, appears to be not strongly associated with wood density (Turner *et al.* 1983). Increased wood density can reduce costs of transporting, handling, and pulping as a smaller volume of wood is required per tonne of dry pulp produced.

Nearly all species of eucalypts grown rapidly in plantations and harvested before much heartwood has formed are suitable for several pulping processes (Higgins 1984). Possible exceptions among the commonly planted species are *E. camaldulensis, E. citriodora,* and *E. tereticornis*, which usually have thick-walled fibres and consequently high density. Young trees of even these three species are successfully used for certain grades of pulp and paper. The pulpwood quality of old eucalypts is inferior because the wood is about 100 kg m^{-3} denser than that of young trees of the same species, kino veins are more common, and the heartwood is occluded with extractives (Hillis 1972). Further, the fibres in the central core of wood of old trees have often failed under compression to form a weak and useless 'brittle heart' (Hillis 1984).

20.1.2 Fuelwood

Fuelwood, like pulpwood, must also be abundant and cheap, and it must be grown close to the place where it will be used so that transport cost is not great. Although the calorific value of all dry eucalypt wood, and almost all other dry wood (Harker *et al.* 1982), is between 19 and 20 MJ kg^{-1}, high density is still an advantage in reducing bulk for transport, reducing maximum moisture content, conferring better burning properties, and producing better charcoal. For fuelwood and charcoal production, growth should be assessed in terms of biomass production (tonnes of dry wood per hectare) rather than volume.

20.1.3 Poles

The principal requirements for poles are straightness, high strength which is enhanced by greater wood density, and freedom from splitting. Log-end splitting has a high commercial significance in South Africa. Selection for *E. grandis* genotypes which were less prone to log-end splitting of poles some time after felling was practiced in South Africa with considerable success, both by simple selection of seed trees in the plantation (Nel 1965) and through selection of plus trees at age 18 years for this trait and their inclusion in grafted seed orchards (Van Wyk and Hodgson 1983). Seed or scions for grafting had been collected at the time of felling. As there was no relation between wood density and log-end splitting, the simpler procedure of assessing wood density from a core extracted from a living tree could not be used. If a clear relation were established between log-end splitting and longitudinal growth stress, it would be possible to select standing trees with low stress using the stress measurement technique developed by Nicholson (1971). Meanwhile, it appears the only way to assess log-end splitting is to fell the tree.

20.1.4 Sawn timber

In many cases adequate sawn recovery is only obtained from trees with a breast height diameter over bark of some 80 cm or more. The sawn recovery from logs of young eucalypts less than about 40 cm diameter under bark at the small end is usually very low since much of the wood splits or bends even with the best sawing and seasoning techniques (Haslett 1988; Waugh and Rozsa 1991). This is due to extreme growth stresses causing tension in the outer layers of wood (Jacobs 1955; Hillis 1984). Evidence for genetic control of growth stress (Nicholson 1973; G. Waugh personal communication) suggests the prospect for reducing growth stress by selection and breeding. Indeed the success of selection for reduced log-end splitting, a related phenomenon, in *E. grandis* in South Africa is one of the notable achievements of eucalypt breeding (Nel 1965; Van Wyk and Roeder 1978). Experience from New Zealand and elsewhere suggests that stresses are worse in tall, slender trees in dense stands than when stand density is low and the trees rapidly reach a large diameter. The gradient in growth stress across eucalypt stems is reduced by failure under compression of wood in the centre of the stem when eucalypts reach a large diameter, about 80 cm or more. In the natural eucalypt forests of Australia, and on a smaller scale in Papua New Guinea, Indonesia, and the Philippines, high-quality sawn timber is obtained from large logs of old trees.

20.1.5 Growth and form

Growth rate is the result of several other adaptive traits, such as tolerance to specific environmental situations. For example, a tree which becomes outstandingly large compared with its neighbours may do so partly because of greater resistance to a defoliating fungal disease, or greater capacity for regeneration of fine roots after their death from waterlogging. Selection for growth rate and volume is usually carried out by measuring diameter with tapes or calipers and height with poles, until height exceeds about 15 m, after which it is too difficult and expensive to measure the heights of large numbers of trees in dense stands.

Survival is also a selection criterion, especially in the early phase of species and provenance testing. Early survival may reflect more the site preparation and planting methods, or meteorological conditions than the genetic make-up. Genetic survival ability is better evaluated at least one year after planting.

Form is usually evaluated subjectively. Each field worker responsible for selection may have a personal criterion of what constitutes a straight or crooked tree, and this may change with time. Nevertheless, it is always possible to mark the straightest and best branched trees, and the worst, and use them as standards for the subjective scoring system. The common selection method is to grade trees according to an arbitrary scale, for example from 1 to 6, 6 being the best (Raymond and Cotterill 1990).

20.1.6 Wood density

Density of eucalypt wood is a complex characteristic since the tissue is made up of varying proportions of different kinds of cells which in turn vary in wall thickness, width, length, and extractives and other materials (for example Davidson 1975). The extent of these variations depends on the species and environment. The average thickness of the fibre wall and the anatomy of its layers are characteristics which contribute to the strength of the wood and its machinability and to various properties of pulp and paper (Bamber 1985).

Wood density influences many properties and conversion processes, including cutting, gluing, finishing, rate of drying, burning, pulping, and papermaking. It provides a good though not always direct indication of the strength, stiffness, and toughness of the timber. Shrinkage is less severe in higher density woods. A strong association was found between basic density and natural durability in *E. grandis* (Nelson and Heather 1972).

Considerable variation in wood density occurs within a species. Very large variations in density were found between 639 trees of *E. regnans* and the means of the 37 families to which they belonged (Doran 1974). These differences were large enough to have had an appreciable effect on the properties of paper made from kraft pulps. For 3000 'plus' trees of *E. grandis*, *E. urophylla*, and their hybrids examined at Aracruz, the density varied from 450 to 650 kg m^{-3} (E. Campinhos personal communication).

A significant relationship was found between the basic density and the seed source for five-year-old trees of *E. grandis* (Bamber and Humphreys 1963), and a high heritability of density was found in young seedlings of *E. deglupta* (Davidson 1972) and *E. regnans* (Higgs 1969). A heritability estimate of 0.45 was obtained for eight-year-old control-pollinated progeny of *E. grandis* by Malan (1988) (see Table 18.4).

Accordingly, because of the significant variations in wood density in unselected trees and because of its high heritability, changes could be expected in wood density by selection and breeding, although the strong adverse genetic correlation of wood density and growth rate found by Malan (1988) in *E. grandis* would require special care in planning a breeding strategy.

20.1.7 Fibre length

Longer fibre length is desirable in eucalypts to increase the strength of pulp and paper (Higgins 1984). There are several cautions in comparing fibre length between trees when planning to select for longer fibres. The differences in fibre length between wood formed at different ages (i.e. different distances out from the pith) are usually greater than the differences between trees and between species. This range of length can be very great, for example 0.45 to 1.6 mm with a mean of 0.99 mm in one group of 19-year-old trees of *E. deglupta* (Davidson 1972).

A high heritability of 0.54 was estimated by Malan (1988) for fibre length of eight-year-old *E. grandis* (Table 18.4). In earlier studies, fibre length in five-year-old *E. grandis* in Australia was found to be unrelated to seed source (Bamber and Humphreys 1963), while significant differences were noted between the fibre lengths of clones of one-year-old seedlings of *E. camaldulensis* (Rudman 1970). Even though fibre length may, in some cases, be more heritable than wood density (for example Higgs 1969), the improvement of mean fibre length even by as much as say 30 per cent by selection may have little effect on pulp and paper quality (with an increase of 0.3 mm the fibres would still be short compared with conifer fibres).

20.2 Selection methods

For selection to achieve substantial genetic gains in any trait (selection criterion) there must be substantial variation in the trait in the population and a high heritability (see Section 18.3.7). If selection is applied to a large number of traits at once none of them will be improved much and there is a danger of adverse genetic correlations operating to devalue one trait at the same time another is improved (Section 18.3.3). It is necessary to decide on the most important traits, perhaps as few as two or three, and concentrate on them. Decisions on which of several different traits to concentrate on and what intensity of selection to use are made for the particular economic circumstances. By far the most efficient method of selection is index selection, taking into account several kinds of information about the trees being considered.

In a detailed account of efficient selection in tree breeding, Cotterill and Dean (1990) noted that in 'the first generation when trees were selected in native stands or genetically unimproved plantations...there may be little choice other than selecting by eye. However, in advanced generations it is possible to implement far more efficient methods of selection'. Efficiency in selection has been a somewhat overlooked aspect of tree breeding. Cotterill (1986b) has pointed out that many advanced generation breeding programs have had elaborate mating designs for controlled pollination and complex statistical designs for comparing large numbers of entries in field trials, and yet did not give sufficient attention to techniques for choosing the best individuals for mass propagation or for the next generation of breeding.

Tandem selection improves several traits one at a time over several generations, one trait in each generation. Independent culling first assesses an easy-to-measure trait (such as diameter) for all the trees and then a difficult-to-measure trait (such as wood density) for perhaps 10 per cent of trees with greatest diameter. Index selection integrates information from many sources into one index value so that all the individual trees can be ranked according to the one scoring system, rather than having to examine the rank of the same individuals separately for diameter, density, straightness, and so on. All three methods of selection, tandem selection, independent culling, and index selection, may be employed in some breeding programs. Index selection becomes an essential tool of the tree breeder in advanced generations of a breeding program as large amounts of data accumulate.

Index selection uses genetic, economic, and pedigree information to maximize the probability of selecting trees with the best genes. The reader is referred to Cotterill and Dean (1990) for a clear explanation of both theory and practice, for methods from a simple index which makes a score combining the measurements of several traits with equal weighting, to a complex computer program for restricted index selection (RESI) using genetic and economic parameters of the selected population and measurements of trees in a series of progeny tests.

20.3 Progeny testing

Progeny testing refers to the evaluation of the performance of offspring from parents with full or partially known identity. Progeny tests of eucalypts are established to serve all or most of the following five objectives.

(1) Estimation of the breeding value of selected parents: by growing the offspring of such parents in family groups in field trials with sound statistical designs, the genetic worth of the parents can be demonstrated. Selecting parents by the performance of their progeny is sometimes called 'backwards' selection, in contrast to selection of the best individuals among the progeny, called 'forwards' selection (Shelbourne *et al.* 1989).

(2) A source of selected trees for the next generation: 'forwards' selection is an important function of progeny testing because normally a large number of unrelated individuals (more than 100) must be selected for each successive generation to avoid inbreeding and to enable a wide choice of trait combinations in the breeding population. Plantations established from seed orchards are less variable because they have a more restricted genetic base than large family trials. Because the purpose of seed orchards is to effect immediate gains, they are therefore composed of only the best breeding material identified through progeny testing (often fewer than 30 selections). Progeny testing on the other hand has a longer-term objective in maintaining and improving a broad breeding base.

(3) Estimation of genetic parameters: this mainly involves estimates of genetic variance and covariance components for estimating heritability and genetic correlations. By combining these parameters with some other statistical parameters, genetic gain predictions can be made (Section 18.3.7). The magnitude of genotype-environment interaction could also be determined from well-designed progeny tests, replicated at two or more sites.

(4) Demonstration purposes: demonstration plantings are perhaps of minor importance scientifically but politically often of major importance for sustained funding by demonstration to decision makers and management that large and cumulative gains in yield and quality are possible. The special requirements for design of trials demonstrating genetic gain are outlined in Section 18.5.

(5) Seedling seed orchards: many eucalypt species with fast growth, early seed production and coppicing ability are very well suited for converting progeny tests to seed producing stands. This is especially important in tropical and subtropical countries where eucalypts are grown on short rotations of less than 10 years.

20.4 Mating designs for progeny testing

A mating design refers to the pattern in which female and male parents are crossed to produce offspring that will be evaluated later to achieve one or more of the objectives discussed above under selection criteria (Section 20.1). Crossing of the parents may be natural (by insects or birds) or artificial (controlled crosses made by hand pollination). The bisexual nature of eucalypt flowers makes artificial crossing more tedious than with pines, for instance. Therefore, the mating designs adopted will depend on availability of resources (money, time, people) and the skills of breeders and field workers.

Many mating designs have been proposed by forest geneticists. General descriptions of designs are given by Zobel and Talbert (1984) who grouped them into incomplete pedigree designs (open-pollinated and polycross) and complete pedigree designs (nested, factorial, bi-parent, diallel). Statistical interpretation of results from progeny tests has been described by several authors including Namkoong *et al.* (1988) and White and Hodge (1989). The concept of heritability as determined from progeny test results has been dealt with, among others, by Hanson (1963). Practical examples of the use of genetic parameters for genetic gain predictions have been provided by Shelbourne (1969, 1991) and Cotterill (1986a). Thus, no detailed description will be given here on mating designs and interpretation of test results. Rather, two mating designs found to be of practical use for eucalypts will be considered: firstly, open-pollinated designs for trials which can yield fast results at relatively low cost (Cotterill 1986a), and, secondly, factorial and partial diallel designs which could form the basis for integrating vegetative propagation with the breeding programs. The use of these designs in breeding strategies is illustated in Section 18.2.3.

20.4.1 Open-pollinated designs

The major advantages of the open-pollinated (OP) progeny tests are that they are inexpensive, simple to perform, yield information rapidly, and can satisfy all the objectives of progeny testing listed in Section 20.3. For this design, seed is obtained from natural pollination of parent trees. Such seed lots are planted in replicated family trials which will yield the information and material needed for a practical breeding program within five years for tropical eucalypts and within ten years for temperate species.

A simplified version of this design is not to retain family identity, but to practise mass selection in large commercial block plantings derived from a broad genetic base, preferrably with at least 200 unrelated individual trees represented in the bulk seedlot used to establish the plantings (Shelbourne 1969).

Concern has been raised that OP progeny tests of eucalypts using seed from natural forests might not give reliable results because of the high, and highly variable, degree of inbreeding among parent trees providing the seeds for the tests (Eldridge and Griffin 1983; Griffin and Cotterill 1988). This is due to a combination of inbreeding due to mating of closely related neighbours and selfing within the massive crowns of large old trees (Section 19.6.2). Therefore care should be taken with interpretation of OP progeny test results from seed collected from natural forests, or from plantations of a very narrow genetic base, or from isolated trees subject to selfing. Heritability will be overestimated and selection will not be accurate (Eldridge 1976; Griffin and Cotterill 1988).

Among a set of families under test, those which perform poorly may have done so because of an extreme level of inbreeding and inbreeding depression, although their value as parents when outcrossed may be high. In other words, their high breeding value may be masked by inbreeding. Families with the highest ranking may be assumed to be among the best, though this could be due to minimal effect of inbreeding so that they are not necessarily higher in rank than some families whose performance is masked by inbreeding (Griffin and Cotterill 1988). Thus the results of open-pollinated progeny tests from seed collected in natural stands are of less value than we might wish, but still useful.

However, in plantations which have a broad enough genetic base the neighbourhood inbreeding has been eliminated because each tree in such a plantation is surrounded by numerous unrelated trees (Fig. 19.7, p. 196). A higher rate of outcrossing (more than 90 per cent outcrossing) can be expected in seed collected in plantations with a broad genetic base than in natural forests (about 70–75 per cent) (Moran *et al.* 1989), provided pollen vectors are abundant and the trees are not too widely spaced. Thus eucalypt progeny tests established with open-pollinated seed from second and later generation plantations with a reasonably wide genetic base can be expected to allow more precise selection, and better estimates of genetic parameters, than first generation progeny tests established with wild seed from natural forests.

Converting an open-pollinated progeny test to a seedling seed orchard is a very attractive option for species that flower profusely at an early age. A prerequisite for easy management of such an orchard is that the species must have a high coppicing ability. Otherwise tall trees would need to be climbed for seed harvesting instead of just felling the trees for the same purpose. By dividing the orchard into blocks or having several different seedling seed orchards, a constant seed crop is ensured by felling and coppicing on a rotational basis (see Section 21.2.6). The young coppice shoots have been used in South Africa as a simple and cheap source of improved material for raising cuttings.

20.4.2 Factorial and diallel designs

A diallel mating design involves n parents crossed in $n \times n$ combinations (full diallel) or more often $(n \times n)-n$ (full diallel without selfs). The half diallel $(n(n-1)/2)$ and other variations reduce the number of crosses (families) to a level that will still give reliable estimates of combining ability of the parents and other genetic parameters. Two useful modifications of the diallel are:

(1) a partial diallel design in which $3n$ crosses are produced so that each parent is crossed 3 times as a male and 3 times as a female in such a pattern that the maximum number of unrelated crosses is obtained (Zobel and Talbert 1984, p. 241);

(2) a pattern in which $3n$ crosses also result from crossing each parent three times as a female but which would not give the maximum number of unrelated crosses. This design has also been referred to as a disconnected factorial mating design (Zobel and Talbert 1984, p. 238).

Both designs will satisfy objectives (1), (3), and (4) of Section 20.3 and to a lesser extent objective (2) depending whether the selected material is used for immediate mass propagation (by seed or cuttings) or for further breeding in recurrent cycles of selection. For further breeding, factorial and diallel designs are restrictive in that related matings cannot easily be avoided in successive generations unless the number of parents is extremely large. A large population of parents, however, means a large number of control-pollinated crosses which are tedious and costly to make with eucalypts.

One way to reduce relatedness in later generations and keep the number of crosses down to a manageable number is to use the mating design known as bi-parental crosses. Each parent appears in only one cross, in the pattern A × B, C × D, E × F,... This design may be combined with open-pollinated mating in tree breeding strategy, so that crosses are only made between parents which performed well in open-pollinated tests.

Control-pollinated designs such as the partial diallel involve a large amount of painstaking and expensive work to produce crosses with known parentage. This is especially true for eucalypts for which the bisexual flowers must be emasculated and bagged before cross-pollinations are done. With due care, uncontaminated crosses can be obtained easily. Successful capsules may eventually contain 30 seeds in *E. grandis* and up to 100 in *E. globulus*. Many controlled crosses have been made with success in Australia, Brazil, Congo, France, Portugal, South Africa, and other countries for producing both intra- and inter-specific crosses.

The value of control-pollinated crosses for production of full-sib families lies in the information to be gained. Full pedigree-control can be maintained (and thus inbreeding avoided), and information on non-additive genetic variation can be obtained. The magnitude of non-additive variance in relation to additive variance will help to determine the breeding strategies to be followed. If large non-additive variance exists, large genetic gain can be obtained by exploiting this variance, especially by means of vegetative propagation. Material for vegetative propagation can be taken directly from the best individual trees from the best full-sib families. Moreover, subsequent crossing of such individuals will aid in generating new and better clonal material faster than would be the case with open-pollinated designs.

20.5 Controlled pollination

Pedigree control requires a successful controlled pollination program which in turn requires a knowledge of the breeding system, flower phenology and morphology as set out in Chapter 19, as well as the necessary technical skills as set out in the following sections. Without detailed biological knowledge, a breeding strategy involving controlled pollination, such as partial diallel mating designs, should not be undertaken. The more complete the knowledge on the reproductive biology of the particular species is, the better the technical aspects of the breeding program can be planned.

20.5.1 Controlled pollination techniques

By training a number of field workers in controlled pollination techniques the breeder can ensure that a large number of crosses can be made in one flowering season. While controlled pollination techniques will differ with different species, the basic principles will remain the same. Since literature on this topic is scarce, and because of the importance of this aspect of eucalypt breeding, most of the publication of Van Wyk (1977) describing controlled pollination of *E. grandis* is reproduced here.

Pollen handling and pollination

As related in Sections 19.1 and 19.2, the flower bud consists of a receptacle which carries an operculum on its rim. The operculum covering the pistil and stamens of a bisexual flower falls away when the flower blooms, enabling the stamens to unfold. The filaments are attached to a staminal ring which forms the rim of the floral cup. A distinct ring is visible at the line of junction between the receptacle and the operculum (Fig. 19.2(f), 19.5(b)).

When the flower bud reaches maturity, it gradually turns from green to yellow, and a split develops at the junction of the receptacle and the operculum indicating it is about to drop. Sometimes pollen is shed inside mature flower buds before the operculum drops. This pollen is viable and can be present in sufficient quantities for cross pollination and can give appreciable amounts of fertilized seed. Collecting pure pollen, separated from the anthers, is not essential and pollen can be applied 'by anther' where the stamens themselves are used to brush pollen onto the stigma.

When mature flower buds are isolated by bagging (Fig. 20.1), just before the opercula are shed, uncontaminated pollen will be available as a result of maximum pollen shed occurring within two days. It is important that flowers are isolated before anthesis because insect visits will contaminate and also reduce the amount of available pollen. The isolation of the pollen flowers, therefore, should be planned in such a way that enough pollen will be available when the flowers to be pollinated are at maximum receptivity.

Another method of pollen collection using mature flower buds is quite feasible. This is useful when pollen shed is hindered by rainy conditions, provided one can be sure of finding the necessary flower buds. Shoots with mature buds can be placed in a jar of

Fig. 20.1. Flowers bagged for controlled pollination on a two-year-old graft of *E. deglupta* in a PICOP seed orchard in the Philippines (photo J. Davidson).

water in a sunny room until the opercula are about to drop.

For pollination by anther no pollen storage is necessary since fresh flowers are used for pollination. Using tweezers, a single flower is picked and the anthers brushed upon the stigma of the flower to be pollinated. A piece of black paper is useful for detecting pollen shed, by tapping the flowers against the paper. The flower is handled carefully to avoid nectar contaminating the pollen and thereby reducing the efficiency of pollination, although this has not yet been confirmed experimentally. Alternatively, a section of staminal ring with a few stamens attached can be broken off the flower cup and used in a similar way.

Pollination by brush involves the collection of mature, opening flower buds from which pollen is extracted two days after anthesis. Excess nectar is removed from open flowers with a piece of blotting paper and the rim of the receptacle is severed, obtaining the staminal ring with the stamens still intact. After this ring has been slit the stamens may be flattened out on a smooth surface and the anthers scraped off the filaments with a sharp razor blade. Alternatively, the anthers can be cut from the intact flower with surgical scissors. The anthers are placed in a container, dried in a dessicator for a few hours, and crushed to get maximum release of pollen grains. After mixing, the pollen is brushed onto the stigmas with a small camelhair paintbrush.

The brush method is appropriate for special purposes, such as experiments requiring mixtures of equal amounts of different pollens or for receptivity tests. One advantage is that it can ensure a mixture of pollen in case non-viable pollen is collected from some flowers. Pollen from one male flower will be sufficient for 10 female flowers.

Pollen should be as fresh as possible. It may be stored from a few days under low humidity and temperatures of about 2° to 4°C, but pollen germination drops considerably after 10 days of storage, especially if it is not well dried. Since flowers and pollen of *E. grandis* are usually available in abundance there is usually no need for storage. However, well-dried pollen of other eucalypts has been stored successfully at –18°C (Boden 1958a).

Flower emasculation and isolation

Yellow flower buds are selected, and open flowers and leaves surrounding the buds are removed. All flowers on a twig do not normally mature at the same time. Immature flower buds are also removed, leaving up to a dozen buds to be enclosed in the isolation bag (Fig. 20.2). A large number of flower buds and umbels are sometimes removed in this process, but there is no point in emasculating immature flowers since their receptivity is very low or zero and they will only exhaust the pollen supply. When studies are done on number of seeds obtained per pollinated flower, it is important that the flowers are at maximum receptivity.

The selected flower buds are then emasculated to avoid pollen contamination, by simply cutting through the tissue of the flower cup slightly below the staminal ring with an emasculating tool of the type described by Meskimen (1965) or a scalpel (Fig. 20.3). During this operation the flower is supported between the thumb and the forefinger to minimize twisting of the pedicel. The first cut usually removes the whole staminal ring, with stamens and operculum attached, but some anthers may remain at the base of the floral cup. A pointed instrument, a jet of water, a blower brush used for cleaning camera lenses, or simply blowing on the flower will remove residual pollen grains likely to be deposited on the stigma during emasculation.

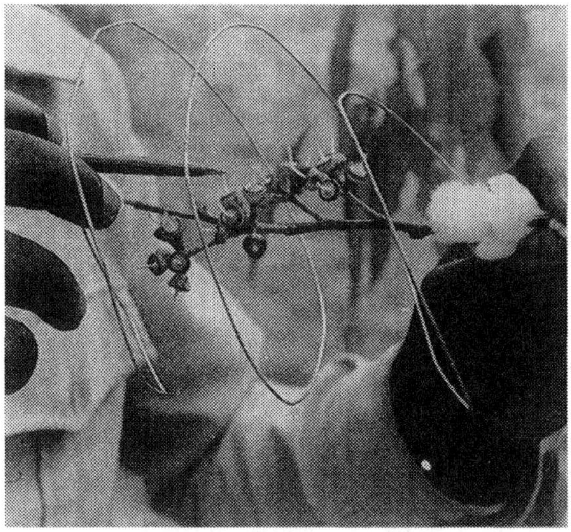

Fig. 20.2. Controlled pollination of *E. tereticornis* at Pointe Noire, Congo. Pollen on a sharpened stick is touched onto receptive stigmas of emasculated flowers. The wire spiral prevents the bag from touching the stigmas (photo P. Maillard, CTFT).

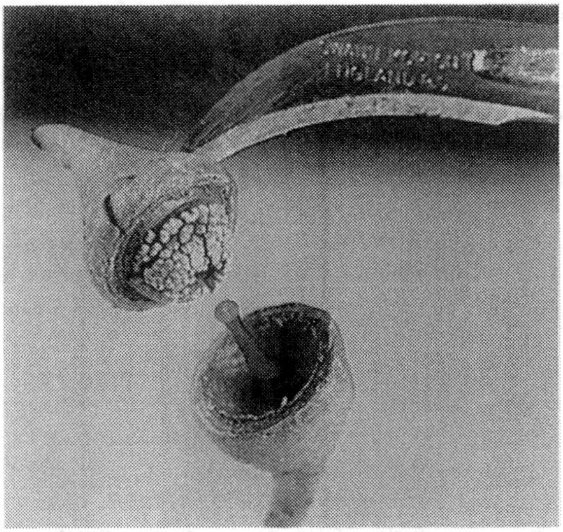

Fig. 20.3. Flower of *E. leucoxylon* emasculated with a curved scalpel (photo A.G. Edward).

After emasculation a wire spiral is drawn over the branch to enclose the flowers, and to support a loosely woven cotton tube or sleeve about 27 × 10 cm (usually referred to as a bag). The spiral is kept in place by tying its ends to the branch, using cotton wool to limit abrasion (Fig. 20.2). The bag or wire should not touch the stigmas. If a branch is too short to support the wire spiral, it can be extended with a twig to enable fixing of both ends of the spiral. The bag is drawn over the spiral and closed at both ends. Bags are then labelled. Flowers of *E. grandis* reach maximum receptivity 4–6 days after fall of the inner operculum. At this stage the stigma is swollen and secretes a sticky substance which facilitates adherence of pollen grains and may act as a medium for pollen germination. Since flowers are very receptive, special precautions against contamination should be taken by removing open flowers in the vicinity of the bag, and by keeping hands and instruments clean. Bags are opened at one end and the pollen applied, either by brush or by anther. After closing, bags may be covered on the upper side with a transparent sheet of plastic material for one or two days as protection against possible rain which will wash pollen off freshly pollinated stigmas. The bags are kept in place until styles have fallen, which may be 2–4 weeks after pollination.

Controlled pollination requires careful planning and record-keeping, and, of course, well trained and conscientious personnel. Emasculation and bagging are done before pollen flowers are isolated to ensure that stigma receptivity and pollen availability coincide. A convenient schedule is emasculation and bagging on Wednesday and Thursday and pollen isolation on Friday, leaving the weekend for the development of the pollen and the emasculated flower. Recording on field sheets includes dates of emasculation, bagging, harvesting, number of flowers emasculated, number pollinated, number of capsules harvested, and other information depending on the requirements of the project.

Observe developing flowers and capsules so as to record losses due to mechanical damage and delete such flowers from the record. Note flowers lost through abscission as this may result in over-estimates in tests of number of seeds per capsule. Cutting of hygroscopic tissue in the emasculation process might lead to early opening of valves and loss of some seed.

It is very important to cut back competing branches around the bag one month after controlled pollination because branches with emasculated flowers could be suppressed causing death and loss of seed.

Methods of drying and extracting eucalypt seed have been described by Boland *et al.* (1980a). The arrangement of seeds in the capsule has significance in the harvest of the products of controlled pollination since great care has to be taken in drying and extraction of capsules to ensure all the viable seed produced is obtained. Extraction may have to be completed using a dissecting needle.

20.5.2 Production of interspecific hybrids

Controlled pollination techniques for interspecific hybridization of eucalypts are similar to those described for *E. grandis* in Section 20.5.1. However, there are several matters to consider before starting the pollinations: complementary characteristics, hybrid vigour, taxonomic affinity, repicrocal recurrent selection, and mass propagation.

One attraction of interspecific hybridization in eucalypts is the opportunity to provide new plantation trees with combinations of characteristics which cannot be obtained in a pure species. For example,

E. grandis × *urophylla* at Aracruz in Brazil has greater coppicing ability and resistance to *Diaporthe* stem canker than could be obtained by selection in *E. grandis*. A second attraction is hybrid vigour or heterosis, when the hybrid grows faster than either parent.

Hybrid vigour has been achieved in spectacular fashion with eucalypt hybrids on sites which are marginal for both parent species, as with *E. grandis* × *tereticornis* at Pointe Noire in the Congo. Using this example, Martin (1989) has drawn attention to the notion of environmentally dependent hybrid vigour. On sites favourable to both parent species, hybrids cannot be expected to be superior, except if the parents were much inbred. Then interspecific hybridization provides an escape from inbreeding depression (Section 19.6.2).

Having decided that interspecific hybridization offers the type of trees required, an orderly plan for controlled pollination can be developed. It would have been agreed that fortuitous natural hybrids are not adequate and that the planting site is marginal for pure species. When the desired characteristics have been enumerated, pairs of species can be chosen with the complementary characteristics, provided these species are not separated by too great a genetic distance.

Taxonomic grouping (based on morphological characters) is a reasonably good basis for predicting pairs of eucalypt species which have genetic affinity and are readily crossable (Griffin *et al.* 1988). Too distant a cross, such as *E. alba* × *deglupta*, if successful, is likely to give heterogeneous, abnormal progeny including dwarfs (Martin 1987).

Among several hundred interspecific crosses at Pointe Noire were 31 among 12 parents of 6 eucalypt species in an incomplete factorial mating pattern (Vigneron 1989b). They were *E. grandis*, *E. pellita*, *E. resinifera*, *E. saligna*, and *E. urophylla* in section *Transversaria* and *E. brassiana* in section *Exsertaria*. The resulting trial of 10 000 hybrid progeny covered 12 ha. (Note: *E. urophylla* was placed in *Exsertaria* by Pryor and Johnson (1971) but Pryor and Johnson (1981) considered *E. urophylla* to be a close relative of *E. pellita* and therefore they and M.I.H. Brooker (personal communication) prefer to place it in *Transversaria*.)

Although the very best combinations could only be found by repeated trial crosses followed by progeny testing, the scale of operations could soon become impractical. The crosses attempted must be limited in number and confined to a few of the best trees in the best provenances of each species (Martin 1989). The crossing program also is confined to F_1 crosses and to combinations which are likely to be easily propagated vegetatively. The principal combinations are *E. urophylla* × *grandis* and *E. urophylla* × *pellita*, for which programs of reciprocal recurrent selection have been adopted (Vigneron 1991).

Barriers to crossing due to the flowering biology of particular species must be included in plans for interspecific hybridization. For example, for species which do not flower at the same time it is necessary to store pollen for several months, as for *E. sieberi* × *regnans* (Eldridge 1963b). Another barrier is unilateral cross-incompatibility between species: the pollen of *E. globulus* will fertilize *E. nitens*, but not the reverse because the style of *E. globulus* is much longer than the maximum growth of a pollen tube of *E. nitens* (Gore *et al.* 1990). Detailed knowledge of the flowering biology of each species is clearly essential when planning to make eucalypt hybrids.

Propagation of a select F_1 individual by seed is likely to produce an unacceptably heterogeneous plantation in the F_2 generation and so F_1 hybrids have been mass propagated by cuttings (Chapter 22). However, the prospect of economic mass production of uniform F_1 hybrid seed should not be dismissed (Martin 1989). It may be possible to devise appropriate controlled-pollination systems for certain species grown in areas where flowering is abundant and labour costs are low. Technological advances which could assist include the chemical stimulation of flowering of grafts with paclobutrazol (Griffin *et al.* 1992) and chemical suppression of pollen production on the female parent.

In future generations the continued production of increasingly improved F_1 hybrids could be made by reciprocal recurrent selection as proposed by Vigneron (1991), each pure species being separately improved by recurrent selection for general combining ability (Section 18.2.3). In each generation interspecific F_1 hybrids would be made between selected individuals of different lines. The hybrid progeny would be propagated vegetatively and clone tests conducted before releasing selected clones for mass propagation (Matheson 1990).

20.6 Experimental design

Two important objectives of progeny testing are evaluation of breeding values of parents and estimation of genetic parameters (Section 20.3). To do this scientifically one cannot just plant blocks or rows of trees and make casual comparisons with regard to growth performances. Populations (families of trees) must be compared with the least possible error. The comparisons in turn will enable selection of the best families and best individuals within families as accurately as possible.

Progeny tests are therefore laid out according to the well-established statistical principles employed by breeders of grain crops since the 1930s. The principles of experimental design and the requirements for management of field trials are basically the same for progeny tests as for the species and provenance trials discussed in detail in Sections 5.3 to 5.7. In progeny testing there are usually very large numbers of entries (families), which means that the convenient and simple randomized complete block design (RCB) is rarely appropriate and some form of incomplete block design is necessary.

20.6.1 Field trials

The shape of the plot in progeny tests varies from rows to square or rectangular patterns. The smaller the plot, the greater the number of replicates planted. In progeny testing, plot size is often smaller than in species and provenance testing (Section 5.6.2). Tree breeders normally aim at 20 to 60 trees per family, depending upon the variability of the experimental site. Cotterill (1984) has pointed out that 20 trees per family are usually adequate to obtain reliable estimates of family characteristics with *Pinus radiata*. The designs with small plots and large numbers of replicates give the highest statistical precision but are more difficult to establish because the identity of each seedling must be maintained throughout. Single-tree plots, where the number of replicates equals the number of trees per family, should be employed only when very high survival can be expected, more than 95 per cent. Therefore, a plot shape and size of a single row of 6 trees (1 × 6-tree plot) with 6 to 10 replicates has become very popular for progeny testing. However, single-tree plots were used with great success in Florida and 2 × 2 and 1 × 4-tree plots are commonly used in South Africa.

Small plots are highly effective in short rotation experiments, i.e. in those experiments that can be concluded by age five years. In long rotation trials, i.e. those aimed at developing breeding material for sawtimber products, small plots are undesirable because silvicultural thinnings will cause missing plots in some replicates. For these trials larger plots (commonly 25 or 36 trees per plot) are planted to allow for at least two thinnings before the final assessment of the trial is made. It is common to make final assessments at about half rotation age. Hence, for long rotation trials, enough trees should remain per family to make reliable assessments of sawntimber variables and defects like log-end splitting.

20.6.2 Evaluation and data processing

Proper evaluation and interpretation of progeny tests form the backbone for development of a long-term breeding program. Progeny tests should be well established and maintained so that mortality figures due to neglect are not part of the first major assessments. Where severe climatic conditions such as frost or drought influence progeny performance, survival figures naturally form a major constituent of trial evaluations.

In addition to height, diameter, stem straightness and branching characteristics, assessment of progeny tests often includes wood properties and timber defects (Section 20.1). Many of the wood and growth characteristics are related, and once the genetic correlations among different traits are determined (from progeny tests), assessment procedures could be simplified or reduced by making use of favourably correlated responses between pairs of traits.

Simple but reliable procedures are best for assessing performance of offspring. For instance, wood density is determined by many fibre characteristics. However, they interact to such an extent that an easily measured composite trait, wood density, is obtained and detailed anatomical measurements are unnecessary (Davidson 1975). Similarly, splitting of eucalypt logs is caused by a variety of factors, not all of which are yet known. However, log-end splitting itself is relatively easy to assess in the field three days after the tree is felled (Van Wyk and Hodgson 1983).

The frequency of evaluation of offspring depends on the variables to be assessed and the breeding strategy followed. Frost damage might be assessed once, or it may be followed with an assessment of resilience (recovery), and another frost damage evaluation later. Growth performance does not need to be assessed more than twice, typically $\frac{1}{4}$ and $\frac{1}{2}$ rotation age, during the effective life of a progeny test unless the researcher has other reasons for measuring trials more frequently. Some trials may serve as pilot trials and be measured earlier, more frequently, and even up to clearfelling age. Information from such trials will be invaluable to determine juvenile–mature correlations for use in defining breeding strategies.

Every effort should be made to keep experimental data free of mistakes. The practice of recording data in field books first and then transcribing it onto computer encoding sheets should be discouraged because it leads to faulty transcriptions and is expensive. One alternative is partially prepared encoding sheets filled in with identifiers like replication, block, plot, and treatment numbers. In the field, a regular check is needed to ascertain that field plot locations match the information on the coding sheets.

When working under wet weather conditions data sheets and clipboards can be wrapped in a transparent plastic bag. More recently, electronic data capturing machines with solid-state memories have been introduced. These machines can be programed to prompt the user for the next entry, have enough memory for a day's work, are weatherproof, and can be connected to micro-computers or even telephone lines for transmitting data.

Data processing is greatly facilitated through the use of computers. Whereas large data sets used to be handled by large mainframe computers in the 1970s, the advent of powerful microcomputers has made the large computers unnecessary for most analysis of progeny data. Many data processing software packages exist for micro-computers so that analyses of large data sets are now possible within field office environments. Turn-around time has been reduced greatly, enabling the tree breeder to produce and interpret breeding results rapidly and efficiently.

Although most computer programs will handle large data sets consisting of the assessments of each individual tree, it is advisable to analyse progeny test data on a plot mean basis. Within plot variation is obtained in a second, separate analysis (Becker 1985). Genetic parameters can still be calculated on an individual tree basis with this approach. The advantages of this approach are: (a) it is cheaper, (b) it uses less computer memory (useful for small computers), and (c) genetic correlations are not subject to error for repeated measurements. Errors can very easily happen unless field workers have a very strict checking system whereby they ensure that trees are matched one to one at the repeat measurement (say matching $\frac{1}{4}$ rotation age with $\frac{1}{2}$ rotation age). Thorough checking to be quite sure the data set does not contain errors is essential before starting analyses.

21

Seed production

The end product of any breeding program is the material to be used in propagating plants for afforestation or reforestation programs. Before the recent progress with mass vegetative propagation, all new eucalypt plantations originated from seed. Although vegetative propagation will decrease the demand for seed, we consider that seed will still play a major role in plantation establishment for many years. Some species are very difficult to propagate vegetatively and even for more favourable species many problems are still being experienced with rooted cuttings. This chapter deals with mass production of genetically improved eucalypt seed with specific discussion of seed production areas, seedling seed orchards, and clonal seed orchards.

21.1 Seed production areas

Seed production areas (Fig. 21.1) are stands specifically managed for seed production, often only as an interim measure until a more advanced seed production program has been established. Such seed production areas can be stands specially planted for seed production or existing stands specially managed for seed production, provided their genetic origin is appropriate. They are a very effective way of making available a seed supply of somewhat improved genetic quality, and they provide an inexpensive supplement to more elaborate breeding activities.

Planted seed production areas should be established at close spacing to allow heavy and early selective thinning so that the trees of best form and vigour can be retained while at the same time a wide spacing is opened up to allow full crowns and good seed crops to develop.

An example of a successful seed production area for *E. grandis* is one in Brazil established by Champion Papel e Celulose S.A. A considerable quantity (10 kg) of seed was collected from more than 50 trees near Coffs Harbour, New South Wales (M. de Freitas, personal communication). Champion planted more than 250 ha at 1600 trees ha^{-1} with this seed at a site near Mogi Guaçu in the State of São Paulo, in 1968. At 4 years 100 ha was thinned to 500 trees ha^{-1}. At $7\frac{1}{2}$ years a second thinning was made to 160 trees ha^{-1} (Freitas *et al.* 1983). Capsules were collected by climbers who used spurs and a cutting blade on a pole to harvest the seed-bearing branches. Seed yield was more than 10 kg ha^{-1} after 10 years (Zani and Kageyama 1984).

Enough experience has been gained to show that systems of this kind work well, though the precise details of spacing, time of thinning, and methods of seed collection from trees which tend to become very large will have to be developed according to the species and in line with facilities and expertise available where the seed production area is established.

The quickest and most common way to make a seed production area is to convert existing mature stands of good growth and genetic quality. If a seed production area is to be planted it should be with a seed mixture derived from a large number of unrelated parents of appropriate provenance. If, however, family identity is known, the tree breeder should rather consider planting a seedling seed orchard (family identity retained) due to the greater genetic gain which can be expected from the orchard.

Seed production areas developed from special plantings require the same attention to isolation from contaminating pollen as seed orchards (Section 21.4.2). Seed production areas developed from existing stands often will have the disadvantage of poor isolation. Usually such stands are selected within mature plantations of the same species. With

Fig. 21.1. Seed production area of *E. grandis* owned by Florestal Rio Doce S.A. near Itabira, Brazil. This eight-year-old plantation originated from a 23 kg bulk collection of seed from near Coffs Harbour, New South Wales. It had been thinned to 150 trees ha^{-1} (Eldridge 1981) (photo K.G. Eldridge).

the co-operation of forest managers the mature stands surrounding the seed production area can either be thinned heavily or replanted with a different species that will not hybridize with the one in the seed production area.

Care should be taken with eucalypt seed production areas not to leave too few seed trees per hectare for seed production. Pollen vectors on plantation eucalypts, mostly insects, have a localized foraging behaviour and if trees are far apart, the frequency of self-pollinations might be increased considerably at the expense of the more desirable cross-pollinations. Zani and Kageyama (1984) studied the effect of tree spacing and harvesting regime on seed crops of *E. grandis* in the Champion seed production area in Brazil. At first a quarter of the trees scattered throughout the 100 ha stand were harvested each year by cutting off the seed-bearing branches. Recovery of these damaged crowns was delayed by overshadowing from adjacent trees harvested in the previous one to three years. Subsequently the stand was divided into four and all the trees in one quarter were harvested each year, by cutting off seed-bearing branches as before. There was a considerable increase in seed viability (910 000 to 1 180 000 viable seed kg^{-1}) and also in seed yield per tree and per hectare.

Another good source of seed for small afforestation programmes is to collect seed from outstanding phenotypes in good-quality stands. Seed collections from good stands, sometimes called 'seed collection areas' or 'seed stands', are considered less improved than those from a seed production area if no selective thinning has been done. However, in many eucalypt stands left past rotation age, competition allows only the dominant trees to flower and produce seed. Thus some seed collection areas are comparable to a seed production area from the standpoint of genetic improvement.

Size of eucalypt seed production areas varies with seed requirements and an area as small as half a hectare may be sufficient, due to prolific seed production.

The amount of genetic improvement from a seed production area is expected to be small. Selection to retain the best trees (about 1 in 10) is one source of improvement. In addition, the neighbourhood inbreeding effect (slow growth from inbreeding among neighbours) of seed collected from natural stands is overcome in the first generation of plantings (see Section 19.6.2). Thus due to the effects of both mild selection and increased outcrossing, seed collected from a local seed production area can be expected to grow into better trees than seed from the same original provenance. However, we are not aware of any progeny tests which evaluate genetic improvement achieved with eucalypts by this cheap and simple procedure.

A seed production area may also be set up by thinning a progeny or provenance trial. Such a seed production area may contain a large amount of genetic variability, and could also function as a genetic base population for future selection (Brune and Zobel 1981).

A special case of poor-quality seed production is the Rio Claro seed source in the state of São Paulo, Brazil (Andrade 1961; Gurgel and Cavalcanti 1978). Many species grow side by side in square plots of about 400 trees, allowing hybridization to occur. Most often the exact source was not known, although some species probably originated from seed taken from only one or two trees. Thus Rio Claro seed is said to suffer from three evils: hybridization, inbreeding, and narrow genetic base. However, *E. citriodora* has not hybridized and the plots of this species at Rio Claro have proved to yield heavy crops of excellent seed, producing extensive plantations of vigorous trees, apparently with a substantial amount of genetic variability. *E. citriodora* (subgenus *Corymbia*), is reproductively isolated from the other species in the arboretum, nearly all of which are in the subgenus *Symphyomyrtus*.

The large, diverse forest planting programs in Brazil prompted the Brazilian Forest Service (IBDF) to certify seed production areas in that country. In the 1970s only imported seed of approved quality could be used in afforestation or reforestation programs if they were to enjoy fiscal incentives. By the 1980s seed of local origin was required, if available from adequately certified areas. Certification of a seed production area required inspection by authorized foresters (usually from universities) who would verify species purity, sanitation, isolation, management practices, and, if possible, wide genetic base (Section 21.5).

Between 1980 and 1985 the IBDF certified 51 seed production areas, provisionally known as seed collection areas until better areas come into production (information from A. Brune). These areas have been established for 13 species and a total area of 1136 ha, as can be seen in Table 21.1.

Other species for which certified areas are expected, as they come into production, are *E. camaldulensis*, *E. pyrocarpa*, and *E. robusta*. Similar certification schemes have been adopted in Brazil for seed production areas of trees from other genera.

It is expected that many seed production areas will be obsolete when seed of higher genetic quality of some species becomes available from seed orchards. Still, the genetic variability in plantations derived from sources of wide genetic base, such as the Champion seed production area at Mogi Guaçu, can be used for future selection. Furthermore, any adequately certified seed production area is valuable as a reserve in case of loss of seed orchards by fire or other causes.

Table 21.1

Eucalypt seed production areas and seed collection areas certified as of August 1985 by the Brazilian Forest Service (IBDF) (compiled by A. Brune.)

Species	Area (ha)	Origin
E. citriodora	10.4	Australia
	8.1	Rio Claro, S.P., Brazil
	5.0	Rio Claro
	155.4	Zimbabwe
E. cloeziana	30.0	Mtao Forest, Zimbabwe
	3.8	S.W. Kennedy + Gympie Qld.
E. deanei	1.3	N. Windsor + Cessnock NSW
E. dunnii	50.4	Moleton, Acacia Creek, north coast, NSW
	49.0	Australia and a Brazilian origin
E. grandis	53.4	Coffs Harbour, NSW
	43.8	Coffs Harbour, NSW
	22.3	Coffs Harbour, NSW
	45.7	Mtao Forest, Zimbabwe
	15.0	Mtao Forest, Zimbabwe
	44.0	Atherton, Qld
	91.9	Mtao Forest, Zimbabwe

Table 21.1
(Continued).

Species	Area (ha)	Origin
	35.0	Mtao Forest, Zimbabwe
	2.4	Rio Claro
	15.5	Coffs Harbour, NSW
	1.2	Coffs Harbour, NSW
	2.5	Coffs Harbour, NSW
	3.3	Coffs Harbour, Rio Claro, S. Africa
	1.5	East of Atherton, Qld
	31.8	Kyogle, Kempsey, Coffs Harbour, S. Africa
	1.3	Australia + Brazil
	4.0	Rathmore, Zimbabwe
E. maculata	25.0	Rio Claro
E. microcorys	4.6	Rio Claro
	2.7	Gympie, W. Coopernook, Wooton, Coffs Harbour
	1.0	Rio Claro
	1.2	Rockhampton, Qld
E. paniculata	10.2	Australia
	10.4	Rio Claro
	22.5	Rio Claro
E. pilularis	1.8	Queensland
	1.9	Near Murgon, Qld
E. saligna	37.2	Itatinga, Brazil
	2.3	NSW (various)
	9.0	Itatinga
	3.0	NSW (various)
	181.3	Itatinga
	1.5	Australia
	8.8	Mt Scanzi + Cessnock, NSW
	4.1	Australia
E. tereticornis	4.0	Australia
E. urophylla	10.8	Bessi-Lau, Timor
	0.5	Dili, Timor
	35.9	Timor
	8.0	Java
E. viminalis	3.4	Canela, Brazil
	215.3	Canela, Brazil

21.2 Seedling seed orchards

A seedling seed orchard consists of a special plantation raised from seedlings, with family identity retained, and managed for maximum seed production. This population of trees consists of open-pollinated families of selected trees (plus trees) planted in a replicated, statistical design. The seed orchard starts off as a progeny test and then is gradually converted into a seed orchard first by thinning the poorest trees

Fig. 21.2. Seedling seed orchard of *E. camaldulensis* in Zimbabwe just after thinning at age three years. The orchard was established as a progeny test of 25 open-pollinated families from a seed collection near Katherine, Northern Territory (photo J.C. Doran).

within each family and later culling some whole families until only the seed-producing trees remain (Fig. 21.2).

Seedling seed orchards are the most common orchard type used in eucalypt breeding programs because the biology and cultivation of nearly all eucalypts of warmer climates allows seed production more easily and cheaply from orchards raised from seedlings than from orchards made from grafts or cuttings. For eucalypts of cooler climates the long delay before abundant seed production in seedling seed orchards may make them less appealing than grafted seed orchards which flower many years earlier.

There has been much discussion since the 1960s about the relative merit of seedling seed orchards and clonal seed orchards (for example Toda 1964). Zobel and Talbert (1984) give a concise description of the pros and cons of both types of orchard and point out that both types have a place in their own right for particular situations. They also acknowledge that seedling seed orchards are very suitable for eucalypts because of their early flowering and rapid growth. In this book we do not wish to try to promote one type of orchard above the other, but rather collate the information from experiences that led to success or failure of different seed orchard types and leave it to the forest manager and tree breeder to make a well-informed decision.

21.2.1 Strategy and planning

Seedling seed orchards must be planned so that advantage can be taken of fast generation turnover. This entails a large number of families in the initial seed orchard to allow for selection and culling in one or two subsequent generations without the danger of inbreeding. This danger, of course, will depend on the degree of culling, but it may be advantageous not to cull too heavily in each generation, retaining perhaps two-thirds of the families (Shelbourne 1991), to maintain large genetic diversity while at the same time obtaining improvement in the traits under consideration.

Advantage should also be taken of the progeny test phase of a seedling seed orchard as a convenient way of assessing the breeding value of each family, before culling to promote seed production on the best trees. For short rotations of six to ten years, reasonably reliable selections should be possible in three- to five-year old trials.

Where good survival is anticipated, as for tropical or subtropical sites, a procedure may be followed where only one replication in time is necessary with a further possible replication in space only if genotype–environment interaction exists. However, for harsher sites, such as sites subject to frost damage, replication in time and space may be important. If a species is tested for cold tolerance, by chance it may be subject to a mild first winter and by the second winter be tall enough to survive ground frosts which would have killed the seedlings in the first year. Although laboratory testing for cold tolerance may partly overcome the problem, it is important to repeat the trial to test the young seedlings a second time under possibly colder conditions. If the trials are also replicated in space, it may not be necessary to repeat them a second time at each site.

21.2.2 Design

Naturally any trial must be statistically designed with a large enough number of replicates. If the number of families is small, the site is uniform, and plot size is small a randomized complete block design may be adequate. More commonly many families are involved (100 or more) and an incomplete block design will be statistically more efficient. A total of 20 trees per family may be sufficient for ranking families (Cotterill and James 1984), but often a larger number is required to make the area of orchard large enough. Plot size should preferably be small for the following reasons.

(1) Small plots will result in a more random distribution of the remaining trees after thinning and culling of some families. This should also effect more efficient cross pollination.

(2) Small plots allow larger numbers of replicates, providing more precise comparisons among families.

(3) Only one tree per plot should remain once seed production commences, to eliminate matings between adjacent siblings.

The shape of the plot can be square but, if not more than 4 trees are planted per plot, it does not matter whether they are planted as 1×4 or 2×2 trees. In South Africa nine replicates of 2×2 tree plots have been planted for each *E. grandis* family. Plots of fewer than four trees substantially increase the amount of labelling and record-keeping and can lead to problems in statistical analysis due to missing plots.

In southern Florida, U.S.A., seedling seed orchards with single tree plots have been used with great success (Section 21.2.6). One disadvantage is that each tree must be labelled and very strict records must be kept of the whereabouts of each tree from nursery, through the planting stage up to the time of each assessment. If four-tree plots are planted, selection for a first 50 per cent thinning within each family can be done very cheaply on each plot without any measurement by simply comparing the four trees in each plot visually, a situation not possible with single tree plots where an assessment of the whole trial is necessary every time a reliable thinning is required.

21.2.3 Spacing

Wide initial spacing is not needed if trials can be assessed early (say from 3–5 years of age). However, too close spacing may inhibit seed production as it will be difficult to open up the stand fast enough by thinning. Thinnings are usually done in at least two stages followed by a final culling, as in the *E. grandis* seedling seed orchard described in Section 21.2.6

Closer spacing is likely to identify best families earlier and more reliably. This is important for short rotation forestry where no thinnings would be practised. Eucalypts are very sensitive to competition, and it is wise to try to select families for their performance under normal plantation conditions of intense competition. If this cannot be done in the seedling seed orchard itself, because of the need for very early thinning to promote seed production, then information on the performance of the same open-pollinated families in separate progeny tests is desirable.

21.2.4 Assessment and culling

The frequency and type of assessments will depend greatly on the management objectives of future plantations, as in progeny testing (Section 20.6.2). In South Africa, where short rotation eucalypt forestry is practised for the production of pulpwood, mining timber, poles and fuel, the following assessments have been done in the series of seedling seed orchards: stem volume, stem straightness, log-end splitting, stem taper, wood density, wood preservation (amount of preservative absorbed), coppicing ability after felling, seed and capsule mass, and finally rootability of cuttings on a family as well as a clonal basis. Stem volume and straightness have been assessed at first and second thinnings as well as at clearfelling. The other features are assessed at a lower intensity after clear-felling of the stand.

21.2.5 Management

Management of seedling seed orchards is a simple matter because the testing stages are closely monitored and thinnings are done at intervals of 1 to 2 years. Thinning too heavily will make the stand prone to wind damage, but if 50 per cent thinnings start early and are well spread over time the remaining trees will develop stability. Coppice sprouts must be killed by spraying the stumps with appropriate herbicides.

Seed harvesting is often done by clearfelling the seedling seed orchard, although repeated harvesting by climbing or use of a truck-mounted hydraulic platform are also common practices. At clear felling the final assessments may be done. If seed is kept separate by family, seed from the poorest families could be thrown away based upon results from the final assessments. At least 50 per cent of the genetic quality (the female contribution) from such poor families is then eliminated from the seed stock, although of course the pollen from these poorer families contributed to the quality of the rest of the seed harvested.

A continuous seed supply may be ensured by planting several such seedling seed orchards and felling them in rotation. If a five-year rotation is required, five such seed orchards could be established, one per year. After five years one seed orchard is felled, bulk seed harvested, and seed is retained from the best selections. A second series of five orchards is started with new improved families. If the species coppices well the first series could be managed to continue seed production from the first coppice crop trees. These trees could be thinned further to eliminate those families found to be inferior in the final assessments. The number of coppice shoots per stump could be reduced if preferred, but it is doubtful whether this will have any merit. Although multi-stemmed trees will develop from the coppice, their sole purpose is to produce seed.

21.2.6 Experience from different programs

Seedling seed orchards have been established in several countries for production of seed for establishment of commercial plantations (Fig. 21.3). Each of these seed production programs has had its own degree of success, and experience has shown that a breeding strategy involving seedling seed orchards can be very successful.

Fig. 21.3. A well thinned seedling seed orchard of *E. grandis* about six years old at Harare, Zimbabwe (photo R.D. Barnes).

Florida, U.S.A. (E. grandis)

A good example of the development of a breeding program through the use of seedling seed orchards is the program for *E. grandis* in southern Florida, U.S.A. Due to limited resources and an increased demand for improved yields from eucalypt plantations early in the 1970s, a strategy was chosen which combined a provenance trial, a genetic base population, a progeny test, and a seedling seed orchard in a single plantation (Geary *et al.* 1983; Meskimen 1983). This strategy is similar to the breeding seedling orchards proposed by Barnes (1984b, 1987) for developing tree breeding programs that suffer from limited resources. An essential element of the success of the Florida program was the rapid turnover of generations achieved, only four years.

The first trials established between 1961 and 1965 provided seed from select trees which was supplemented with imported seedlots to establish the first comprehensive genetic base population in 1973. This *E. grandis* base population, known as G-POP73, consisted of 296 families. Up to 60 seedlings were planted as single tree plots in a completely randomized design for a total of 14 195 seedlings at 1500 stems ha^{-1} on 9.5 ha. The origins of the families (Franklin and Meskimen 1984) were as follows:

First-generation importations:	
New South Wales and southern Queensland, Australia: mostly from selected trees	76
South Africa: mostly from controlled pollinations among 19 seed orchard clones	118
Miscellaneous non-Australian importations	16
From gardens in Florida	22
Second-generation selections:	
from mother trees selected in Florida tests	50
Third-generation selections:	
from grandmother and mother trees selected in Florida tests	14

Two-stage selection was planned for the 1973 plantation. In the first stage the test was assessed for height, diameter and survival at age 2.4 years and culled very heavily leaving only 431 trees (about 50 stems ha^{-1}) at age 2.9 years. Seed was collected from the 431 trees at age 4 years to establish the 1977 genetic base population (G-POP77), the early assessment of which would provide data for the second stage of selection in the 1973 plantation.

In the second stage of selection in G-POP73 trees were culled from 431 trees to 80 trees at age 5 years. Trees retained were chosen on the superior performance of their progeny at age 6 months in the 1977 plantation (G-POP77). These 80 trees (which were much too far apart at about 9 stems ha^{-1}) produced the commercial seed crop of the G-POP73 orchard from 1978. Seed collection was from a truck with a hydraulic platform which reached to 15 m. The same machine was used to cut the trees back to that height.

The 1977 *E. grandis* base population (G-POP77) was established on 17.3 ha with 31 725 trees representing 529 families, about 60 trees per family (Meskimen 1983). Of these families 431 came from G-POP73 and 98 were new accessions. The layout was a completely randomized design with single tree plots (Meskimen 1983). Stocking of G-POP77 was 1900 stems ha^{-1} at an espacement of 1.8 m within paired rows 2.3 m apart and 3.5 m spacing between pairs. The 3.5 m spacing allows access to any tree by machinery for cultivation or seed collection.

The most serious deficiency of *E. grandis* for growing in Florida is its susceptibility to freezing damage in winter. Although the area is at latitude 27°N and the summer and much of the winter are favourable to tree growth, severe cold is often experienced. A mass of freezing air can blow from the cold continent to the north with devastating results for the growing, unhardened eucalypts.

Two of the most important selection criteria for the breeding program in Florida are freeze resistance and freeze resilience (Section 11.4.2). Radiation freezes which occur on still clear nights as thermal inversions are more common than advective (wind-blown) freezes. Although for radiation freezes genetic adaptation (cold tolerance) plays a role in survival, rapid early growth seems to contribute most to avoiding damage. Fast-growing genotypes elevate tender crown tissue above lethal temperatures while thick, insulative bark is developed at levels close to the ground. Advective freezes, however, challenge physiological cold hardiness of the trees because turbulent cold air blows through the crowns of tall, mature trees (Franklin and Meskimen 1984; Geary *et al.* 1983).

Different genotypes recover differently after frost damage. The crowns of cold-resistant trees are lightly damaged, if at all, by severe freezes. Other selections are only freeze-resilient: their stems may be killed to the ground by a hard freeze in their first winter, but they resprout rapidly. In subsequent winters they suffer only superficial damage unless exceptional freezing weather is experienced.

One possible criticism of the Florida program is that too much effort went into establishing and testing of the base population at a single site. Also the use of a single tree plot design placed an extra burden on the managers of the program. It may be argued that more sites should have been tested through the use of different seedling seed orchards in a multiple population approach (Namkoong *et al.* 1980) or a strategy of breeding seedling orchards (Barnes 1984b) over a range of sites. Furthermore, the full range of natural provenances of *E. grandis* was not tested in the early years of the project. Later it was found that some newly introduced north Queensland provenances suffered less cold damage than improved seed from Florida seed orchards (Franklin and Meskimen 1984, p. 352).

Another criticism is that the mixing of improved material from advanced generations with first-generation selections, theoretically at least, would have slowed down progress in genetic gain. Also, it turned out that mixing of species in the very early populations caused a degree of hybridity which may be marginally detrimental in a long term breeding program based on seedling regeneration.

Despite these relatively minor criticisms, it is clear the Florida seedling seed orchard program has adequately demonstrated that it has been successful and that substantial genetic gains can be obtained quickly with such a strategy (Section 18.5). With the ever increasing interest and progress being experienced with vegetative propagation, even larger genetic gains can be expected using this relatively simple breeding approach to select individuals to be mass propagated by cuttings (Meskimen *et al.* 1987).

Other species being dealt with along similar lines in Florida but on a smaller scale are *E. amplifolia* (Rockwood *et al.* 1987), *E. robusta* (Dvorak *et al.* 1981) and *E. tereticornis*.

South Africa (E. grandis)

In contrast to the Florida program, the *E. grandis* seedling seed orchards in South Africa are based on a large population of trees that have formed commercial plantations for many decades. Many of these plantations were established from 82 kg of seed imported from unknown origin between 1897 and 1915 (Poynton 1979). The original importations were under the name *E. saligna* and although stands of mixtures of *E. grandis* and *E. saligna* have existed, the present population is almost exclusively *E. grandis* (Burgess *et al.* 1985). The total area planted to *E. grandis* in 1989 was more than 300 000 ha (Anon. 1990).

A series of *E. grandis* seedling seed orchards was started in 1976 to supplement the existing clonal orchards. Whereas the clonal orchards had been established originally to supply seed for the long term sawn timber program, the seedling seed orchards (SSO) were to supply seed for short rotation forestry programs. Rapid generation turnover was aimed at with selections at about half-rotation age within each SSO providing the seed for the next generation SSO. These SSOs very closely serve the functions listed by Barnes (1984b, 1987) who refers to them as breeding seedling orchards.

From 1976 to 1983 six first-generation seedling seed orchards were established in South Africa. Each of these orchards started out as a progeny test consisting of 99 open-pollinated families. The parent trees of these families had been selected for vigour, stem straightness, and desirable branch formation, i.e. relatively thin, horizontal branches in a well-shaped crown. Timber defects and wood properties were to be dealt with once the progeny of the parents were evaluated.

The families were planted as 4-tree plots (2×2) in a randomized complete block design with 9 replicates. Espacement was 2.1×2.1 m. Growth evaluations were done at 2 and 3.5 years respectively, prior to thinning. At each thinning 50 per cent of the current stocking was removed. A further culling of 40 per cent of the families was done based on both evaluations. Thus, a stand of 3564 trees was reduced to approximately 500 trees leaving approximately 30 m^2 for each seed tree. This stand was clearfelled for seed harvesting at the age of about 5.5 years. At this stage all the remaining trees were assessed for growth rate, stem straightness, log-end splitting, grain spirality, stem taper, and capsule and seed production. Timber samples were collected for wood-density determination and timber preservation tests. Coppice shoots from the stumps were collected three months after clearfelling to test for rootability of cuttings. A further culling and thinning of families and trees was done based on all these results before the coppiced stand was finally left for seed production. Lately, with the growing interest in clonal forestry, the information and material from these combined progeny tests and seed orchards have been used intensively to start clone banks as source material for commercial programs utilizing rooted cuttings.

Seed from the best tree in each family (or more than one tree in the most superior families) has been used to establish the next-generation seedling seed orchards. These orchards are usually established at more than one site to guard against possible family–site interactions with respect to growth performance. Four such second generation seedling seed orchards had been established by 1987.

Seedling seed orchards have also been planted in South Africa for *E. fastigata* (1981), *E. macarthurii* (1984), and *E. elata* (1985). These included 100, 90 and 81 families planted as four-tree plots in 20, 12 and 10 replicates respectively. All families came from parents selected in South Africa, but many families derived from recent seed importations from Australia have been used to supplement breeding material in seedling seed orchards of these species because they have a much more restricted genetic base than *E. grandis* in South Africa.

Zimbabwe (E. grandis)

In 1962, before the skills had been acquired to graft eucalypt trees and to avoid the delay in developing grafting techniques, the breeding program in Zimbabwe adopted a strategy of mass selection supplemented by progeny testing in seedling seed orchards (Mullin *et al.* 1981). Superior trees were selected of *E. grandis*, *E. paniculata*, *E. resinifera* and *E.* 'saligna', the latter being of uncertain identity.

For each of the species, seed from selected trees was used to raise plants for open-pollinated seedling seed orchards. A layout based on 300 planting spots at 3.7 metres square spacing was adopted. To improve the opportunity for selection within families, a form of group planting was used in which ten seedlings, selected within each family on appearance in the nursery, were established about 1 m apart in a circle around each of the 300 planting spots. In the course of the next three years the orchards were subjected to four assessments and thinnings that reduced them, on the basis of growth rate and stem form, to the best individual trees of the best families (Mullin *et al.* 1981). By the completion of the fourth thinning the orchards had been reduced from their initial 3000 stems per species to:

E. grandis: 74 stems comprising all 25 of the original families;
E. paniculata: 31 stems comprising 20 of the original 60 families;
E. resinifera: 88 stems comprising 61 of the original 100 families;
E. 'saligna': 82 stems comprising all 25 of the original families.

The 1962 seedling seed orchards at Mtao Forest Reserve were used successfully for some years for *E. grandis* needs until it was suspected that pollen of *E. tereticornis* was reaching the trees in the orchards and creating problems of hybridization (Mullin *et al.* 1981).

Seedling seed orchards were also started for *E. camaldulensis* but the initial selection sources were limited and had to be supplemented with provenance trials. For *E. grandis*, subsequent strategies include the breeding seedling orchards (BSO) described by Barnes (1984b, 1987). The BSOs function as selection, progeny test, seed production areas, and breeding archives. They are planted at close spacing and thinned early and frequently on a selective basis to leave equal numbers of trees per family up to the time of assessment, which will be when the trees reproduce sufficiently well to permit the start of the next selection cycle.

Australia (E. regnans and E. globulus)

The first eucalypt seed orchard in Australia was established by APM Forests Pty Ltd at Silver Creek in south-eastern Victoria between 1970 and 1973 on an area of 11.5 ha. *E. regnans* seedlings were raised from open-pollinated seed collected from plus trees selected in even-aged natural forests aged 30 years (Richmond 1971; Eldridge 1971). The local provenance (Traralgon Creek and Mirboo East, Table 14.1, p. 135 and Fig. 14.4, p. 134) was the source of 120 of these families, 5 were from other parts of Victoria, and 40 from southern Tasmania (Cameron and Kube 1983). Subsequent provenance trials showed the local provenance was indeed the best for the intended planting area in the Strzelecki Ranges of south-eastern Victoria (Griffin *et al.* 1982a).

To obtain a flat enough site the orchard was located at the drier extreme of the range of the species. It was well isolated by surrounding plantations of *Pinus radiata*. Four contiguous orchards were established in successive years, each with about 40 families and 16 replicates in randomized complete blocks. The plot of each family was planted in a line of five with 0.6 m between seedlings and a 6 × 6 m spacing between family plots (planting spots). Over the next 3 years each plot was thinned to the single most vigorous tree. This type of design was first used in Brazil in 1952 for *E. maculata* by Pasztor (1962). A seedling seed orchard was preferred to a grafted clonal orchard because grafting was considered difficult to achieve on a large scale with the technical resources available (Cameron and Kube 1983).

Selection between families was carried out at nine years when 10 out of 40 families in each section of the orchard were culled on the basis of growth at the orchard site and from progeny tests of the same families grown at normal plantation spacing on more typical *E. regnans* sites which are steeper, wetter, and more fertile.

Seed collections have been made with a 15 m or 20 m hydraulic platform on a truck and were largely confined to picking clusters of capsules by hand in an effort to retain developing crops for subsequent harvests. The considerable expense of this method was preferred to cutting large branches or felling a whole section of the orchard, both of which would have required several times the area of orchard for the same sustained annual amount of seed. Note that *E. regnans* does not coppice and a felled section would have to be replanted.

Seed production in the *E. regnans* seed orchard has been abundant only in some years (every fourth year in natural forests (Ashton 1975a)) and only after about age 10 years. In one of the best years, 1988, the harvest was 8 kg ha^{-1} from an orchard aged 11.5 years, established in 1976 close to the 1970–73 Silver Creek orchard and managed in the same way (Davis and O'Connor 1988). At age 11 to 13 years seed harvests averaged about 5 kg ha^{-1}, sufficient for operational seed requirements, but in some years there was so little seed on the trees that no collection was made.

A seedling seed orchard of *E. globulus* of similar design in north-western Tasmania, established by APPM Ltd in 1980, also has had a rather small and late yield of seed : 3 kg ha^{-1} at 9 years and 6 kg ha^{-1} at 10 years (W.N. Tibbits personal communication). Evidently seed orchard planners cannot expect temperate eucalypts to produce seed as early or abundantly as *E. grandis*.

21.3 Clonal seed orchards in South Africa

The term clonal seed orchard refers to a collection of selected trees, vegetatively propagated, planted in isolation from genetically inferior trees of the same species, and managed so that the maximum amount of seed can be obtained in the most economic way. The orchard trees originate from selected trees (plus trees) chosen for a variety of desirable characteristics that normally include growth rate, stem straightness, and wood properties. Propagation of the trees is vegetative and may be by means of grafts, cuttings, air-layering, or even by tissue culture techniques. The most common clonal orchards are grafted seed orchards; however, the recent successes with rooted cuttings will encourage establishment of clonal orchards through the use of cuttings, especially for species that flower within a few years after planting.

One of the main reasons for use of relatively expensive grafted plants for seed orchards rather than seedlings is to speed up seed production. Grafting material (scions) taken from mature parts of the parent tree (ortet) is physiologically mature and normally retains this physiological condition and tends to flower within one or two seasons after grafting. Hence, cross-pollination among grafted trees (ramets that had originated from selected parents) will ensure a relatively rapid supply of genetically improved seed, especially for temperate eucalypts which may take ten years to flower abundantly when planted as seedlings. The clonal seed orchards of *E. grandis* in South Africa described in detail in this section are good examples of the techniques required for eucalypt clonal orchards.

21.3.1 Background

In South Africa, the first clonal seed orchard of *E. grandis* was established in 1965 at De Hoek State Forest, Northern Transvaal. Although 8 ha was set aside, the area was almost never fully stocked because of slow rate of establishment and scion–stock incompatability problems. Therefore, to obtain potential seed production per unit area, average figures per ramet should be translated to an area basis (148 ramets = 1 ha). In 1988 the first seed orchard contained 1015 ramets from 74 clones and the second orchard, established in 1969 at J D M Keet Forest Research Station (formerly Zomerkomst) contained 1676 ramets from 78 clones.

Both orchard sites are near the town of Tzaneen at latitude 23°55'S. The De Hoek orchard is at 1300 m altitude and 1730 mm average annual rainfall and the Zomerkomst orchard is at 750 m altitude with 930 mm rainfall. As will be apparent from the high seed yields described in Section 21.3.6 these two sites have environments close to the optimum requirements for seed production of *E. grandis*.

21.3.2 Selection

Parent trees were assessed for volume production, stem straightness, desirable crown characteristics, and freedom from log-end splitting after felling and cross-cutting. Initially, selection intensity was very high with only one tree in 10 to 50 ha being accepted. Later, a dual purpose selection program yielded more and younger plus trees with somewhat lower selection intensity. However, standards are not necessarily lowered because short rotation selections go through one phase of screening on volume production and stemform before being included in the sawn timber program.

21.3.3 Methods of producing grafts

First attempts (in the late 1960s) with *E. grandis* cuttings in South Africa were not very promising. Even though some rooting was obtained with cuttings from coppice shoots, lack of facilities and other practical considerations compelled attempting various grafting techniques.

Initially, unselected stock plants were raised for grafting. However, in an investigation on scion–rootstock relationships, Van Wyk (1977b) obtained a much higher rate of success in grafting when rootstocks were grown from seed collected from the ortet before scions were taken from the same tree for grafting.

Raising eucalypt seedlings for rootstocks needs careful planning because seed from selected ortets must be harvested well in advance of the grafting season. In South Africa, grafting is usually done when the rootstocks are 10 months old. Recently, success has also been obtained with younger rootstocks, as is experienced in France and New Zealand. In South Africa, grafting can be done throughout the year but best results have been obtained in spring. This work has been in an area with temperatures mainly within the range from 15° to 25°C. Large deviations from these temperatures in distinctly different climatic types will probably dictate a more exact definition of the grafting season for eucalypts, and there may be variation in requirements of different species.

In South Africa rootstocks are normally grown in plastic containers of about 5 litre capacity. After grafting, the successful grafts are usually kept in partial shade in the nursery for several weeks before planting. The large containers require large amounts of nursery soil which increases costs. Therefore, a preliminary test of field grafting was carried out by making 129 grafts at two sites to investigate the technique (Hodgson 1977b). The 88 per cent success obtained after 3 months was comparable with nursery grafting indicating that field grafting could be adopted as a regular procedure.

After trials of several grafting techniques (Van Wyk 1977b) the procedure finally adopted was a method similar to the rind graft described by Garner (1968). By this method an intial 'take' of more than 90 per cent can be obtained, but thereafter, as commonly experienced, incompatibility sets in and this is liable to reduce success to less than 50 per cent with some clones.

21.3.4 Design

A spacing of 8.2 × 8.2 m was adopted for the first clonal seed orchards of *E. grandis* in South Africa. This spacing facilitates mechanical maintenance of grass swards in the orchards but may be too wide for efficient cross-pollination. Therefore planning has been carried out for close planting of a group of clones. The clones to plant in each group were selected for their high breeding value from the results of control-pollinated progeny tests. They were the parents which produced outstanding full-sib families with regard to volume production, for example clones 4, 6, 15, 50 in Table 18.1, p. 163. Grafts from these parents were then planted close to each other in groups of 2, 3, or 4 per planting site, in other words, one planting spot would have several clones that were to grow into a 'single' crown. These groups were placed 14 m apart while grafts within a group were only 2 m apart. With this spacing it is hoped that bees, while concentrating on the flowers of the 'single' crown, would enhance cross-pollination and at the same time increase the frequency of offspring from outstanding specific combinations.

A prerequisite for this scheme is a knowledge of flowering times of each clone so that pairs or groups that synchronize can be planted together. Also graft incompatibility often causes one or more individual grafts (ramets) in a group to fall behind the development of others. This leads to rapid suppression of some ramets and careful pruning must be done, or even replacement of grafts to have groups develop equally. A better approach may be to plant at least two grafts per clone per site and adopt a hedgerow approach where the groups are planted close to each other along a hedgerow. If a graft then failed another graft of the same clone could fill the gap. Different clones would touch each other within the hedgerow. The hedgerows could then be spaced 15 m apart to encourage bees to work up and down such rows before switching to another hedge.

21.3.5 Management

Originally the branches of ramets were tied down or weighted down in an effort to keep seed-producing capsules within reach. Trees were also pollarded to encourage growth of side branches (Fig. 21.4). Although this practice was effective, it proved to be very expensive and resulted in a great loss of flowers and capsules. With this practice the crowns of the trees did expand rapidly in diameter and it was found that a 8.2×8.2 m espacement was too close for ramets 5–6 years of age.

Fig. 21.4. Grafted *E. grandis* in Zomerkomst seed orchard, South Africa. Ramets are pollarded every four years to encourage growth of side branches (photo K.G. Eldridge).

The strong apical growth of *E. grandis* means that each annual crop of seed is 2–3 m higher up the tree than the previous crop. Some means of crown control was required. Accordingly, Hodgson (1977a) investigated control methods by using four treatments, one of these being to allow growth to continue for six seasons after grafting, and when a height of about 15 m had been obtained, felling was done at a height of about one metre above the graft union as in Fig. 21.5.

In this case the intention was that after felling, coppice shoots should be allowed to grow with a view to rotational felling thereafter. The other three treatments involved pollarding at various heights and treating the resulting coppice shoots by either tying them down so that seed collection could be done at ground level, or cutting them back at the most appropriate stage.

The timing of cutting trees to produce coppice shoots could be important in relation to the initiation of flower buds. If pruning is done well before spring new coppice shoots will be produced early enough to attain development of flower buds on these shoots. However, seed from the previous flowering season only ripens a few months before bud initiation. Since there is a large range in flowering time among clones, and it is preferred to allow 4 to 5 months for seed to ripen, it was found that harvesting operations were too close to flower initiation to obtain good flower production on ensuing coppice shoots. Therefore, very little flower production was obtained the first year after harvesting. With *E. grandis*, capsules remain unopened on the trees as long as branches stay alive. Considering all this, it was found that a rotation of 4 years should be adopted for harvesting by clearfelling. Some lower branches may die off in this period but usually they can be reached with pruning shears on long poles for harvesting of capsules on such branches before they die and shed their seed.

Fig. 21.5. A graft of *E. globulus* about 10 years old cut one metre above the graft union and sprouting vigorously, in a CELBI seed orchard in Portugal (photo K.G. Eldridge).

Grafted *E. grandis* trees flower within 2 years after grafting if scions are taken from trees that had flowered before. Scions taken from young coppice shoots were found to flower a year or more later. For *E. nitens*, it seems as if flowering is further delayed even on grafted trees because no flowers had been produced on such trees 5 years after grafting in South Africa.

Bee-hives have also been introduced into the orchard, although it is not known how many of these should be provided per unit area. The ideal would be to arrange that the bees in the hives need not go elsewhere for eucalypt nectar and pollen, but this still leaves the question of bees entering the area from outside. In fruit orchards 5 bee colonies per 2 hectares are considered to be enough for sufficient fruit (Free 1970). Since older eucalypt trees flower profusely, possibly 3 colonies per hectare will be most efficient. Bees from different colonies in competition range more widely than bees in single colonies which normally forage constantly on small areas. This behaviour of competing bees, would be beneficial for promoting cross-pollination, because non-competing bees may tend to work on a single ramet.

21.3.6 Seed yield

The two South African seed orchards of *E. grandis* described in Section 21.3 have produced very high yields of seed, up to 2 kg of seed from a single graft. For the three years 1977–1979 the overall average seed production was 21 kg ha^{-1} yr^{-1} from a total of 16.8 ha (Van Wyk and Hodgson 1983). Only 20 per cent of the trees in the two orchards were harvested in these years due largely to the cycle of very severe top-pruning being out of phase at that time. With top-pruning of a quarter of the orchard each year, the potential production of a fully stocked *E. grandis* orchard in this favourable area is probably of the order of 40 kg ha^{-1} yr^{-1}. A single hectare could provide enough seed to plant 10 000 ha per year!

21.4 Management for seed production

Some of the accumulated South African experience of managing eucalypts for seed production has been outlined in previous sections. As for seed orchards of other forest trees (Faulkner 1975), it is clear that the special biological characteristics of each species and the local economic and environmental conditions determine what can be done, and there is no substitute for on-the-spot experience and knowledge.

21.4.1 Sites for maximum seed production

Selection of a site where the species will produce abundant seed is essential and, at present, can only be done by trial and error, by observing seed production of the species on a range of sites. Three cases illustrate the point.

(1) In Portugal four grafted seed orchards of *E. globulus* were planted by CELBI between 1970 and 1975 (CELBI 1979). Only one of these orchards could be said to be moderately successful in producing seed (3 kg ha^{-1} at 10 years). The other three, some distance inland from the mild conditions of the coastal area, failed to produce seed in commercial quantities (L. Leal, personal communication).

(2) In Western Australia seedling seed orchards of *E. diversicolor* were established in 1972 and 1979 on sites favourable for growth within the natural range of the species. Despite heavy thinning they failed to produce much seed. At a drier place further inland and outside the natural range, a small stand of planted trees subject to partial flooding in winter and drought in summer has produced much heavier crops of seed than have ever been observed on young trees within the natural range (R. Breidahl, personal communication).

(3) In Papua New Guinea *E. deglupta* flowers more profusely at Bulolo (inland at 700 m on the main island of PNG) than in its natural range so the seed orchards were located there rather than in New Britain (J. Davidson information).

Another aspect of choice of site is distance from the orchard manager's office. A distant orchard is likely to receive much less care and attention, and produce less seed, than one that can be seen through the office window.

In planning the size of a eucalypt seed production area or seed orchard the seed size of the particular species must be taken into account. It is better to plan for a number of viable seeds than simply a number of kilograms. For example, 1 kg of *E. globulus* seed contains about 70 000 viable seed while for *E. camaldulensis* and *E. grandis* the number would be about 700 000 (Boland *et al.* 1980a). Thus, production of enough seed to establish the same plantation area would require ten times the orchard area for *E. globulus*, assuming the same annual harvest of, say, 10 kg ha^{-1} for each species.

21.4.2 Isolation

An important question is isolation of the seed production area from potentially contaminating genetic sources. The maximum distance from another stand with which a given stand can easily cross is dependent for the most part on the distance that pollen vectors travel to effect pollination.

A study of pollen dispersion by bees (Pacheco *et al.* 1986) using ^{32}P-labelled pollen in a seed orchard of *E. saligna* in Brazil revealed that the greatest activity of the bees was up to 100 m from the hives and decreasing gradually up to the studied distance of 300 m. It appears that a buffer zone of 100 to 200 m planted with non-interbreeding trees would be realistic for insect pollinated species (Van Wyk 1981; Griffin 1989b).

Seed production areas and seed orchards can be isolated very effectively by planting them in plantations of another genus or another eucalypt species with which there is a complete barrier to interbreeding. This is quite often feasible especially where more than one species is used in a silvicultural program. For example if the seed orchard is to be either *E. grandis* or *E. saligna*, isolation can be provided by *E. citriodora*. For certain species many such arrangements can be planned using knowledge of breeding barriers between subgenera (Pryor and Johnson 1971).

21.4.3 Managing the trees

Managing eucalypts for seed production is still a matter of practical experience with little scientific basis. For free-growing trees a wide enough spacing is required so that low branches remain green and bear seed, there is little competition with neighbouring trees, and there is plenty of light over the whole crown. The repeated thinnings required for this purpose can be achieved as part of genetic culling, as described in Section 21.2.6 for South African seedling seed orchards. Eventually the trees grow too tall for access by climbers or hydraulic towers. In the Florida *E. grandis* orchards, trees were lopped back at 15 m every few years and in the Australian *E. regnans* orchards at 20 m.

For coppicing species, the trees can be cut and seed collected on a cycle of about 4 to 8 years, depending on species. The most intense management of eucalypts for seed production has been in the South African orchards of *E. grandis* (Section 21.3.5; Hodgson 1977a) where grafts were severely cut back every four years (Fig. 21.4). However, the extra seed obtained from tying down branches proved to be too little for the great effort involved, and the practice ceased after a few years.

The prospects for greatly increased production as a result of new techniques for intensive management of eucalypts for seed production are very encouraging. Future possibilities for managed seed production include the use of paclobutrazol (sold as 'Cultar' and 'Clipper'), a chemical which inhibits the action of the gibberellin group of growth hormones, reduces growth of branches, and in some cases promotes early and abundant flowering (Griffin *et al.* 1992). Also, particular cycles of moisture and drought, heat and cold are being developed empirically to promote flowering of eucalypts grown in large pots, opening up the prospect of small, intensively managed seed orchards in controlled environment facilities. Another possibility is the use of improved pollen management in orchards where open pollination prevails, to increase the numbers of flower buds becoming ripe capsules and the numbers of seeds per capsule. When research on the physiological processes involved in eucalypt flowering is further advanced it may be possible to control flowering with precision and produce seed in control-pollinated clonal orchards at a price low enough to compete with vegetative propagules in some eucalypt species (Martin 1987; Tibbits *et al.* 1990).

21.4.4 Seed collection

Seed collection from eucalypts presents some difficulties as the crops of capsules are on long thin branches which are difficult to reach. The technique used in seed production areas in Brazil is for climbers with knives to chop off seed-bearing branches (Section 21.1). This has two disadvantages: destruction of developing crops for future years and increasing danger to climbers working several metres higher on each visit. A hydraulic platform on a truck solves the problem only where sufficient funds are available and the topography is flat. An alternative is to establish large enough areas to fell a whole block for seed each year (as in South African seed production areas, Section 21.2.6).

After felling, the coppice regrowth produces another good crop of seed within a few years. For a non-coppicing species such as *E. regnans* it may be cheaper to use a truck-mounted hydraulic platform despite the expense of this equipment.

A detailed study of costs showed the best technique for collecting in an orchard of *E. regnans* with a good crop at age 11.5 years, was to use a truck with a 20 m boom. The two operators in the bucket picked capsules literally by hand to ensure the future value of developing crops was not diminished (Davis and O'Connor 1988). Considering the great value of the orchard seed in terms of the expected improved growth and form of new plantations, the high cost of collecting the *E. regnans* seed was considered well-justified.

21.5 Seed certification

Seed buyers require some assurance that their purchases are of the quality claimed since genetically improved seed is more valuable than unselected, routine seed. There are seed certification schemes operating in Western Europe and North America, usually based on the OECD (Organisation for Economic Cooperation and Development) scheme for the control of forest reproductive material moving in international trade (Anon. 1971). The OECD scheme, whether for trade within a country or for international trade, puts seed (or other reproductive material such as cuttings) into three broad categories: source identified, selected, and certified:

(1) Source identified: the minimum standard, requiring a definition of the locality of collection, a description of the stand, and supervision of collection and processing by a Designated Authority.

(2) Selected: the Designated Authority must be satisfied that the stand providing the seed has been selected for certain criteria. A seed production area of known and suitable provenance and a reasonably wide genetic base would meet these criteria.

(3) Certified: the superiority of this seed source must be demonstrable in field trials. A well-conducted seed orchard would satisfy the Designated Authority for this category.

In Brazil seed certification is conducted by the Federal Government forestry authority (IBDF) through its Commission of Control of Forest Seed (Sampaio and Suiter 1983). This scheme effectively controls the quality of seed used in plantations financed under the fiscal incentive scheme (Galvão and Couto 1984). A company planning a plantation will not receive money from IBDF unless the seed they propose to use is of suitable quality. IBDF recognizes three categories, similar to those of the OECD:

(1) Seed collection area: seed trees selected at the rate of not less than 1 in 10 in a suitable stand managed for wood production.

(2) Seed production area: a selected stand heavily thinned to promote seed production, and somewhat isolated from inferior pollen.

(3) Seed orchard: the orchard must contain at least 50 intensely selected clones and be well isolated.

Other eucalypt seed traded in Brazil is not subject to IBDF control, and few other countries growing eucalypts have any formal control over quality of eucalypt seed traded. In Australia, where a great many small seedlots are collected for export each year, there is no formal certification scheme. However, the quality of seed and of detailed information provided for many years by CSIRO's Australian Tree Seed Centre and by the French Centre Technique Forestier Tropical in France gives assurance of authentic place of origin and botanical identity, corresponding to OECD 'source identified'.

Prudent buyers protect their own interests by asking for details of number and quality of seed trees and exact location and genetic history of each seed stand. Also, they find out as much as possible about the credentials of seed collectors and merchants. Plantation managers who seek the cheapest seed regardless of source should be reminded that the Latin motto *caveat emptor* (let the buyer beware) is as true today as it was 2000 years ago.

An obvious solution to doubts about the genetic quality of purchased eucalypt seed is to grow your own genetically improved seed (Eldridge 1981), and much of this book is devoted to explaining how to do so.

22

Mass vegetative propagation

22.1 Advent of eucalypt cuttings for plantations

Before 1960 the only genera of forest trees mass propagated for wood production by making rooted cuttings of superior individuals were *Populus* and *Salix* in temperate areas and *Cryptomeria* and *Cunninghamia* in subtropical areas. Since that time progress in cloning techniques and a better, though still far from complete, understanding of the maturation process and ways of reversing it, have led to other genera of forest trees such as *Picea, Pinus, Sequoiadendron*, and eventually *Eucalyptus* joining the list of trees able to be mass propagated by rooted cuttings.

Vegetative propagation is a technique widely used in tree breeding to manage breeding populations more efficiently. Grafts, cuttings, or micropropagules of selected trees are brought together in a clone bank where the genotypes are conserved and pollinations are made near ground level, in safety. Clonal seed orchards are established using the same techniques. Vegetative propagation for clone banks and seed orchards is on a small scale compared with the mass propagation described in the following sections.

Canberra, Australia, appears to have been the scene for the first success in producing rooted cuttings of eucalypts (Fielding 1948; L.D. Pryor, personal communication). Both Fielding and Pryor set shoots cut from seedlings of several species with mist sprays and bottom heat in greenhouses.

A French forester, Bouvier, working in Morocco in 1950, discovered by chance the possibility of mass vegetative propagation of juvenile eucalypts. It was usual in North Africa to prune *E. camaldulensis* seedlings to 25–30 cm height to provide more uniform and hardy planting stock. The off-cuts, consisting of the terminal shoot, plus usually a few pairs of leaves, fell between the seedlings and on the ground in conditions of high humidity and high soil-moisture levels. It was observed that some of these discarded shoots formed roots after a few days.

This observation led to more formal trials being undertaken in the following year at two other nurseries, using seedlings of *E. camaldulensis, E. gomphocephala*, and hybrids of *E. robusta*. By the mid-1950s several thousand rooted cuttings had been planted in the forests nearby before it became more generally recognized that many eucalypt species could be rooted from cuttings taken from seedling material, up to about node 15 (Franclet 1956, 1963; Giordano 1956; Pryor and Willing 1963; Bachelard and Stowe 1963). The first trials on the rooting of cuttings of mature eucalypts were carried out in North Africa in 1954 and investigations continued there for almost 20 years (Franclet 1956, 1963, 1969, 1970). However, it was clear that decline in capacity to root with ageing of the plant was an important problem to be addressed (Pryor and Willing 1963).

Both *E. deglupta* in Papua New Guinea (Davidson 1974a) and mature *E. robusta* growing as an exotic in Hawaii (Pryor and Willing 1963) were noted to sometimes produce aerial roots from the lower trunk (up to a height of 2–3 m and to 10 m above ground respectively). It has often been observed that windthrown *E. camaldulensis* can produce adventitious roots from stems when uprooted trees fall into swampy ground (Fig. 22.1). These observations pointed to the possibility that these species might be easily rooted as cuttings.

E. deglupta appears to be by far the easiest of the eucalypts to propagate by rooted cuttings. In 1967 Pryor and Willing successfully rooted cuttings from a mature tree of *E. deglupta* in a 15-year-old plantation near Port Moresby, Papua New Guinea (L.D. Pryor, personal communication). The shoots, which were sent by air to Canberra, had sprouted on a broken branch 6 m above the ground.

J. Davidson began extensive experiments on vegetative propagation of *E. deglupta* from 1967 in Papua New Guinea and from 1969 under controlled

Fig. 22.1. Adventitious roots on windthrown *E. camaldulensis* at the edge of Lake Coorong, Victoria (photo K.G. Eldridge).

environmental conditions in a phytotron in Canberra, Australia (Davidson 1972, 1974a,b). In Papua New Guinea he started mass propagation of *E. deglupta* firstly in a mist chamber, then in a small glasshouse, and then as the numbers of cuttings increased he moved operations to nursery beds with overhead mist nozzles and hessian covers for shade and protection against heavy rain. Several thousand cuttings were planted in the field at Keravat in 1968 and 1969, all derived from trees up to five years of age. (*E. deglupta* seems unique among the eucalypts in that cuttings can be taken successfully from crowns of trees five years and older, without rejuvenation.) The cuttings were raised from the beginning in the one container (peat pots), so no transfers were required prior to planting in the field. Sprouts arising from small patch scions taken from trees up to 25 years old and grafted on to seedling stocks also rooted well. Coppice shoots growing on stumps of trees more than 100 years old could be rooted but rooting was less vigorous than from trees less than 25 years old (Davidson 1974a, b, and unpublished information).

E. grandis cuttings were raised successfully in nurseries in the Coffs Harbour area of New South Wales in the early 1970s, but less than one hectare of plantation was established. The nursery techniques developed by L.D. Pryor, R.R. Willing, and I.P. Burgess consisted of collecting cuttings from young coppice shoots on stumps of selected trees felled at age 3 to 7 years, and from seedlings kept in large pots and repeatedly cut back. Cuttings were set in vermiculite in plastic pots under mist spray in shaded, plastic-covered beds in summer (Burgess 1974; Clarke 1975).

Between 1964 and 1973, French researchers of the Centre Technique Forestier Tropical (CTFT) succeeded in the transfer of the rapidly-developing technology of cloning of old trees to the Congo (Martin and Quillet 1974) where continuous further progress has been made (Jean and Delwaulle 1981; Delwaulle 1983, 1985a, b; Delwaulle and Martin 1984).

In the 1970s mean annual increment of eucalypts planted near Pointe Noire in the Congo doubled with the use of cuttings of the best interspecific hybrids which occurred fortuitously in plantations (Delwaulle and Martin 1984). Just as impressive are the results of a similar program carried out by Aracruz Florestal S.A. in Espirito Santo state, Brazil (Brandão *et al.* 1984). In both cases the change was from not very well adapted pure species or inappropriate species hybrids occurring by chance in plantations. Zobel (1981) attributed the bulk of the increased yield at Aracruz to genetic gain, whilst acknowledging that improved establishment techniques were also an important contributing factor.

Cuttings are increasingly the preferred planting stock for eucalypt plantations, as a result of the successful programs at Aracruz and Pointe Noire becoming well known to plantation managers throughout the world. For example, in South Africa more than one million cuttings of selected *E. grandis* were produced annually in the mid-1980s (Denison and Quaile 1987). At the same time controlled pollinations were made of *E. grandis* with other species to produce hybrids which, after testing and selection, were propagated for marginal areas which are either too dry or too cold for pure *E. grandis* (Van Wyk 1985a). In France AFOCEL have selected high productivity, cold-tolerant clones of hybrids (*E. gunnii* × *dalrympleana*, *E. gunnii* × *viminalis*, and *E. gunnii* × *globulus*) in preference to pure species which have poorer growth rate and form (*E. coccifera*, *E. gunnii*, and *E. pauciflora*) or inadequate cold tolerance (*E. globulus* and *E. viminalis*) (Chaperon 1983; Destremau *et al.* 1983; Potts and Potts 1986). Another example, without hybrids, is in the Terai region of Nepal where well-conducted provenance and progeny trials of tropical *E. camaldulensis* and *E. tereticornis* established since 1981 were searched for outstanding individuals. The selected trees were felled and cuttings were taken from basal coppice shoots. By 1986 several thousand cuttings per year were produced for fuelwood plantations (White 1986).

22.2 Advantages of mass vegetative propagation

The main characteristics and advantages of plantations of vegetatively propagated eucalypts, namely uniformity, adaptation, cost, and wood production are discussed in this section. Libby (1985) gave an account of the major potential advantages of mass vegetative propagation and clonal forestry for plantations of forest trees in general.

The term 'clonal forestry', strictly speaking, refers to use of a number of tested selected, identified clones, as at Aracruz and Pointe Noire. However, the term is sometimes used loosely to include another scheme for vegetative multiplication of small quantities of very valuable seed. Genetic gain may be achieved more quickly in some breeding plans by mass vegetative propagation of seedlings of selected pedigreed families without testing and identification of individual clones. In this book we use the term 'clonal forestry' in the strict sense, 'vegetative propagation of families' for the second case, and mass vegetative propagation to cover both.

22.2.1 Uniformity

Excellent uniformity may be obtained from rooted cuttings with all the trees within one clone at a given age on the same site having about the same diameter, height, and form (Fig. 22.2). Aracruz provides a good example of uniformity in wood quality.

Fig. 22.2. Uniformity of clones within rows of a three-year-old trial of *E. grandis* cuttings at Aracruz, Brazil (photo E. Campinhos).

Seedling populations derived directly from eucalypt seed collected in the natural habitats are often highly heterogeneous in growth rate and form. Seed from eucalypt seed orchards and seed production areas usually produces heterogeneity in plantations due to recombination and a certain amount of selfing. Under intensive nursery management inbred eucalypt seedlings are likely to grow well at first, escape culling in the nursery, and reach the plantation where they grow poorly and become suppressed. Controlled pollination of unrelated parents eliminates heterogeneity due to inbreeding. Of course heterogeneity is completely eliminated by using single clones in large blocks. The Aracruz and Pointe Noire plantations are established in a mosaic of monoclonal blocks of about 20 ha of each clone.

In addition to being more even in growth and form, clonal plantations of eucalypts appear to have more favourable adult crown characteristics (long and narrow). It is not known whether this is a selection effect, a consequence of equal growth rate of all plants without assertion of dominance, or perhaps a physiological attribute of clones (topophysis or cyclophysis, see Section 22.3).

22.2.2 Adaptation

Vegetative propagation provides the opportunity for very close matching of a selected clone to the special characteristics of a particular site. Hybridization between species has enabled the selection of individual clones capable of growing very rapidly on sites to which these clones are specifically adapted. The sites include marginal or degraded sites such as dry areas of Morocco and cold winter environments of southern France. However, the first large plantations of hybrid eucalypt clones were on sites with high growth potential but for which no single pure eucalypt species had been found to be well adapted (Section 20.5.2). Examples include Aracruz in Brazil, Pointe Noire in the Congo, and more recently, parts of the eastern Transvaal in South Africa.

Deployment of eucalypt clones has been restricted to the sites on which they were developed and on which they are known to do well. It seems to us unlikely, and undesirable, that particular eucalypt clones will be grown in large plantations on a wide range of sites worldwide as certain poplar clones have been (Mohrdiek 1983). With the unit of management being the individual clone, clones can be deployed in sets to specific sites. In contrast, a seed orchard produces the same general product for a wide range of plantation sites.

In each case of successful clonal eucalypt plantations the selected well-adapted planting stock, combined with improved cultural practices, has dramatically increased biomass production, making greater use of the water, sunlight, and nutrients available to but not used efficiently by the previous unimproved and poorly adapted planting stock. Many millions of hectares of degraded tropical forest land no longer carry commercial forest and many areas are denuded or covered with weeds or *Imperata* grass and require rehabilitation (Lovejoy 1985). The use of well-adapted eucalypt clones offers a means to rehabilitate some of this denuded and degraded land, bring it into tree production, and take pressure off the world's remaining intact tropical moist forests.

22.2.3 Cost

A distinct cost advantage of vegetative propagation is the short-term ability to quickly capture a greater proportion of additive and non-additive genetic variation than can be done using seed orchards. In this way the breeding program can greatly reduce the number of years, and hence the cost, before a newly selected, improved product of breeding can be established in large areas of plantation (Matheson and Lindgren 1985). In other words, cuttings can provide better plantations more quickly than seed.

The nursery costs of production of cuttings have decreased with increased rates of success for rooting and with rooting of the cuttings in the final container in which they are taken to the field.

Mass vegetative propagation of valuable and expensive genotypes (selected natural or artificial hybrids, or pedigreed controlled crosses) differs from the use of seed and seedlings in that vegetative propagation requires only a small specialized seed production facility with a large propagule manufacturing operation. In contrast, propagation by seedlings requires a large and usually expensive seed production facility, a seed orchard, in addition to the usual nursery facilities.

22.2.4 Wood production

The outstanding advantage of the use of cuttings in Brazil and the Congo is that the growth rate of eucalypt plantations established from cuttings in the 1980s was about double that of plantations established from the unimproved seed available before 1975, after making allowance for improved establishment practices.

In the Congo, where the pulpwood plantations of eucalypt cuttings at Pointe Noire exceeded 15 000 ha by 1984, the clones were derived from hybrids of several eucalypt species, the most recent hybrids developed by CTFT being from controlled pollination. The earlier clonal plantations were of clones of selected individuals of the hybrids known as *E.* 'PF_1' and *E.* '12ABL × *saligna*'. Wood samples of 65 clones from six-year-old Congo plantations were of satisfactory kraft pulp quality, variable from clone to clone, and nearly uniform within one clone. There were good prospects for improvement in pulp yield and strength by selection of the best clones (Petroff and Tissot 1983).

The hybrid *E.* 'PF_1' appeared in 1959 in a plantation of *E. alba* from Indonesia which was designated *E.* 'platyphylla', a synonym of *E.* alba. Thus arose the name *E.* 'platyphylla F_1' (PF_1) given to the progeny. Although the other parent is uncertain it was probably Brazilian *E.* 'alba', itself of partly hybrid origin, a descendent of *E. urophylla* and possibly including *E. tereticornis* and *E. grandis*. Thus the combination is probably *E. alba × urophylla* (Delwaulle and Martin 1984).

E. '12ABL' is an inbred land race of *E. tereticornis* which is thought to be descended from a single tree in Madagascar (Martin 1971). *E.* 'saligna' in the mass propagated Congo hybrid *E.* '12ABL × saligna' is thought to be *E. grandis* from South Africa (Delwaulle 1985b). Thus the combination is probably *E. tereticornis × grandis*.

The growth rate of monoclonal plantations of the better clones of the two hybrids on the savanna sands at Pointe Noire in 1983 was of the order of 30–35 $m^3 ha^{-1} yr^{-1}$, mean annual increment at 6 or 7 years (Delwaulle and Martin 1984). Recently selected clones from control-pollinated hybrids are growing even faster. This is indeed a remarkable improvement over the 12 $m^3 ha^{-1} yr^{-1}$ obtained there in the 1960s with *E. tereticornis* '12ABL' (Chaperon 1984b) and 20 $m^3 ha^{-1} yr^{-1}$ from stands of selected provenances of *E. urophylla* (Delwaulle and Martin 1984) and *E. cloeziana* (Bouvet and Delwaulle 1983).

In Brazil Aracruz Celulose S.A. has used *E. grandis × urophylla* to obtain similar gains in production and provide a solution to replacement of eucalypt plantations attacked by *Diaporthe* canker (*Cryphonectria cubensis*) (Brandão *et al.* 1984). As in the Congo, the first hybrids arose by chance in seedling plantations. The seed had a mixed origin with *E. grandis* predominating but hybrids with *E. urophylla, E. saligna* and other species were common. Planting commenced in 1967 and by 1978, when more than 30 000 ha of seedlings had been planted and felling commenced for the newly opened kraft pulp mill, the production was of the order of 33 $m^3 ha^{-1} yr^{-1}$ over bark (Brandão *et al.* 1984). In 1979 the first commercial-scale planting of cuttings was made only four years after first attempts by Aracruz to make cuttings with methods developed earlier in the Congo, Australia and Papua New Guinea. By 1990 the company was raising about 50 million rooted cuttings per year. The wood production in test plantations aged 5.5 years in 1982 averaged 70 $m^3 ha^{-1} yr^{-1}$ over bark (Zobel *et al.* 1983) and some individual plots have produced more than 100 $m^3 ha^{-1} yr^{-1}$. Wood density of the ortets of selected clones at Aracruz was 25 per cent greater than that of the original seedling population, and a similar increase in density and improved pulp yield per hectare was predicted in the clonal plantations.

22.3 Effects of maturation state

This section deals with the question of why meristems of young eucalypts, and other plants, can be more easily propagated than those of old trees.

Difficulties encountered when cloning eucalypts are many:

(a) rooting shoots from young seedlings or from coppice from the base of a tree is relatively easy, but it becomes difficult or impossible from increasingly older parts of a tree (cyclophysis, or maturation of the apical meristem (Olesen 1978));

(b) the position in the crown where a shoot is collected influences rooting ability and subsequent vegetative vigour; for example, a graft or air layer from a branch of some species may grow horizontally and slowly instead of upright (known as plagiotropy, an example of topophysis (Olesen 1978));

(c) reduction in uniformity, rooting ability, and subsequent vegetative vigour of some ramets even when cuttings were thought to have been taken from the most favourable positions in the tree and sophisticated techniques used;

(d) apparent long-lasting transmission and imprinting of this heterogeneity in successive generations of rooted cuttings.

For other hardwoods many attempts have been made to associate these problems with a number of anatomical, biochemical, or physical criteria (Robinson and Wareing 1969; Bonga 1982). Early attention was paid in particular to the onset of flowering as an indication of the mature phase being reached. Thus onset of flowering may indicate the end of the juvenile period for rooting cuttings because of the close association which has been shown to exist between juvenility and rooting ability for several species by Shaffalitsky de Muckadell (1959) and others. Among the eucalypts, many species occasionally flower in the nursery. *E. deglupta* regularly flowers in plantations as early as nine months of age and species such as *E. brassiana, E. camaldulensis, E. grandis,* and *E. tereticornis* when grown as exotics often start to flower within two years.

A detailed examination of the leaf morphology of eucalypts, where one can regularly observe three and often five leaf types during the development of a number of species (Jacobs 1955; Penfold and Willis 1961), indicated restoration of juvenile function was possible particularly where damage by fire or severe insect attack caused epicormic growth leading to new, leafy twigs.

The existence of several stages from juvenile to mature in forest trees has been recognized for many years. A wide spectrum of ideas has been put forward ranging from Passecker's (1947) conclusion that the part of the tree near the root always stayed juvenile while the distal parts become more mature, to more complex hypotheses such as that of Krenke (1940) who proposed a cycle of senescence and rejuvenation in plants which would explain why juvenile zones could be found in the crowns of mature trees. The duration of the juvenile stage became shorter the greater the distance from the roots that each cyclical flush occurred. In the very mature stages of some species the juvenile stage might be absent, resembling the situation described by Passecker (1947). Assaf (1966) mentions a zone favourable to rooting in fruit trees in tissues formed at the time the apical meristems reactivate after dormancy.

22.3.1 Rejuvenation in eucalypts

It has been recognized for many years that most eucalypt species can be propagated by rooted cuttings when these cuttings are taken from seedlings (Fielding 1948; Franclet 1956, 1963; Giordano 1956; Bachelard and Stowe 1963). Pryor and Willing (1963) pointed out that the capacity for juvenile material to root was generally limited to the first 15 nodes formed after the cotyledons. Adult tissues of eucalypts have always been impossible to propagate by cuttings, except for *E. deglupta*.

Adult tissue of *E. grandis* was shown by Paton *et al.* (1970, 1981) to contain a rooting inhibitor. At about this time Davidson (1974a) described work carried out in 1968 where he made methanol extracts of growing shoots and leaves of material of *E. deglupta* of various ages. By using young seedling cuttings of *E. deglupta* and germinating cress seeds to bioassay the extracts he was able to show qualitatively the presence of a chemical rooting inhibitor in adult material (more than 5 years old).

This limitation of rooting ability to a zone near the base of the trunk close to the root system can be exploited in producing rooted cuttings from older eucalypt trees because of the dormant bud strands which occur in eucalypts. These originate from the base of every leaf formed by the eucalypt tree (Cremer 1972; Paton 1983a).

The tissue at the outer end of the bud strands remains meristematic and retains the physiological age attained when the bud tissue formed during differentation of the leaf. It follows that, in trees with two leaves per node, the first 15 sets (sometimes a few more) of these dormant bud clusters are capable of producing coppice shoots which can be propagated readily by rooted cuttings. For this we use the term 'apparent rejuvenation' since no reversal of the physiological maturation state is involved.

Dormant bud strands also occur in the upper trunk of mature trees and in the branches. All leaves still attached to the tree have an axillary bud and one or more accessory buds. All actively growing terminal shoots have meristems. It is possible, as demonstrated in the following sections, to apply various treatments to yield rooted material from dormant bud strands, axillary buds, accessory buds, and meristems. This requires a reversal of the maturation state of those tissues before rooting will occur. For this we use the term 'true rejuvenation' with respect to rooting ability.

22.3.2 Rejuvenation by coppicing

In 1954–55, in the forest of Lalla Ito, Morocco, French foresters were carrying out some of the earliest known breeding work on eucalypts. Twenty-four plus trees of *E. camaldulensis* were selected from among several thousands in plantations about 40 years old and with an average height of 47 m. Many of these individuals were probably hybrids. The best individual emerging after tests of splitting, peeling, and other wood qualities was 'Lalla Ito No. 23', probably a hybrid between *E. tereticornis* and *E. camaldulensis* (A. Franclet, personal communication).

At various times after felling, sprouts were collected from the stump of tree No. 23. Results of rejuvenation are shown in Table 22.1 reproduced from an early paper by Franclet (1956). Results were poor by present-day standards with some other species but, nevertheless, one of the better attempts was the successful rooting of 8 per cent of 180 cuttings made from six-month old coppice sprouts treated with 0.01 per cent auxin solution.

22.3.3 Rejuvenation by serial production of cuttings over several generations

Initial work on this aspect followed on the previously mentioned trials on coppice shoots carried out at Lalla Ito, Morocco in 1954–55. A second generation of cuttings (Table 22.2) was prepared from the first

Table 22.1
Relative success in striking cuttings prepared from coppice sprouts collected at various times following felling of a single eucalypt (from Franclet 1956).*

Type of source material	Hormonal treatment**	Number of cuttings prepared	Per cent of cuttings rooted
New coppice sprouts from stump	0.1% IBA in talcum powder	15	7
Herbaceous coppice sprouts 3 months old	None	170	0
Lignified coppice sprouts 6 months old	24 hours basal dip in 0.01% IBA solution	180	8
	24 hours basal dip in 0.01% NAA solution	65	0
Lignified coppice	Cowdung coating (no hormone)	200	0
	0.1% IBA in talcum powder	22	10
	0.1% NAA and IBA in talcum powder	8	0
Tall coppice shoots	0.3% NAA in talcum powder	15	7

* Probably *E. tereticornis* × *camaldulensis*, cuttings taken from 4/8/55 to 19/9/56

** NAA napthalene acetic acid, IBA indole butyric acid

Table 22.2
Performance of second generation cuttings from coppice shoots of a single clone of E. camaldulensis *(from Franclet 1956).*

	Hormone treatment	Number of cuttings prepared	Per cent of cuttings rooted
(a) Effect of different hormone treatments on rooting of uniform cutting material			
	0.3% NAA in talcum powder	50	24
	0.3% IBA in talcum powder	50	48
	0.1% IBA and NAA in talcum powder	50	46
	Nil	50	12
(b) Rooting of different types of second generation cuttings with the same hormone treatment			
Well-established mother plants	0.1% NAA and IBA in talcum powder	100	65
Recently-established mother plants	0.1% NAA and IBA in talcum powder	100	42
Sub-terminal shoot	0.1% NAA and IBA in talcum powder	24	58
Terminal shoot	0.1% NAA and IBA in talcum powder	24	20

generation (described in Table 22.1) (Franclet 1956). The rooting success in this case reached as much as 65 per cent for one group of 100 cuttings and demonstrated, for the first time in eucalypts, the rejuvenating effect of making cuttings from cuttings over two or more generations.

22.3.4 Rejuvenation by repeated grafting

In 1956, while grafting scions from tree 'Lalla Ito No. 23' in Morocco to plant the first seed orchard, French foresters noticed that the grafts were producing juvenile leaves. Some shoots taken from these grafts and rooted as cuttings produced even more juvenile leaves. This discovery led in 1962 to the adoption of grafting to rejuvenate eucalypt twigs taken from the crown of mature trees.

Amongst the numerous experiments conducted in Tunisia on this aspect (Franclet 1970), one involved grafting scions taken from an 83-year-old *E. camaldulensis* on to seedling rootstocks 4–6 months old raised from half-sib seed of the mother tree (to reduce graft incompatibility). When shoots sprouted on the scions they were removed and grafted onto more seedling rootstocks. Three experimental treatments were applied, each one having a shorter interval between graftings (one, two, and four to six graftings per year) respectively. Only the more frequent graftings rejuvenated this particular clone to the stage where further propagation by rooted cuttings could take place. The cuttings had a growth habit and appearance similar to seedlings raised from the same mother tree.

The behaviour of large numbers of grafts of *E. deglupta* made by top-cleft, side veneer, bottle-grafting, and patch-grafting techniques was studied in work carried out between 1967 and 1972 by Davidson (1974b). He noted that with bottle-grafts of mature scions on to seedling stocks the scions sometimes lost their leaves and roots developed on the base of the scion in the container of water (Davidson 1974b). In bottle-grafts of *E. deglupta* on to juvenile rootstocks, graft incompatibility was more common with adult scions than with juvenile scions. This effect was reduced in patch-grafts of adult scions on to seedling stocks (patch-grafting gave about 85 per cent initial take). Evidence was mounting that rooting inhibitors found in adult leaves and bark (Paton *et al.* 1970; Paton and Willing 1974) could be involved in both graft incompatibility and rejuvenation. Paton and Willing (1974) also found bottle-grafts of adult *E. grandis* scions on to adult rootstocks of the same species were rarely successful but grafts of adult scions on to juvenile rootstocks gave 90 per cent success. In *E. deglupta* the new shoots coming from the accessory buds of the adult scion patches had juvenile leaves and it seems the absence of adult leaves and the small size of the scion patch affected the degree of rejuvenation and improved also the chance of making a successful graft.

The rejuvenation process induced by grafting a mature scion on to a juvenile stock was explained for eucalypts by Paton *et al.* (1981) in terms of the action of 'inhibitor G' (terminology of Paton *et al.* 1970). These workers took scions from the crown of a 15-year-old tree of *E. grandis*. Leaves of scions were shown to have a very high level of inhibitor G (7.5 mg g^{-1} fresh weight or more) and such scions were impossible to root directly. The scions were grafted on to seedling rootstocks above node 3 (above the third leaf pair). When diameter growth of the scions exceeded 1–2 cm (6 months to 2 years after grafting), they were defoliated and strangulated resulting in growth of epicormic shoots. Epicormic shoots were collected at various distances from the point of grafting. Juvenility was estimated in two ways: ability of cuttings made from the epicormics to form roots and the level of inhibitor G in the new tissues. The authors summarized their results in tabular form (Table 22.3). These results can be compared with values obtained using adult material collected directly from a 10 m high tree and seedling *E. grandis* about 40 cm high (Tables 22.4, 22.5).

The G level in epicormic shoots taken from a mature scion near the graft union (Table 22.3) is comparable to that of basal epicormic sprouts taken from an adult tree (Table 22.4) or axillary shoots taken from the lower nodes on seedlings (Table 22.5). This rejuvenation effect is 'true rejuvenation' and depends on the distance between the site where the epicormic buds arise on the mature scion and the root of the juvenile stock. Neither the act of grafting itself nor the first four nodes of the seedling stock appear to influence rejuvenation. Rejuvenation, however, requires the initials of future epicormic shoots to remain under the influence of the roots for a minimum of six months, at least in *E. grandis*. This effect is slowly reversible over a longer period of time as the epicormic tissues themselves begin to age (Paton 1984b).

Table 22.3

Levels of inhibitor G and rooting ability of epicormic shoots induced on adult scions of E. grandis *grafted on juvenile rootstocks (based on Paton et al. 1981).*

Maturation criterion	Distance from roots of juvenile stock	
	Near (basal)	Far (distal)
Rooting percentage	30–50	0
G level (mg g^{-1} fresh wt.)	0.28	2.3

Table 22.4

Levels of inhibitor G and rooting ability of epicormic and primary shoots taken from the base of an adult tree of E. grandis *(based on Paton et al. 1981).*

Type of shoot	Maturation criterion	Distance from roots	
		Near (basal)	Far (distal)
Primary shoots	Rooting percentage	–	0
	G level (mg g^{-1} fresh wt.)	–	7.5
Epicormic shoots	Rooting percentage	10–80	0
	G level (mg g^{-1} fresh wt.)	0.3	7.5

Table 22.5

Levels of inhibitor G and rooting ability of primary shoots and axillary shoots taken from seedling E. grandis *(based on Paton et al. 1981).*

Type of shoot	Maturation criterion	Distance from roots	
		Near (basal)	Far (distal)
Main axis	Rooting percentage	80–100	0
	G level (mg g^{-1} fresh wt.)	0.02	1.3
Axillary	Rooting percentage	40–80	0
	G level (mg g^{-1} fresh wt.)	0.25	2.5

22.3.5 Rejuvenation by micropropagation

Micropropagation is the propagation of differentiated organs, such as buds, embryos, or hypocotyls in sterile culture. The term tissue culture is sometimes used in a broad sense to describe the techniques and laboratory facilities required in micropropagation, but tissue culture strictly speaking is confined to the propagation of cells or callus tissue in sterile culture.

In a comprehensive review of literature on vegetative propagation of the eucalypts, Hartney (1980) recorded successful attempts at callus culture (tissue culture) of several eucalypt species, but only one case of eucalypt plants, with both shoots and roots, being produced in culture from callus cells. Callus cultures were established successfully from hypocotyls and young seedlings of *E. citriodora* by Lakshmi Sita (1979) and these eventually formed plantlets but similar cultures derived from leaves and shoots of one-year-old seedlings failed to differentiate. Aneja and Atal (1969) found they could regenerate plants from lignotuber tissue of *E. citriodora*, but not from root-tip or main stem tissue.

Nodal segments, hypocotyls, and leaves have been used as explants in organ culture of eucalypts, with successful formation of both shoots and roots. The nodal segments contained at least one axillary, accessory, or terminal bud. There are several published records (reviewed by Hartney 1980) of the success of R.A. de Fossard and his associates in propagation from nodal explants of *E. grandis* seedlings and epicormic shoots, from *E. grandis* nodes above node 50 and up to node 80, and from the crowns of 25-year-old *E. ficifolia*.

What is notable about the work since 1975 is that significant breakthroughs have been made in propagation by callus and organ culture of adult material. Success in propagating material taken from a height above that at which rooting is normally possible from unrejuvenated cuttings (about 15 nodes for several species, Paton *et al.* 1970), means that these methods of culture can be used for rejuvenation of older material. More effective media have become available for multiplication of shoots and to improve rooting frequency (for example Barker *et al.* 1977; Boulay 1983; Hartney and Svensson 1992), and multiplication rates potentially are high, enabling transfers (re-culturing) as frequently as every three to four weeks. After five to six transfers the clone is considered sufficiently rejuvenated or reactivated to enable subsequent propagation as cuttings under glasshouse or open nursery conditions (Franclet 1984).

The relatively high cost of micropropagation is due to the high content of skilled labour employed in the procedure. This cost must be amortized over as many plantlets as possible. For species such as *E. nitens* which cannot be propagated by rooting cuttings even from seedlings (Tibbits *et al.* 1990), a commercial-scale procedure for micropropagation will be required. Micropropagation is the preferred method in a number of Australian nurseries producing thousands of ramets per year of salt-tolerant clones of *E. camaldulensis* (for example, Hartney and Kabay 1984) even though this species is relatively easy to propagate by rooted cuttings. In southern China the million micropropagated eucalypt plantlets produced in 1991 were used mainly as foundation stock in cuttings nurseries (Yan Muqin, personal communication). However, for much larger numbers of micropropagated plantlets, several million per year, it would be necessary to have either automated transfers for sub-culturing in the micropropagation laboratory or simpler and cheaper methods of micropropagation.

22.4 Mass propagation of cuttings for industrial plantations

Mass vegetative propagation is only recommended when there is something worthwhile to propagate, such as outstanding individuals in plantations or products of a breeding program.

22.4.1 Setting up clone banks for mass propagation

Selection of ortets

Ideally, the original tree from each clone, the ortet, would be an outstanding individual selected towards the end of a series of species elimination, provenance, and progeny trials carried out on sites representative of the area to be planted, or from progeny tests in later generations of an advanced breeding program. (Methods are given in Chapters 5 and 20). These outstanding individuals (plus trees) are selected to become ortets of clones at an age when the upper crowns of all eucalypts (except *E. deglupta*) have long lost the ability to root directly from cuttings.

Scion material for grafting can be collected by climbing, shooting down of branches with a rifle, or by means of various devices employing ropes (sling-shots, bow and arrow, weighted line, commando saw, etc.). These are the same techniques which are used in seed collection (Boland *et al.* 1980a, p. 109).

Transfer of ramets to the clone bank

In the first place it is necessary to transfer ramets of the ortets from the place of selection to a site near the nursery and other facilities where rejuvenation and multiplication will take place.

This initial transfer theoretically is possible using most of the techniques of vegetative propagation such as grafting, air-layering, or cuttings. The simplest and most commonly used method is to cut down the selected tree and propagate coppice shoots as cuttings provided the species is easy to root, but of course this is not possible with non-coppicing species except perhaps in the first or second year. We will consider grafting and air-layering first.

Risk avoidance

Since the selected ortet is a valuable part of the breeding program, it is wise to reduce the risk of losing the genotype by using grafting or air-layering for placing it in a clone bank. First, open-pollinated seed should be collected from the plus tree for growing related seedlings for grafting stock to reduce graft incompatibility. Next, a few grafts should be made, and only when successful grafts are established should the tree be cut down for coppicing. The main reason for taking these precautions is that not all cut stumps will produce coppice shoots, even with the most outstanding plus trees of coppicing species in humid tropical areas.

Grafting

Much of the success of grafting eucalypts depends on experience gained locally for a particular species since success depends on the species, seasonal environmental conditions, the technique used, and the patience and skill of the grafter.

The best stock is a vigorously growing seedling of the same species and as closely related as possible to the potential scion parent, preferably from seed of the same tree. The stock is usually decapitated at about the fourth or fifth node for a top-cleft graft, but the height needs to be varied about this norm in order to match the diameters of the stock and scion at the union. The rind graft is preferred for *E. grandis* in South Africa (Van Wyk 1977a). Non-adhesive grafting tape is used.

In a temperate climate the graft is usually placed in a heated glasshouse, while in warmer subtropical and tropical areas, the grafts are usually kept in a nursery with high shade provided. Union is normally effected within about two weeks and the scion begins to grow about one month after grafting.

In species which prove difficult to graft by the top cleft method, patch grafting offers a better chance of success (Davidson 1974b). Several other grafting techniques, including bottle grafts and side veneer grafts, have also been used successfully with several species of eucalypts (Pryor and Willing 1963).

The scion from the adult tree (the ortet) selected in the field is in the mature or adult phase of maturation. For a short time after growth recommences from the grafted scion a more juvenile phase is retrieved at which time further scions for grafting should be collected to repeat the process of rejuvenation. With each successive grafting, rejuvenation lasts for a longer time. The rejuvenating influence of the stock is more effective when the level of the graft is near the root collar (Franclet 1984).

One advantage of patch grafting for the initial graft is that several bud patches can be grafted even on the first seedling node. Usually cleft grafting is adopted for the second and following grafts in the rejuvenation series since the new scions (shoots from the first top-cleft or patch graft) are much smaller in diameter than the adult material and are grafted on to small seedlings. These small seedling rootstocks must be growing vigorously. The small size and softer nature of the scion material makes it very difficult or impossible to manipulate bud patches. Stocks and scions often as small as 2 mm in diameter at the union are prepared using a very sharp scalpel and the union bound with non-adhesive cellophane tape about 5 mm in width.

Air-layering

Successful air-layering (Fig. 22.3) requires girdling of the shoot on the ortet, application of auxin powder and then a humid dressing of peatmoss or other similar material covered with polythene sheet. Although 20 or more species of eucalypts have been successfully propagated by air-layering (Hartney 1980) the technique remains an uncertain one and of course it is very difficult to apply to crowns of tall trees. However both air-layering and grafting are

Fig. 22.3. Successful air-layer (marcot) of *E. deglupta* ready for planting at PICOP, Philippines (photo J. Davidson).

non-destructive practices which can be used in cases where permission cannot be obtained to fell the ortet, or for non-coppicing eucalypts.

Cuttings

Two types of treatment of the ortet are possible to enable rooted cuttings to be obtained from a mature selected plus tree. One is by stimulation of epicormic shoot growth by ringbarking or wounding the tree with an axe or chainsaw. This technique has been used successfully with some species, for example *E. grandis* in Florida (Meskimen 1987), but unless it can be applied very low on the trunk the epicormic shoots do not make very good cuttings.

The second is a more destructive technique where the tree is felled leaving a short stump (less than 20 cm high) which later usually produces numerous coppice sprouts (sometimes more than 100 per stump). In the equatorial tropics, this can be carried out at almost any time of the year. In temperate climates spring is the best time (Clarke 1975; Heth *et al.* 1986), while in monsoon climates it is best carried out following the monsoon. In all cases the aim is to cut back the ortet at the beginning of an active period of vegetative growth. Of course the technique is only applicable to species in which coppice can be obtained from trees old enough for selection, usually 5 to 10 years in fast growing eucalypts. However, even 'non-coppicing' species such as *E. regnans* and *E. delegatensis* will coppice if cut at the age of one or two years, a suitable age for selection if the material for propagation is from a selected control-pollinated family, and juvenile–mature correlation is high.

Coppice shoots are normally large enough for making cuttings 6 to 12 weeks after the tree is felled. Again the time taken depends on local environmental conditions and also on species and diameter of the stump; smaller diameter stumps coppice faster. The shoots are suitable for harvest when the stems and leaves have developed an appropriate shade of green and some lignification has taken place, but the shoots are less than 1 m long. Succulent shoots and shoot tips do not make good single-node cuttings in most species. Once the naked axillary buds start to develop the material becomes less suitable for making cuttings.

Individual cuttings are prepared from the base of each sprout upward by cutting single nodes (one pair of leaves), discarding the shoot tip. However, two nodes are preferred when the internodes are short. For certain species the tip makes a good cutting provided one or two nodes are included.

The first generation of coppice shoots on the stump is less suitable (gives a lower success rate) for making cuttings than subsequent harvests from the second and subsequent generations of coppice from the same stump. This is true of both the ortet and the ramets, but more important for the ramets to which one returns frequently for more material. The ortet stump is usually abandoned after the first harvest when the plantation is regenerated.

Precisely what constitutes the ideal shoot to set as a cutting, what pretreatment is required, and exactly what should be done in the nursery depends very much on the species and the local environment. While a prescription of the current technique used at Aracruz or Pointe Noire may be helpful for a new operation, the final details are a matter for local decision based on local experience.

22.4.2 Management of clone banks

The clone bank of mother plants used for production of clonal material for plantation use consists of selected clones replicated many times following rejuvenation and planted on a fertile site in close proximity to the nursery. A large area may be required as there may be as many as 100 clones and hundreds of ramets of each clone (Fig. 22.4). For a plantation program of up to 13 000 ha per year the Aracruz company maintains many hectares of clone bank. This area will decrease as the company harvests coppice shoots from stumps after felling blocks of particular clones.

Rejuvenation of material in the clone bank

The material in the clone bank (or clone archive) will consist initially of first generation cuttings, air-layers, or grafts. This material usually still requires further rejuvenation to improve the rooting success of future cuttings. This can be achieved either by serial grafting, hedging of the original graft, or micropropagation (*in vitro* multiplication).

Grafting. From four to six serial graftings are usually sufficient to rejuvenate a ramet to the point where it strikes cuttings easily (Franclet 1984). This method is time-consuming and costly but can be justified at the initial stage of the program. Alternatively the grafted ramet in the clone bank can be left for one or two years to grow into a tree. The graft can then be cut immediately above the union and allowed to coppice. The rejuvenating influence of the stock has usually been sufficient to allow cuttings to be obtained successfully from such coppice shoots, but care must be taken that only those shoots arising on the remaining part of the scion are used and never those from the stock.

Hedging. The coppice shoots can be harvested completely several times or one or two strong shoots can be kept trimmed back to form a hedge of not more than about 5 to 10 nodes in height. With most species the hedges produce more cuttings in a year than several complete removals of coppice shoots from a stump over the same period. (Several terms are used for ramets which are repeatedly pruned to provide shoots for cuttings. French eucalypt propagators use the term 'pied mère' or 'mother plant', poplar growers use the term 'stool', radiata pine breeders use the term 'hedge' or 'hedge plant'.)

Micropropagation. Micropropagation using explants (Section 22.3.5) is now routine for many species, provided the explants are carefully chosen and the timing of subculturing is closely monitored (Boulay 1983; McComb and Bennett 1985; Hartney and Svensson 1992). The subcultured plantlets make excellent cuttings either *in vitro*, or in the nursery for use as mother plants for further rejuvenation by a

Fig. 22.4. Clone bank of cuttings at Aracruz, before harvesting shoots to set as cuttings in the nursery (photo E. Campinhos).

series of cuttings or directly for mass propagation in the glasshouse or nursery. Any clone that remains difficult to root is removed from the program at this stage.

Harvesting. Early in a clonal program when the number of mother plants must be increased rapidly, harvesting shoots from mother plants can start at 4 to 6 months. However, when a program is well established, one to two years will pass before the clone bank is felled to provide coppice for mass propagation. At this age the stem diameter at the base should be between 8 and 12 cm (or more, but not less than about 8 cm); this size produces abundant coppice shoots. In temperate countries one can achieve less than 6 months of suitable coppice harvest, in tropical climates up to 12 months. Harvests may be almost continuous within the harvesting period, taking about one third of the shoots at a time every 10 to 15 days or complete harvests may be made at longer intervals of 4 to 6 weeks, depending on seasonal conditions. Harvests rotating around the clone bank in a defined pattern enables smoother scheduling of the nursery operations because material arrives at the propagating benches evenly and continuously.

It is essential for the success of the clone bank that the physiological condition of the mother plants is appropriate, both the juvenility and general health and vigour. The ground between the mother plants should be cultivated, fertilized, and irrigated as necessary to promote vigorous growth of coppice shoots.

In France and the Congo experience so far has shown that some shade seems to be needed for optimum development of coppice shoots having the best pigmentation and physiological health for making cuttings. Management of the clone bank on a two-year rotation where one half of the trees (every second row or every second tree alternately in adjacent rows) is cut annually is suggested to provide shade. However, in an experiment conducted by the University of Viçosa in Brazil the best rooting ability was obtained from mother plants in full sunlight, and shade was not beneficial (A. Brune, personal communication). Further study is evidently needed on this subject.

Mother plants are perpetuated by periodically allowing one or two coppice shoots to grow on for one or two years, and then the harvesting process is repeated every few weeks. This whole procedure can be done about five times before the stump deteriorates to an extent where coppicing ability is lost.

At any stage where mother plants begin to produce material which roots poorly, perhaps due to poor health or to ageing, these ramets should be eliminated from the program. New mother plants of the same clone can be produced by recycling through a period of *in vitro* propagation or by repeated grafting.

In large programs, clones may first be centralized and rejuvenated in a single clonal archive. From there rejuvenated material may be dispersed to a number of operational clone banks each situated near a local nursery supplying planting stock for separate parts of a planting program. Sometimes the mother plants are made portable, in large pots.

22.4.3 Nursery techniques for mass propagation of cuttings from clone banks

Preparation of the cuttings

In tropical regions, where most eucalypt cuttings are produced, several cuttings are taken from the basal part of a vigorous juvenile coppice shoot usually with four leaves per cutting (two nodes) (Figs. 22.5, 22.6) but sometimes two leaves (one node) (Fig. 22.7). The end of each leaf is usually cut off to reduce transpiration losses. In temperate regions, such as the south of France, the internodes are much shorter and the type of cutting varies with season. In winter the cutting consists of the entire leafy axillary shoot harvested from the mother plant. No further preparation is given except to remove sufficient leaves from the base of the cutting in order to insert it in the substrate. Any moribund leaves and the shoot tip would be removed at this stage also, leaving usually 4 to 8 leaves on the cutting (2 to 4 nodes).

Cuttings are plunged into a solution of fungicide (6 g Benlate/10 litres of water), then drained before treatment with hormone. Hormone should be used always to obtain better root growth even though most cuttings will eventually root without it. A light dusting of the extreme basal end of the cutting in a commercial preparation 1 per cent indole-butyric acid (IBA) (for example Rhizopon AA 1 per cent) is used in France (Chaperon 1984a). Elsewhere up to 3 per cent IBA (for example Seradix) has been used (Davidson 1974a).

Fig. 22.5. Coppice shoot from a mother plant in a clone bank, to be cut into two-node sections (photo E. Campinhos).

Fig. 22.6. Two-node cutting ready to set in the nursery (photo E. Campinhos).

A dibble stick such as a meat skewer, pencil, or piece of dowel is inserted in the damp substrate to make a hole for inserting the cutting to prevent physical damage to the base of the cutting or dislodging the hormone powder (Fig. 22.7).

Substrate and containers

Experience in Congo, Brazil, and France has demonstrated that cuttings cannot be prepared in a suitable manner on a commercial scale unless they are raised directly in the planting container.

Fig. 22.7. Inserting cuttings after dipping in hormone powder (photo K.G. Eldridge).

The types of containers being used seem to depend on local experience. Containers which have rigid sides and a shape designed to minimize root deformation are best. Usually they are tapered and have internal vertical ridges (Campinhos and Ikemori 1983). Containers with non-woven walls (peat, paper, etc.) are permeable to roots and are often preferred to polythene tubes (Davidson 1974a; Chaperon 1983), and have the advantage that they allow the roots to be pruned by contact with the air. Polythene bags (with holes for drainage) are still widely used but are less suitable for growing cuttings since mechanical root pruning is necessary unless containers are supported off the ground (for example on chicken wire mesh). It is very important to have well-developed lateral and vertical root systems on the plants before they are transferred to the field, but container size should be no bigger than is required to hold the root system for the period in the nursery.

Various media are used for filling the containers. Milled bark, sand, aged sawdust, peat, vermiculite, perlite, and loam have all been used successfully in various combinations and mixtures.

The substrate should be medium textured and free-draining while having at the same time good water retention. For good rooting, cuttings must have ample heat, moisture, and air. 'Ample heat' has not been defined precisely, but a stable, non-fluctuating temperature within the range 20° to 24°C seems appropriate. The texture has an important role in that it affects drainage, air movement, and emerging root architecture and later plays a part in the weaning of young plants.

Organic matter should be low since its presence can increase the risk of rotting of the cutting. A sterile medium is desirable to eliminate weed seeds and pathogens. Steam, heat sterilization, or chemical treatment can be used.

Fertilizer normally is added either as pellets (slow release) or in granular form during potting or later as a solution applied with the normal irrigation water. Fertilizers vary from slow release granules such as 'Osmocote' to more conventional ones such as triple super phosphate and NPK (a mixture of nitrogen, phosphorus, and potassium in various proportions).

In Papua New Guinea weathered pumice was used as the rooting substrate. This was taken from deep in the soil profile and was found to be largely free of weeds and pathogens and did not require sterilization. In the Congo and Brazil cuttings were raised initially in local sandy subsoil in tubes of polythene film but now the trend is towards vermiculite in rigid polythene tubes. Vermiculite proved suitable in the stable, warm, humid climate at Aracruz (Brandão *et al.* 1984). Further inland, where there are larger climatic fluctuations, the best substrate for the rooting process was found to be finely ground charcoal. Since charcoal does not adhere well to the root system, it is used only as a top layer above sterilized soil in the rigid plastic container. In France pulverized pine bark is used (Chaperon 1983).

Root formation

The vigour and vegetative condition of the mother tree and the environmental conditions into which the cuttings are placed determine the time taken for the roots to appear. Rooting is quicker in tropical countries (for example 6 days in the hot season in Brazil, Congo, and Papua New Guinea) than in temperate countries (10 to 12 days in summer in France and South Africa). The rooting system which develops is dependent on the medium used. Differences in root thickness, branching, flexibility, number, and point of origin on the cutting occur in different media.

Rooting success depends on the proper balance of aeration and moisture within the medium; too little or too much moisture (poor drainage) causes roots to emerge near the surface of the medium. Adequate drainage enables roots to emerge from the base of the cutting.

The quality of the emergent root system is inversely proportional to the time taken for the first roots to appear. Cuttings which root rapidly generate roots at the base and also up the stem of the cutting (Fig. 22.8). Roots are therefore numerous. Cuttings which take a long time to root produce some callus at the base of the cutting and only a few roots emerge from this callus.

There are differences in response between species. For example, *E. globulus, E. nitens,* and *E. regnans* have a reputation for being difficult to root, when *E. camaldulensis, E. deglupta,* and *E. robusta* cuttings will root when their bases are immersed in water alone.

Weaning

Between the acquisition of a functional root system and the resumption of active aerial growth the

Fig. 22.8. Numerous roots from base and stem of cuttings of eucalypt hybrids at Pointe Noire, Congo (photo G. Quillet).

cuttings require a careful transition from continuous or intermittent mist and shade (Fig. 22.9) to normal nursery conditions of full sunlight and watering by sprinkler several times a day. This phase is called weaning. Duration of weaning depends on seasonal conditions and the type of cutting used and varies between two and four weeks.

In the Congo cuttings are set under shade and the time between mistings is progressively increased (Delwaulle 1983). Shade is removed completely when the axillary buds develop the first pair of leaves. Normally, it takes from two to three months from rooting of the cutting until the planting stock is ready for the field. For some species such as *E. deglupta* and *E. camaldulensis* in a tropical environment only six weeks are required to produce plants 25 to 30 cm high.

Importance of discarding poor-rooting clones

In the eucalypts, as with other trees, rooting aptitude varies greatly between species, within the same species or hybrid, from one clone to another, and between ramets of a clone according to the maturation stage of the different mother plants. In the Congo clones of the *Eucalyptus* hybrid 'PF_1' show a variation in rooting ability from 1 to 90 per cent with a mean of about 50 per cent. The same problem has been found with *E. globulus* in Portugal and Tasmania. If the poor rooting clones are discarded then a rooting percentage of greater than 80 per cent can be achieved for commercial production. Of course, the number of clones is then greatly reduced and many clones of excellent potential in growth and form are discarded.

Fig. 22.9. Rooting of cuttings requires several weeks under mist spray and shade (photo E. Campinhos).

Care must be taken with propagation techniques otherwise advancing maturation will result in the unnecessary rejection of good clones. Some of the clones in young clonal plantations of *E. globulus* in Portugal showed indications of early maturity with mature leaves at one year. This appeared to have been the result of taking cuttings from too high on the shoots of the mother plants in clone banks, over 3 or 4 generations of repropagation from the ortet. Cutting back the mother plants close to ground level in the clone banks and repropagating only from the lowest nodes would help to arrest unwanted maturation. The problem can be avoided by changing to propagation of cuttings from seedlings of selected control-pollinated families.

Trials in the mid-1980s with *E. globulus* in Portugal lead to around 5 per cent of clones tested having a consistent rooting ability above about 60 per cent for shoots which were collected as coppice from either ortets in plantation or mother plants grown in field clone banks under irrigation. However, for *E. globulus* the environment of the mother plant has proved critically important. Trials were made in the late-1980s under a 'cascade system' of taking cuttings from young rooted cuttings growing in containers inside a greenhouse with appropriate environment control. As a result, 25 per cent of *E. globulus* clones tested had rooting ability above 70 per cent, and with far less seasonal variation (P. Wilson and P.P. Cotterill, CELBI, personal communication). In view of this finding CELBI constructed 2 ha of glasshouse in 1991 to produce 10 million *E. globulus* cuttings annually. A large proportion of the glasshouse area is devoted to raising mother plants under good environment control.

Recent investigations in Portugal with a wide range of eucalypt species and hybrids show that it is unwise to generalize about the rooting ability of different sources of material (P. Wilson, personal communication). This is because species can be very specific in their requirements (treatment and environment) for good rooting success, particularly with regard to mother plant environment and handling. A system which may give good rooting success for *E. grandis* does not work well for *E. globulus* and may give zero rooting for *E. nitens*.

For *E. gunnii* in France the mean rooting percentage of all clones is about 50 per cent and the average of the best clones 80 per cent. For *E. deglupta* the mean of all clones is 85 to 90 per cent and the best clones more than 95 per cent. Many other species yield an average of over 80 per cent on a commercial basis after culling the clones with poor rooting aptitude (for example *E. brassiana, E. camaldulensis, E. grandis, E. gunnii* × *dalrympleana, E. tereticornis, E. urophylla* and *E. viminalis*).

It seems reasonable to expect more than 70 per cent success in rooting cuttings with several eucalypts of importance in commercial operations. This figure assumes proper rejuvenation, discarding clones with poor rooting ability, and setting cuttings directly in the container destined for plantation establishment.

22.4.4 Production of cuttings in temperate climates

In temperate countries (for example France, Chaperon 1983) the procedure is different from that at Aracruz and Pointe Noire and varies with the season so that cuttings can be produced almost all year round (the exception is mid-winter, late November to early January in France).

Winter propagation is practised under a polythene cover inside a plastic greenhouse with bottom heat, thereby retaining a warm, moist atmosphere near the cuttings. The substrate is watered and treated with fungicides, if necessary, once a week (more regularly in mid-winter). Good drainage is essential since the substrate must be damp but not waterlogged.

Temperature of the substrate is kept at about 20°C at the level of the base of the cutting. If temperatures are too low rooting is retarded. Above optimum temperature there is a high risk of developing rot in the cuttings. The ambient temperature under the polythene cover is not normally regulated.

Since light plays an important role in rooting, artificial illumination is usually necessary to supplement natural daylight. A minimum of 200 hours of light a month is required for good results. Four hundred hours a month is even better (approximately 13 hours each day).

Propagation in spring and autumn is carried out under intermittent mist or fog but still inside the greenhouse. The higher temperatures arising in the greenhouse are reduced by air circulation and by whitewashing the plastic cover. Additional shade

(50 per cent) is given just above the plants (this also contributes to breaking up the mist droplets). Temperatures above 30°C have been observed to contribute to rotting of the cuttings and other problems (canker necrosis, wilt disease, foliar callus, etc.). Overheating must be avoided by closely monitoring greenhouse temperature and humidity and applying any necessary controls.

Summer propagation can be carried on outside under intermittent mist sprays. Shade cloth (50 per cent) is placed just above the cuttings to exclude the burning rays of the sun and maintain a humid atmosphere about the aerial parts of the cuttings. It also plays a role in keeping the new shoots (next generation of cuttings) in a better physiological condition for rooting.

22.5 Production of cuttings on a small scale

The preceding sections described the practices used in propagating eucalypts by cuttings for large-scale industrial plantations. The same principles can be applied to achieve improvement in growth rate and quality, on a much smaller scale.

A small program such as planting woodlots for poles and posts, fuelwood, or oil production will not be able to carry the high cost of initial investment in rejuvenation and relocation of large numbers of clones.

Clonal propagation with the procedures now in use in Brazil and the Congo requires a large, very-well-managed nursery, continuity of employment of highly motivated key staff, good record keeping in the office, and accurate labelling in the nursery and field. To apply such procedures for social forestry projects in rural villages may lead to a prohibitively high price per plant. However, if village people wished to grow their own cuttings they could obtain from a central research and development nursery several selected clones as a small number of potted, physiologically juvenile ramets (mother plants) which could be multiplied in a small, well-organized nursery (Eldridge 1984).

A nursery which grows eucalypt seedlings reliably in a warm climate would need only minor improvements in its facilities and work standards to produce cuttings, provided good advice and training were available on the new techniques. A start can be made on a very small scale: one square metre of bed, a plastic cover, sticks for shade, a watering can, and a thoroughly reliable attendant.

It is possible to start off with a small number of good-quality ortets or ramets and make cuttings in a repetitive series to achieve a rapid multiplication of plants for the field over 3 to 6 generations in a season.

Costs are reduced because successive generations of mother plants are kept (temporarily) in the nursery itself and cuttings can be taken from them. This plan in which each crop of cuttings is set from the last crop, is described as 'cascading' (bouturage en cascade, Chaperon 1983). Costs can be further reduced by leaving most of the plants in the nursery unlabelled. Production can be spread throughout the year in tropical and subtropical climates, and with a suitable nursery for up to 9 months in temperate climates (Chaperon 1984a). Clones can be changed at any time because they are not immobilized in a clonal archive or clone bank. Successive generations of mother plants are not retained in the nursery but are themselves eventually planted in the field as routine planting stock.

A certain percentage of mother plants will produce cuttings which have poor rooting ability. It is necessary to avoid further multiplication of these because of the risk of creating, even within a clone, a line of poor rooting individuals. Judgement is made more on whether roots are formed or not than on numbers and quality of roots formed on a cutting.

While costs of production are lower and management simpler for this method it is not as free of risk as for the method using clonal archives or clone banks. The latter offer security of long-term production of vegetative material of high quality without the risk of degeneration of mother plant which is likely when they are part of the production cycle and not individually identified.

23

Looking ahead

23.1 Genetics in silviculture

Continued expansion of the planting of fast-growing eucalypts seems inevitable because of their ability to help meet the world-wide demand for wood. Equally inevitable are disputes about the suitability of eucalypts for particular areas, and their banning in some places, in spite of improvements in land-use planning. What is the place of eucalypt breeding in these moral and political dilemmas?

There seems no point in inhibiting eucalypt breeding activities because some, or even many, plantations have been badly sited. Eucalypt breeding can be justified by its contribution to greater production in quantity and value in well-planned plantations. One aspect, agroforestry, deserves much more attention than it has received in the past. We agree with White (1990) that the concept of co-existence of food and wood production as part of ecologically sustainable agriculture should be included in eucalypt breeding objectives and the policy for genetic research on eucalypts. Eucalypt plantations are not only grown as extensive industrial monocultures. Small woodlots and windbreaks produce a large proportion of the world's eucalypt harvest, and, increasingly, eucalypts are grown with compatible intercrops. These food plants carry the costs of site preparation, weeding and fire prevention with a trade off of slightly reduced tree growth but overall a greater return to the farmer than either trees or food crops alone (White 1990). There are examples in Bangladesh (Davidson 1985a), southern China (Ou *et al.* 1989), and Nepal (White 1985).

Whatever the intended application, the general objectives of breeding are likely to remain as increased wood production and good health, with selection for certain quality traits for particular species and circumstances, including straighter stems, branch shedding, coppicing ability, pulp yield, and frost resistance.

Some plantation managers are already asking how much longer they should continue investing in eucalypt breeding. How soon will they be able to say enough improvement has been made? It may be that a goal of reasonable straightness has already been achieved in some plantations of *E. grandis* in Brazil, for example, but that other traits such as branch shedding and pulp yield need further improvement. Looking ahead, there is little doubt that selection and breeding will be required for resistance to new pests and diseases not yet introduced to Brazil. Also there may be new planting sites or cultural practices and new uses for *E. grandis* wood all of which will require selection and breeding. Thus we can predict that the need for eucalypt breeding skills will continue as long as eucalypts are grown in plantations.

23.2 Genetic resources

The emphasis in this book is on making use of the great richness of the genetic resources of the eucalypts by assembling appropriate base populations as the foundation for long-term genetic improvement. The specialized techniques of advanced generation breeding, tissue culture, and genetic engineering have not been dealt with in detail as we consider them to be matters for the future and not of such immediate concern as assembling genetic resources. After all, eucalypts are only in the first stages of domestication unlike grain crops, for example, where more advanced techniques are necessary because natural genetic resources are almost extinct. Flexibility built into eucalypt breeding plans will allow the introduction of such techniques as they become necessary for requirements which cannot be met by 'traditional' methods. We contend that the basic need in eucalypt breeding is well-planned recurrent selection and mating, starting with a suitable base population, and will remain so for many decades for two reasons. Firstly, the efficient use of more 'advanced' techniques, for producing new varieties and for mass propagating them, requires the availability of a pool of well-adapted material of superior standard in many economically important traits. Such material comes from the continually improving breeding populations to which genetic engineering might add a limited number of special traits in future. Secondly, the expense of the advanced techniques may not be justified for the major eucalypt products, fuelwood, poles, and pulpwood, for which improved varieties are required. For such low priced commodities we have suggested ways to provide better trees to meet the goals of growers fairly quickly and at a modest cost through population improvement.

Names of the eucalypts will be a continuing challenge to eucalypt breeders as taxonomic research leads to new species and new groupings. There is little prospect that eucalypt names will ever be fixed once-and-for-all. For example, new names of fast-growing species with potential for tropical plantations are likely to arise from the botanical exploration of the highly variable *E. urophylla* and its close relative *E. pellita*. Naming of natural provenances is simple enough as they are fully described by their exact place of origin, as in the Petford provenance of *E. camaldulensis*. But cultivated varieties and seed sources from plantations of a mixture of provenances may be more difficult to name with confidence. For clones the advent of DNA fingerprinting will bring precision in identification.

Great benefit has been obtained relatively cheaply from eucalypt provenance research by choice of the best geographic seed sources within a species to match particular sites, and further benefits can be expected from continued research. Provenance studies in several species have shown distinct patterns of geographic variation in economically important traits, leading the eucalypt breeder to collect seed in the best regions of provenance, and to avoid the worst. However, provenance studies of some fast-growing eucalypts are far from complete due to deficiencies in sampling the full natural range of the species or inadequate design of field trials, or both. For example, there is at present insufficient evidence to know whether or not there are distinct patterns of geographic variation in three major species, *E. globulus*, *E. tereticornis*, and *E. viminalis*. The very thorough range-wide collections of *E. globulus* in the late 1980s and subsequent numerous and well-designed trials will soon provide a definite answer for that species, but for the other two species continued intensive research is needed. Even in the much-studied *E. grandis* there are contradictory results from some trials.

After 200 years of introduction of eucalypts to many countries and about 30 years of international scientific research on species and provenance trials, one might ask, 'Is there any more out there?' Is there any need to keep looking for a better provenance and for more wild trees to add to existing breeding populations? The answer will vary with species and planting sites and with the genetic quality of existing land races. There is clearly much benefit to be obtained by experimentation to match species and provenances to the site when a new planting project is started with the first handful of eucalypt seed. But even in areas where eucalypts have been grown for many decades the incentives for more wood from less land call for an adventurous approach to trying something new and better. For example, in *E. camaldulensis*, perhaps the most widely planted and the most intensively sampled eucalypt, there are promising areas which have never been sampled. Two extreme sites from which seed might be collected in future for field trials are an extremely hot, dry and saline site on the Rudall River in the Great Sandy Desert of Western Australia and an extremely frost-prone site on the Yass River near Canberra. Similarly, in *E. grandis* the highest altitude southern occurrences west of Coffs Harbour, New South Wales, and the lowest altitude northern occurrences near Cairns, Queensland, have not contributed to overseas base populations of this species.

When it is realized what a great abundance of natural variability there is in all species of fast-growing eucalypts, some of it still not sampled and brought into cultivation, it would be unwise to confine future breeding activities to land races of uncertain origin. In the 12 chapters of this book reviewing provenance results we found evidence in several species that land races had inferior growth in comparison with newly imported wild material. All sources of useful genetic variability, including both land races and selected natural provenances, should be used when assembling genetic resources for a eucalypt breeding program.

The continued availability of unrelated, genetically diverse material is likely to be a major limiting factor in the long-term productivity of plantations. Such new material for infusion into breeding populations will only be available if forest genetic resources are conserved. Good plantation records, which document the genetic history of domestication, can be a major contribution to *ex situ* conservation (Eldridge 1990). Gene conservation *in situ* for several of the fast-growing eucalypts is relatively easy because large tracts of natural forest of the better provenances remain and are protected in national parks and equivalent reserves. As natural forests are eliminated special plantations for *ex situ* conservation become necessary.

23.3 Developments in breeding

Before efficient selection can commence, a base population must be assembled in open-pollinated progeny tests with the best natural provenances well-matched to the planting site. This step remains to be taken in some long established eucalypt plantation projects, as well as in new projects.

Some inaccuracy of selection among open-pollinated families from natural forests means a large number of families is required in the first generation and the progeny of most of them should be carried over to the next generation. In the second and later generation it appears that the relatively high rate of inbreeding, due to selfing, neighbourhood inbreeding, and non-random mating is overcome and selection among open-pollinated families will be more precise.

The amount of inbreeding and inbreeding depression per generation in an open-pollinated breeding strategy is an important subject for further research. If eucalypts in plantations generally have a high outcrossing rate, like conifers, then breeding strategies and methods of statistical analysis and selection developed for conifers will be readily applied to eucalypts. If not, and the mixed mating system of natural forests prevails in plantations, eucalypt breeding in advanced generations may require the development of special procedures, in particular the greater expense of advancing the breeding population from one generation to the next by controlled pollination which will allow complete avoidance of inbreeding.

Hybrid breeding in eucalypts so far has been confined to selecting outstanding F_1 interspecific hybrids, and the accompanying mass propagation by rooted cuttings. Future trends are likely to include reciprocal recurrent selection and perhaps more elaborate strategies with hybrid populations.

Mass propagation of the products of eucalypt breeding is likely to continue to be both by seed and by vegetative propagation, and both subject to great advances in technique and efficiency. Development of commercial-scale procedures for tissue culture and somatic embryogenesis are the subject of intense research activity by several large pulp and paper companies. Research and development also continues with the aim of achieving an understanding of the processes of maturation and finding ways to reverse them. Such breakthroughs would greatly increase the speed and efficiency of putting the products of eucalypt breeding into wood producing plantations.

High-technology enhancement of seed production is also close to being a practical reality, bringing the possibility of very small seed orchards managed for production of particular crosses. Meanwhile, lower technology seed production in seedling seed orchards and seed production areas seems likely to continue as the means of mass propagation for eucalypt plantation projects where funds are short, as they usually are.

23.4 Successful eucalypt breeding

In this book we have shown how genetic gain in the widest sense can be achieved in eucalypt crops at four levels: first, through choice of the best species; second by choice of the best provenances; third, by selection of many trees in the best provenances and mass production of their seed in seed orchards; and fourth, by selection of a few outstanding individual trees including inter-specific hybrids and mass production of them by vegetative propagation. The first two levels are the easiest and cheapest to accomplish but the third and fourth require more expertise, time, and money.

Throughout the book we have emphasized that, although many eucalypt plantation projects are already considered successful, large gains can be made in new plantation projects, and in many existing projects, by choosing well-adapted species and provenances through testing new introductions in well-designed field trials. Large initial gain achievable from choice of species and provenances is a once-only event for each project. Such a step should not be by-passed since gain obtainable in these early stages is usually much greater than from several cycles of improvement later by recurrent selection and mating within well-adapted breeding populations, each of which will give less remarkable, but repeatable, genetic gain in each future generation. What we are advocating is always to take the big leap ahead in the first generation, pause to establish seed production areas, then, as resources permit, proceed to population improvement by recurrent selection over subsequent generations. More advanced breeding methods, such as hybridization or genetic engineering with their accompanying vegetative propagation, can be considered as need and opportunity permit. However, because eucalypt breeding populations are highly heterozygous due to their early stage of domestication, there is no doubt that substantial population improvement will continue to be achieved by well-planned recurrent selection.

Fig. 23.1. The shape of things to come. A genetically transformed clone of *E. camaldulensis* being multiplied by micropropagation. Two genes were inserted using *Agrobacterium*, one for tolerance to the antibiotic kanamycin, and the other for the bacterial enzyme GUS. Using this transformation methodology, economically valuable genes can be inserted into selected clones (photo Cooperative Research Centre for Temperate Hardwood Forestry, Hobart).

Glossary

Most of the entries are based on Allard (1960), Falconer (1989), and Snyder (1972).

Breeding Systematic genetic improvement of a species by selection of individuals for certain characteristics and mating the selected individuals to produce an improved population.

Breeding value The value of a parent judged by the performance of its progeny from several crosses; expressed numerically as twice the mean deviation from the overall mean of all crosses, i.e. twice the general combining ability.

Character Any detectable phenotypic property of an organism; the same meaning as trait.

Clone The genetically uniform individuals (ramets) propagated vegetatively from a single sexually produced individual (the ortet).

Density Basic density of wood is measured as dry weight divided by soaked green volume.

Domestication To bring under human control; wild trees brought into cultivation in plantations become genetically altered over generations of selection.

***Ex situ* conservation** Gene conservation at a distance from the natural stand employing special plantations or the long-term cold storage of seed, pollen, or micropropagated clonal material.

Family The sexually produced individuals raised from a single tree, the progeny of that tree.

General combining ability (GCA) The average performance of the progeny of one parent when it is crossed with a number of other parents; expressed as the mean deviation from the overall mean of all crosses.

Genetic correlation A way of expressing the genetic relationship between two traits and estimating correlated response to selection, which is the change in one trait when we select for another (see Section 18.3.3).

Genetic parameter A numerical quantity which specifies a population in respect to some genetic characteristic, for example additive genetic variance, phenotypic variance, heritability, genetic correlation.

Genotype An individual's hereditary constitution; it interacts with the environment to produce the phenotype.

Graft The cut surface of a shoot (the scion) taken from a selected tree is matched with the cut surface of a seedling (the stock) and bound with tape; the tissues grow together forming a graft union.

Heritability The proportion of total variation in a trait which is due to genetic effects, the rest is due to environmental effects; estimated from statistical analysis of progeny tests.

Inbreeding Mating among closely related trees, including selfing.

Index selection A statistical method for efficient selection using a large amount of data on several traits (such as growth, form and wood density) of individual trees and families, taking into account their heritability and other genetic parameters and the relative economic importance of each trait.

***In situ* conservation** Gene conservation 'in place', in natural forests.

Land race A land race develops when exotic trees are introduced in a new environment; genetic changes take place in the population of trees over one or more generations of selection by natural or human agencies; a land race of poor quality develops when the first planting was from a poorly adapted provenance or, worse, from the seeds of a single tree (see Section 3.4).

Lignotuber A woody tuber containing numerous buds developed at the base of the first few pairs of leaves of seedlings of most eucalypt species; can become very large in some species, particularly the mallees.

Mallee The growth form of many eucalypt species of drier regions; a multi-stemmed shrub in which the stems arise from a large lignotuber.

Operculum The cap of a eucalypt flower bud formed by fusion of sepals or petals; it falls off as the flower opens.

Ortet The original sexually produced plant from which a clone has been derived.

Outcrossing Mating among unrelated trees of different genotypes.

Phenotype The appearance or performance of an individual which results from the interaction of its genotype and the environment.

Plus trees Selected, outstanding, above-average trees.

Progeny test A statistically designed field test of progenies (= families).

Provenance The place where seed comes from, normally used for seed from natural forests not from plantations; because of genetic adaptation to their distinctly different natural sites, seed of different provenances of the one species can differ greatly in performance when planted together at one site (see Section 3.1).

Ramet An individual member of a clone.

Replicate In a statistically designed field trial each seedlot (family or provenance) is repeated (replicated) in several plots at random over the whole trial area; one replicate contains one plot of each seedlot in the trial.

Scion The shoot taken from a tree to make a graft.

Seed orchard A special plantation of highly selected trees, isolated to minimize contamination with pollen from outside sources, and managed for maximum seed production.

Seed production area A plantation of suitable provenance raised from the seed from a reasonably large number of unrelated trees (10 to 20, or more) and thinned for seed production.

Selfing Pollination of an individual tree or clone with its own pollen.

Specific combining ability (SCA) The deviation of a cross from the performance predicted on the basis of the general combining ability of the two parents.

Sublines A method of minimizing the unwanted effects of inbreeding in a well-managed breeding population; there could be a separate subline for each major provenance region; selection and mating is within the subline in each generation of population improvement, eventually leading to some inbreeding; selected individuals are taken from each subline to make wide outcrosses by controlled pollination as foundation stock for clones; alternatively grafts of selected individuals from each subline would form seed orchards where wide outcrossing takes place.

Tissue culture Growing plant cells or tissues or plantlets in sterile culture on artificial media.

Trait The character selected : growth rate, stem straightness, wood density, etc.

Vegetative propagation Not propagated by seed but by rooted cuttings, grafts, or tissue culture.

References

Numbers in square brackets show the Section where the reference occurs.

ABARE (Australian Bureau of Agricultural and Resource Economics) (1990). *Australian forest resources 1989*. Australian Government Publishing Service, Canberra. [1.2.2]

Abbott, I.D. and Pederick, L.A. (1984a). Genetic variation in *Eucalyptus delegatensis*. 1. Provenance variation. *Research Branch Report* 265. Forests Commission of Victoria, Melbourne. [8.4.1]

Abbott, I.D. and Pederick, L.A. (1984b). Genetic variation in *Eucalyptus delegatensis*. 2. Altitudinal variation in growth (two years), seed dormancy and cold hardiness. *Research Branch Report* 266. Forests Commission of Victoria, Melbourne. [8.4.1, 8.4.2]

Abd-Alla, M.F., El-Negoumy, S.I., El-Lakany, M.H., and Saleh, N.A.M. (1980). Flavenoid glycosides and the chemosystematics of *Eucalyptus camaldulensis*. *Phytochemistry* **19**, 2629–2632. [6.3]

Acosta-Solis, M. (1963). Economia, propagacion y rendimiento del eucalipto en el Ecuador. Instituto Ecuadoriano de Ciencias Naturales, Contribucion No. 40, Separata de la *Revista Tecnica* Nos. 14 y 15, pp. 3–8. Quito, Ecuador. [1.3]

Ades, P.K. and Burgess, I.P. (1982). Improvement in early growth rate achieved by phenotypic selection of *Eucalyptus grandis*. *Australian Forest Research* **12**, 169–173. [3.4]

Ades, P.K. and Burgess, I.P. (1983). Provenance trials of Eucalyptus grandis in northern New South Wales. *Silvicultura,* São Paulo **31**, 396–397. [11.4.1]

ADTR (Australian Department of Trade and Resources) (1982a). *Australian farming systems: an introduction*. AGPS (Australian Government Publishing Service), Canberra. [4.3.2; Fig. 4.1]

ADTR (1982b). *Australian farming systems: the tropical region*. AGPS, Canberra. [4.3.2; Fig. 4.2]

ADTR (1982c). *Australian farming systems: the subtropics*. AGPS, Canberra. [4.3.2; Fig. 4.3]

ADTR (1982d). *Australian farming systems: the temperate region*. AGPS, Canberra. [4.3.2; Fig. 4.4]

ADTR (1982e). *Australian farming systems: the arid and semi-arid region*. AGPS, Canberra. [4.3.2; Fig. 4.5]

Allard, R.W. (1960). *Principles of plant breeding*. Wiley, New York. [18.2.3, 18.3]

Anderson, C.A. (1982a). The effect of high pH and P, on the development of lime-chlorosis in two seedling populations of *Eucalyptus obliqua* L'Hérit. *Plant and Soil* **69**, 199–212. [13.3]

Anderson, C.A. (1982b). The effect of calcium on the germination, growth and mineral nutrition of acidic and calcareous populations of *Eucalyptus obliqua* L'Hérit. *Plant and Soil* **69**, 213–223. [13.3]

Anderson, C.A. and Ladiges, P.Y. (1978). A comparison of three populations of *Eucalyptus obliqua* l'Hérit. growing on acid and calcareous soils in southern Victoria. *Australian Journal of Botany* **26**, 93–109. [13.3, 13.4.1; Fig. 13.5]

Anderson, C.A. and Ladiges, P.Y. (1982). Lime-chlorosis and the effect of fire on the growth of three seedling populations of *Eucalyptus obliqua* l'Hérit. *Australian Journal of Botany* **30**, 47–66. [13.3; Table 3.1]

Anderson, E. (1949). *Introgressive hybridization*. Wiley, New York. [1.5, 19.6.3]

Andrade, E.Navarro de (1961). *O eucalipto* (2nd edn). Co. Paulista de Estradas de Ferro, São Paulo. [1.2.1, 1.2.2, 21.1]

Andrew, I. (1986). Simple experimental design for forestry trials (revised edition). *FRI Bulletin* No. 71. Forest Research Institute, Rotorua. [5.6.2]

Aneja, S. and Atal, C.K. (1969). Plantlet formation in tissue cultures from lignotubers of *Eucalyptus citriodora* Hook. *Current Science*, Bangalore **38**, 69. [22.3.5]

Anon. (1964). *Annual report of the Forestry and Timber Bureau for the year 1963*, pp. 47–48. Department of National Development, Canberra. [6.4, 16.3]

Anon. (1966). *Forestry in Western Australia. Bulletin* 63 (2nd edn). Forests Department of Western Australia, Perth. [19.5]

Anon. (1971). *OECD scheme for the control of forest reproductive material moving in international trade*. OECD, Paris. [21.5]

Anon. (1983). *Inventário florestal nacional (sintese dos resultados)*. Instituto Brasileiro de Desenvolvimento Florestal, Brasilia. [1.2.2]

Anon. (1987). *L'arboretum de Ruhande*. Institut des Sciences Agronomiques du Rwanda, Butare. [5.2]

Anon. (1990). *Commercial timber resources and roundwood processing in South Africa 1988–89*. Forestry Branch, Department of Environment Affairs, Pretoria. [21.2.6; Table 1.1]

Appleton, R., Cameron, J.N., Dean, C.A., and Whiteman, P.J. (1988). Combined and multiple trait selection in a 16.5 year old *Eucalyptus globulus* provenance trial (VRD13). *Gippsland Technical Report* 88–17. APM Forests, Morwell. [10.4.1]

Arisman, H. (1985). Indonesia's breeding program for *Eucalyptus deglupta* and *Eucalyptus urophylla*. In *Proceedings ASEAN-Australian forest tree improvement program,* Item 7, Bangkok meeting, August 1985. [16.5]

Ashton, D.H. (1956). Autoecology of *Eucalyptus regnans* F. Muell. Unpublished Ph.D. thesis, University of Melbourne. [19.4.3]

Ashton, D.H. (1958). The ecology of *Eucalyptus regnans*. The species and its frost resistance. *Australian Journal of Botany* **6**, 154–176. [13.5, 14.3.2, 19.6.3]

Ashton, D.H. (1975a). Studies of flowering behaviour in *Eucalyptus regnans* F. Muell. *Australian Journal of Botany* **23**, 399–411. [19.2.2, 19.2.3, 19.3.2, 19.5]

Ashton, D.H. (1975b). The root and shoot development of *Eucalyptus regnans* F. Muell. *Australian Journal of Botany* **23**, 867–887. [14.1, 19.6.1, 19.6.2; Fig. 19.7]

Ashton, D.H. (1976). The development of even-aged stands of *Eucalyptus regnans* in central Victoria. *Australian Journal of Botany* **24**, 397–414. [19.6.2; Fig. 19.7]

Ashton, D.H. (1981). The ecology of the boundary between *Eucalyptus regnans* F. Muell. and *E. obliqua* L'Hérit. in Victoria. *Proceedings of the Ecological Society of Australia* **11**, 75–94. [13.5, 19.6.3]

Ashton, D.H. (1984). Artificial hybrids produced by pollination of *Eucalyptus regnans* F. Muell. by *E. obliqua* l'Hérit. and *E. baxteri* (Benth.) Maiden and Blakely at Foster North, Victoria. *Victorian Naturalist* **101**, 193–199. [13.5]

Ashton D.H. and Sandiford, E.M. (1988). Natural hybridisation between *Eucalyptus regnans* F. Muell. and *E. macrorhyncha* F. Muell. in the Cathedral Range, Victoria. *Australian Journal of Botany* **36**, 1–22. [13.5, 19.3.3, 19.6.3]

Ashton, D.H. and Williams, G. (1973). The occurrence of gum-topped stringybarks in the Trentham forest. *Victorian Naturalist* **90**, 90–92. [13.5]

Assaf, R. (1966). Etude sur la croissance des rameaux de diverses especes fruitieres. These Docteur Ingenieur, Université de Toulouse. [22.3]

Awe, J.O. (1973). Provenance variation in *Eucalyptus camaldulensis* Dehnh. Unpublished M.Sc. thesis, Australian National University, Canberra. [Table 3.1]

Awe, J.O. and Shepherd, K.R. (1975). Provenance variation in frost resistance in *Eucalyptus camaldulensis* Dehnh. *Australian Forestry* **38**, 26–33. [6.4.6]

Awe, J.O., Shepherd, K.R., and Florence, R.G. (1976). Root development in provenances of *Eucalyptus camaldulensis* Dehnh. *Australian Forestry* **39**, 201–209. [6.4.5]

Bachelard, E.P. and Stowe, B.B. (1963). Rooting of cuttings of *Acer rubrum* L. and *Eucalyptus camaldulensis* Dehn. *Australian Journal of Biological Sciences* **16**, 751–767. [22.1, 22.3.1]

Bamber, R.K. (1985). The wood anatomy of eucalypts and papermaking. *Appita* **38**, 210–216. [20.1.6]

Bamber, R.K. and Humphreys, F.R. (1963). A preliminary study of some wood properties of *Eucalyptus grandis* (Hill) Maiden. *Journal of the Institute of Wood Science* **11**, 63–70. [20.1.7]

Banerjee, S.K., Nath, S., and Nandi, A. (1990). Provenance trials of *Eucalyptus tereticornis* in two lateritic soils of West Bengal. *Indian Journal of Forestry* **13**, 1–7. [15.3]

Banks, J.C.G. (1972). Studies on the taxonomy and ecotyptic variation of *Eucalyptus viminalis* Labill. Unpublished M.Sc. thesis, Australian National University, Canberra. [Table 3.1]

Banks, J.C.G. and Hillis, W.E. (1969). The characterisation of populations of *Eucalyptus camaldulensis* by chemical features. *Australian Journal of Botany* **17**, 133–146. [6.3; Table 3.1]

Barber, H.N. (1956). The natural history of natural selection. *Australian Journal of Science* **18**, 148–159. [17.1]

Barber, H.N. (1958). The process of natural selection. In *Proceedings 10th international congress of genetics*, Montreal, Canada, Vol. 2, pp. 13–14. [19.6.1]

Barber, H.N. (1965). Selection in natural populations. *Heredity* **20**, 551–572. [13.3; Fig. 13.2]

Barber, H.N. and Jackson, W.D. (1957). Natural selection in action in *Eucalyptus*. *Nature*, London **179**, 1267–1269. [19.3.2; Table 3.1]

Barker, P.K., de Fossard, R.A., and Bourne, R.A. (1977). Progress towards clonal propagation of *Eucalyptus* species by tissue culture techniques. *International Plant Propagators' Society Combined Proceedings* **27**, 546–556. [22.3.5]

Barlow, B.A. (1981). The Australian flora: its origin and evolution. In *Flora of Australia*. Vol. 1. *Introduction*, pp. 25–75. Australian Government Publishing Service, Canberra. [2.2.1]

Barnes, R.D. (1984a). Genotype-environment interaction in the genetic improvement of fast-growing plantation trees. In *Site and productivity of fast-growing plantations*, Proceedings IUFRO Symposium (ed. D.C. Grey, A.P.G. Schönau, and C.J. Schutz) Vol. 1, pp. 197–213. South African Forestry Research Institute, Pretoria. [21.3.5]

Barnes, R.D. (1984b). A multiple population breeding strategy for Zimbabwe. In *Provenance and genetic improvement strategies in tropical forest trees,* Proceedings IUFRO Conference, Mutare, (ed. R.D. Barnes and G.L. Gibson), pp. 619–632. Commonwealth Forestry Institute, Oxford. [18.1, 18.2.3, 21.2.6]

Barnes, R.D. (1987). The breeding seedling orchard in a multiple population breeding strategy for tropical trees. In *Proceedings Simposio sobre silvicultura y mejoramiento genetico de especies forestales* Vol. 2, pp. 1–18. Centro de Investigaciones y Experiencias Forestales, Buenos Aires. [6.5, 18.2, 21.2.6]

Barnes, R.D. and Mullin, L.J. (1989). The multiple population breeding strategy in Zimbabwe - five year results. In *Breeding tropical trees*, Proceedings IUFRO Conference, Pattaya, Thailand, November 1988, (ed. G.L. Gibson, A.R. Griffin, A.C. Matheson) pp.148–158. Oxford Forestry Institute. [18.2.3]

Barr, N.A. (1983). Sawing qualities of *Eucalyptus saligna* Sm. as a guide to selection of trees for breeding and seed collection. *Silvicultura,* São Paulo **32**, 800–801. [11.4.1]

Barrett, R.L. and Carter, D.T. (1970). *Eucalyptus camaldulensis* provenance trials in Rhodesia (Part 1, early results). *Research Bulletin* 2 (Part 1). Rhodesia Forestry Commission, Salisbury. [6.4.1, 6.4.5]

Barrett, R.L., Carter, D.T., and Seward, B.R.T. (1975). *Eucalyptus grandis* in Rhodesia. *Rhodesian Bulletin of Forestry Research* 6. Rhodesia Forestry Commission, Salisbury. [11.2, 11.4.1]

Barton, A.F.M., Cotterill, P.P., and Brooker, M.I.H. (1991). Heritability of cineole yield in *Eucalyptus kochii. Silvae Genetica* **40**, 37–38. [18.3]

Bateman, W. (1961). Seed collection of the genus *Eucalyptus* in northern Australia. In *Proceedings FAO 2nd world eucalyptus conference*, São Paulo Brazil, Vol. 1, pp. 569–579. [19.4.4]

Beadle, N.C.W. (1981). *The vegetation of Australia.* Cambridge University Press. [2.2]

Becker, W.A. (1985). *Manual of quantitative genetics.* Academic Enterprises, Pullman. [18.3, 20.6.2]

Bellefontaine, R. (1979). Onze annes d'amélioration genetique forestière. *Annales de la Recherche Forestière au Maroc* **19**, 15–48. [6.5]

Bentham, G. (1867). *Flora Australiensis*, Vol. 3 (*Myrtaceae* to *Compositae*). Lovell Reeve, London. [19.1]

Bevege, D.I. (1976). A green revolution in the high yield forest: rational silviculture or new technology? *Australian Forestry* **39**, 40–50. [1.2.3]

Bird, T., Kile, G.A. and Podger, F.D. (1974). The eucalypt crown diebacks — a growing problem for forest managers. *Australian Forestry* **37**, 173–187. [3.5.1]

Blake, S.T. (1953). Botanical contributions of the northern Australia regional survey. I. Studies on northern Australian species of Eucalypts. *Australian Journal of Botany* **1**, 185–352. [16.3]

Blake, S.T. (1977). Four new species of *Eucalyptus. Austrobaileya* **1**, 1–10. [16.3]

Blake, T.J. (1981). Salt tolerance of eucalypt species grown in saline solution culture. *Australian Forest Research* **11**, 179–183. [6.4.7]

Blakely, W.F. (1965). *A key to the eucalypts* (3rd edn). Forestry and Timber Bureau, Canberra. [1st edn 1934, The Worker Trustees, Sydney]. [13.3, 19.1, 19.2.2]

Blume, C.L. (1849). *Eucalyptus deglupta* Bl. *Museum Botanicum Lugd.-Bat.* 83–84. [7.3, 16.3]

Boden, R.W. (1958a). Handling and storage of pollen in *Eucalyptus* breeding. *Australian Forestry* **22**, 73–81. [20.5.1]

Boden, R.W. (1958b). Differential frost resistance within one *Eucalyptus* species. *Australian Journal of Science* **21**, 84–86. [9.3.1; Fig. 9.4]

Boland, D.J. (1977). Abnormal *Eucalyptus leucoxylon* F. Muell. seedlings. *Australian Forest Research* **7**, 151–161. [19.6.1]

Boland, D.J. (1978). Geographic variation in *Eucalyptus leucoxylon* F. Muell. *Australian Forest Research* **8**, 25–46. [Table 3.1]

Boland, D.J. (1981). Eucalypt seed for Indian plantations from better Australian natural seed sources. *Indian Forester* **107**, 125–134. [1.3, 6.5, 15.2, 18.2.3]

Boland, D.J. (1985). Taxonomic revision of *Eucalyptus delegatensis* R.T. Baker (Myrtaceae). *Australian Forest Research* **15**, 173–181. [8.3; Fig. 8.4]

Boland, D.J. and Dunn, A.T. (1985). Geographic variation in alpine ash (*Eucalyptus delegatensis* R.T. Baker). *Australian Forest Research* **15**, 155–171. [8.1, 8.3, 8.4.2; Table 3.1; Fig. 8.1]

Boland, D.J. and Sedgely, M. (1986). Stigma and style morphology in relation to taxonomy and breeding systems in *Eucalyptus* and *Angophora* (*Myrtaceae*). *Australian Journal of Botany* **34**, 569–584. [19.2.3]

Boland, D.J., Brooker, M.I.H., and Turnbull, J.W. (1980a). *Eucalyptus seed.* CSIRO, Canberra. [1.1, 1.2.1, 3.5.3, 5.4, 19.1, 19.2.2, 19.4.4, 19.5, 19.6.3, 20.5.1, 21.4.1, 22.4.1]

Boland, D.J., Moran, G.F., and Kleinig, D.A. (1980b). *Eucalyptus delegatensis* R.T. Bak. provenance seed collections 1977–78. *Seeds Section Report* No. 3. CSIRO Division of Forest Research, Canberra. [8.3]

Boland, D.J. *et al.* (1984). *Forest trees of Australia* (4th edn). Nelson and CSIRO, Adelaide. [1.2.1, 2.1.1, 2.2.4, 9.1, 10.3, 13.1, 14.1, 17.1; Table 2.1]

Boland, D.J., Brophy, J.J., and House, A.P.N. (1991). *Eucalyptus leaf oils: use, chemistry, distillation and marketing.* Inkata Press, Melbourne. [1.2.2]

Bond, H.W. and Brown, W.L. (1979). The exploitation of floral nectar in *Eucalyptus incrassata* by honeyeaters and honeybees. *Oecologia* **44**, 105–111. [19.3.2]

Bonga, J.M. (1982). Vegetative propagation in relation to juvenility, maturity and rejuvenation. In *Tissue culture in forestry*, (ed. J.M. Bonga and D.J. Durzan), pp. 387–412. Martinus Nijhoff/Dr W. Junk, The Hague. [22.3]

Booth, T.H. (1985). A new method for assisting species selection. *Commonwealth Forestry Review* **64**, 241–250. [4.3, 4.3.3]

Booth, T.H. (1990). Mapping regions climatically suitable for particular tree species at the global scale. *Forest Ecology and Management* **36**, 47–60. [4.3.4]

Booth, T.H. (1991). Where in the world? New climatic analysis methods to assist species and provenance selection for trials. *Unasylva* **165**, 51–57. [4.3, 4.3.3]

Booth, T.H. and Pryor, L.D. (1991). Climatic requirements of some commercially important eucalypt species. *Forest Ecology and Management* **43**, 47–60 [4.3.3, 5.1; Table 4.1]

Booth, T.H., Nix, H.A., and Hutchinson, M.F. (1987). Grid matching: a new method for homoclime analysis. *Agricultural and Forest Meteorology* **39**, 241–255. [4.3.3]

Booth, T.H., Nix, H.A., Hutchinson, M.F., and Jovanovic, T. (1988). Niche analysis and tree species introduction. *Forest Ecology and Management* **23**, 47–59. [4.3.3, 4.3.4, 11.1; Fig. 4.6]

Booth, T.H., Stein, J.A., Nix, H.A., and Hutchinson, M.F. (1989). Mapping regions climatically suitable for particular species: an example using Africa. *Forest Ecology and Management* **28**, 19–31. [4.3.3, 11.2; Fig. 4.6]

Booth, T.H., Stein, J.A., Hutchinson, M.F., and Nix, H.A. (1990). Identifying areas within a country climatically suitable for particular tree species: an example using Zimbabwe. *International Tree Crops Journal* **6**, 1–16. [4.3.3]

Borough, C.J., Incoll, W.D., May, J.R., and Bird, T. (1984). Yield statistics. In *Eucalypts for wood production* (ed. W.E. Hillis and A.G. Brown) pp. 201–225. CSIRO, Melbourne. [14.1]

Borralho, N.M.G., Kanowski, P.J., and Cotterill, P.P. (1992). Genetic control of growth of *Eucalyptus globulus* in Portugal. I. Genetic and phenotypic parameters. *Silvae Genetica* **41**, 39–45. [Table 18.4]

Boulay, M. (1983). Micropropagation of frost resistant eucalypts. In *Eucalyptus in California, General Technical Report* PSW-69, (ed. R.B. Standiford and F.T. Ledig), pp. 102–107. USDA Forest Service, Berkeley. [1.6.2, 22.3.5, 22.4.2.]

Bouvet, J.M. and Delwaulle, J.C. (1983). Introduction d'*Eucalyptus cloeziana* au Congo, Pointe Noire — Parcelle 77.13. *Bois et Forêts des Tropiques* **200**, 7–20. [22.2.4]

Brandão, L.G., Campinhos, E., and Ikemori, Y.K. (1984). Brazil's new forest soars to success. *Pulp and Paper International* **26**(12), 38–40. [1.6.1, 18.2.2, 22.1, 22.2.4, 22.4.3]

Brewbaker, J.L. (1960). *Agricultural genetics*. Prentice-Hall, New Jersey. [18.3]

Briggs, J.D. and Leigh, J.H. (1988). *Rare or threatened Australian plants* (1988 revised edition). Australian National Parks and Wildlife Service Special Publication No. 14. ANPWS, Canberra. [3.5.1]

Brooker, M.I.H. (1986). An introduction to the study of characters in *Eucalyptus*. In *Proceedings of a workshop on seed handling and eucalypt taxonomy*, Harare July 1985 (ed. R.D. Ayling and B.R.T. Seward), pp. 119–142. International Development Research Centre, Ottawa. [2.2.2]

Brooker, M.I.H. and Kleinig, D.A. (1983). *Field guide to eucalypts*. Vol. 1. *South-eastern Australia*. Inkata Press, Melbourne. [2.1.1, 10.3, 11.1, 17.1, 19.1]

Brooker, M.I.H. and Kleinig, D.A. (1990). *Field guide to eucalypts*. Vol. 2. *South-western and southern Australia*. Inkata Press, Melbourne. [2.1.1, 19.1]

Brown, A.G. and Turnbull, J.W. (1986). The Australian environment. In *Multipurpose Australian trees and shrubs: lesser-known species for fuelwood and agroforestry* (ed. J.W. Turnbull), pp. 11–27. Australian Centre for International Agricultural Research, Canberra. [2.2, 4.3.1]

Brown, A.G., Eldridge, K.G., Green, J.W., and Matheson, A.C. (1976). Genetic variation of *Eucalyptus obliqua* in field trials. *New Phytologist* **77**, 193–203. [13.4.1, 13.6; Tables 3.1, 13.1]

Brown, A.G., Wolf, L.J., Ryan, R.A., and Voller, P. (1989). Tredat: a tree crop data base. *Australian Forestry* **52**, 23–29. [4.5]

Brown, A.H.D. (1978). Isozymes, plant population genetic structure and genetic conservation. *Theoretical and Applied Genetics* **52**, 145–157. [19.6.1]

Brown, A.H.D. (1979). Enzyme polymorphism in plant populations. *Theoretical Population Biology* **15**, 1–42. [19.6.1]

Brown, A.H.D., Matheson, A.C., and Eldridge, K.G. (1975). Estimation of the mating system of *Eucalyptus obliqua* l'Hérit. by using allozyme polymorphisms. *Australian Journal of Botany* **23**, 931–949. [19.6.1, 19.6.2; Table 19.2]

Brune, A. (1983). Methods of establishing genetic base populations for selection-designs for long-term, *ex-situ* maintenance of *Eucalyptus* gene pools and breeding populations. *Silvicultura,* São Paulo **31**, 527–528. [1.3, 1.4.3]

Brune, A. and Zobel, B. (1981). Genetic base populations, gene pools and breeding populations in *Eucalyptus* in Brazil. *Silvae Genetica* **30**, 146–149. [21.1]

Burdon, R.D. and Namkoong, G. (1983). Multiple populations and sub-lines. *Silvae Genetica* **32**, 221–222. [18.2.3]

Burdon, R.D., Shelbourne, C.J.A., and Wilcox, M.D. (1978). Advanced selection strategies. In *Documents FAO third world consultation on forest tree breeding* Vol. 2, pp. 1133–1147. CSIRO, Canberra. [18.1]

Bureau of Meteorology (1988). *Climatic averages Australia*. Australian Government Publishing Service, Canberra. [4.3.2]

Burgess, I.P. (1974). Vegetative propagation of *Eucalyptus grandis*. *New Zealand Journal of Forestry Science* **4**, 181–184. [22.1]

Burgess, I.P. (1975). A provenance trial with blackbutt: 9 year results. *Australian Forest Research* **7**(1), 1–9. [18.3.5]

Burgess, I.P. (1983). The natural occurrence of *Eucalyptus grandis*, its distribution patterns in natural forests, its characteristics and conservation. *Silvicultura,* São Paulo **31**, 397–399. [11.1]

Burgess, I.P. (1984). Tolerance of two years old *Eucalyptus grandis* and *Eucalyptus saligna* to winter frost in Canberra. In *Proceedings IUFRO Colloque international sur les eucalyptus resistants au froid,* September 1983, pp. 358–366. AFOCEL, Paris. [11.4.2]

Burgess, I.P. (1988). Provenance trials of *Eucalyptus grandis* and *E. saligna* in Australia. *Silvae Genetica* **37**, 221–227. [11.2, 11.4.1; Tables 11.2, 11.3]

Burgess, I.P. and Bell, J.C. (1983). Comparative morphology and allozyme frequencies of *Eucalyptus grandis* Hill ex Maiden and *E. saligna* Sm. *Australian Forest Research* **13**, 133–149. [11.3, 11.4.2; Table 3.1]

Burgess, I.P. and Griffin, A.R. (1991). Flowering time of *Eucalyptus* species in their natural habitat. *Division of*

Forestry User Series No. 1. CSIRO Division of Forestry, Canberra. [19.2.2]

Burgess, I.P., Bell, J.C., and van Wyk, G. (1985). The identity of trees currently known as *Eucalyptus grandis* in the Republic of South Africa based on isozyme frequencies. *South African Forestry Journal* **135**, 24–30. [11.2, 21.2.6]

Burley, J. and Wood, P.J. (1976). A manual on species and provenance research with particular reference to the tropics. *Tropical Forestry Papers* 10. Commonwealth Forestry Institute, Oxford. [5.2, 5.4, 5.6.2, 5.6.3, 5.6.4]

Butcher, T.B. (1986). *Eucalyptus globulus* provenance trials in Western Australia. In *Proceedings ninth meeting (Burnie) Research Working Group No. 1, Forest Genetics,* pp. 95–96. Australian Forestry Council, Canberra. [10.4.1]

Butcher, T.B. (1988). Co-operative breeding strategy for *Eucalyptus globulus* in Western Australia. In *Proceedings tenth meeting (Gympie) research working group No. 1, Forest Genetics,* pp. 254–257. Australian Forestry Council, Canberra. [10.5]

Cameron, J.N. and Kube, P.D. (1983). Management of seedling seed orchards of *Eucalyptus regnans* — selection strategy and flowering studies. *Silvicultura*, São Paulo **32**, 765–768. [21.2.6]

Cameron, J.N., Cotterill, P.P., and Whiteman, P.H. (1989). Key elements of a breeding plan for temperate eucalypts in Australia. In *Breeding tropical trees* Proceedings IUFRO Conference, Pattaya, Thailand, November 1988 (ed. G.L. Gibson, A.R. Griffin, A.C. Matheson) pp. 159–168. Oxford Forestry Institute. [18.2.2]

Campbell, A.J. (1888). Field Naturalists' Club of Victoria expedition to King Island, November, 1887. Official report. *Victorian Naturalist* **4**(9), 128–162. [10.4.1]

Campinhos, E. and Ikemori, Y.K. (1978). Tree improvement program of *Eucalyptus* spp.: preliminary results. In *Documents FAO third world consultation on forest tree breeding*, Vol. 2, pp. 717–735. CSIRO, Canberra. [11.4.3]

Campinhos, E. and Ikemori, Y.K. (1983). Nova técnica para a produção de mudas de essências florestais. *IPEF*, Piracicaba **23**, 47–52. [11.4.3, 22.4.3]

Camphinos, E. and Ikemori, Y.K. (1989). Selection and management of the basic population *Eucalyptus grandis* and *E. urophylla* established at Aracruz for the long term breeding programme. In *Breeding tropical trees*, Proceedings IUFRO Conference, Pattaya, Thailand, November 1988 (ed. G.L. Gibson, A.R. Griffin, A.C. Matheson) pp. 169–174. Oxford Forestry Institute. [1.5, 11.5, 18.2.2, 18.2.3]

Cannon, P.G. (1982). Growth of eucalyptus in six species and provenance trials in the Department of Cauca: results after three years. *Research Report* 81. Carton de Colombia S.A., Cali. [10.4.1]

Carpenter, R.A. (ed.) (1981). *Assessing tropical forest lands: their suitability for sustainable uses.* Tycooly International, Dublin. [4.2.1]

Carr, D.J. and Carr, S.G.M. (1959). Floral morphology and the taxonomy of *Eucalyptus*. *Nature*, London **184**, 1549–1552. [2.1.2]

Carr, D.J. and Carr, S.G.M. (1962). Natural groups within the genus *Eucalyptus*. In *The evolution of living organisms* (ed. G.W. Leeper) pp. 426–445. Melbourne University Press. [1.2.1, 2.1.2]

Carr, S.G.M. (1972). Problems of the geography of tropical eucalypts. In *Bridge and barrier: the natural and cultural history of Torres Strait* (ed. D. Walker) ANU publication no. BG/3, pp. 153–181. Australian National University, Canberra. [2.2]

Carr, S.G.M. and Carr, D.J. (1972). *Eucalyptus calophylla* var. *maideniana* Hochr.: a male tree. *Proceedings of the Royal Society of Victoria* **85**, 49–50. [19.3.5, 19.6.1]

Carr, S.G.M., Carr, D.J., and Ross, F.L. (1971). Male flowers in eucalypts. *Australian Journal of Botany* **19**, 73–83. [19.6.1]

Carter, E.J. (1987). From seed to trial establishment: a handbook giving practical guidelines in nursery practice and the establishment of simple species and/or provenance trials. *DFR User Series* No. 2. CSIRO Division of Forest Research, Canberra. [5.4, 5.6.2, 5.6.3; Fig. 5.1]

Cass Smith, W.P. (1970). Stem canker disease of red flowering gums. *Journal of the Department of Agriculture*, Western Australia **11**, 33–39. [19.6.1]

CELBI (1979). *Eucalyptus globulus* seed orchards within CELBI. In *Proceedings technical consultations on fast-growing plantation broadleaved trees for Mediterranean and temperate zones*, FAO/IUFRO Lisbon October 1979, pp. 195–207. (FAO paper no. FO:FGB-7-3/6.) [21.4.1]

CELBI (1982). *Propagação vegetativa do* Eucalyptus globulus *Labill., enraizamento de estacas.* CELBI, Figueira da Foz, Portugal. [1.6.1]

Chaperon, H. (1978). Particularités de l'amélioration génétique des *Eucalyptus* au Congo Brazzaville. In *Documents FAO third world consultation on forest tree breeding* Vol. 2, pp. 579–591. CSIRO, Canberra. [19.3.5]

Chaperon, H. (1983). Clonal propagation of *Eucalyptus* by cuttings in France. In *Eucalyptus in California, General Technical Report* PSW-69, (ed. R.B. Standiford and F.T. Ledig), pp. 108–114. USDA Forest Service, Berkeley. [1.6.1, 22.1, 22.4.3, 22.4.4, 22.5]

Chaperon, H. (1984a). Selection individuelle et clonage de l'*Eucalyptus*. In *Proceedings IUFRO Colloque international sur les Eucalyptus resistants au froid*, Bordeaux September 1983 pp. 535–558. AFOCEL, Paris. [22.4.3, 22.5]

Chaperon, H. (1984b). Influence of propagation by cuttings on the breeding strategy of forest trees. In *Provenance and genetic improvement strategies in tropical forest trees*, Proceedings IUFRO Conference, Mutare, (ed. R.D. Barnes and G.L. Gibson), pp. 135–148. Commonwealth Forestry Institute, Oxford. [22.2.4]

Chaturvedi, A.N. (1973). Rotation in *Eucalyptus* hybrid plantations. *Indian Forester* **99**, 205–214. [15.2]

Chaturvedi, A.N. (1976). Eucalypts in India. *Indian Forester* **102**, 57–63. [15.2]

Chaturvedi, A.N., Sivaji, P., and Jayaram Prasad, D.V. (1989). *Eucalyptus* provenance trials in Andhra Pradesh. *Indian Forester* **115**, 445–454. [15.3]

Chippendale, G.M. (1976). *Eucalyptus* nomenclature. *Australian Forest Research* **7**, 69–107. [10.3]

Chippendale, G.M. (1981). Distribution density of *Eucalyptus* species in Australia. *Search* **12**, 131–133. [2.2.4]

Chippendale, G.M. (1988). *Eucalyptus, Angophora (Myrtaceae). Flora of Australia* 19. Australian Government Publishing Service, Canberra. [1.2.1, 2.1.2, 2.2.4, 10.3; Table 2.1]

Chippendale, G.M. and Wolf, L. (1981). *The natural distribution of Eucalyptus in Australia.* Australian National Parks and Wildlife Service, Canberra. [1.2.1, 2.2.4, 3.5.1, 7.3, 10.3, 17.1]

Chippendale, G.M. and Wolf, L.J. (1984). EUCALIST: computerised data retrieval system for *Eucalyptus (Myrtaceae). Australian Forest Research* **14**, 147–152. [2.2.4, 4.3.3]

Christensen, P.S. (1971). The purple crowned lorikeet and eucalypt pollination. *Australian Forestry* **35**, 263–270. [19.3.2]

Churchill, D.M. and Christensen, P. (1970). Observation on pollen harvesting by brush tongued lorikeets. *Australian Journal of Zoology* **18**, 427–437. [19.3.2]

Clarke, B. (1975). Establishment of eucalypt plantations, Coffs Harbour, N.S.W. *Science and Technology*, Melbourne **12**(6), 10–13. [22.1, 22.4.1]

Claudot, J. (1963). Contribution of Morocco on the genetic improvement of introduced eucalypts in the Mediterranean Basin. In *Proceedings of the world consultation on forest genetics and tree improvement*, Stockholm Vol. 1, FAO/FORGEN 63-4/9, pp. 1–15. FAO, Rome. [6.5]

Coates, D.J. and Sokolowski, R.E. (1989). Geographic patterns of genetic diversity in karri (*Eucalyptus diversicolor* F. Muell.). *Australian Journal of Botany* **37**, 145–156. [Table 3.1]

Cook, I.O. and Ladiges, P.Y. (1991). Morphological variation within *Eucalyptus nitens s. lat.* and recognition of a new species, *E. denticulata. Australian Systematic Botany* **4**, 375–390. [12.3]

Cossalter, C. (1977). Location and ecological data of some provenances of *Eucalyptus deglupta* Blume in the Celebes and Ceram Islands - characteristics of the natural stands. *Forest Genetic Resources Information*, FAO Rome **6**, 16–23. [Fig. 7.2]

Costin, A.B. (1954). *A study of the ecosystems of the Monaro region of New South Wales with special reference to soil erosion.* Government Printer, Sydney. [17.1]

Cotterill, P.P. (1984). A plan for breeding radiata pine. *Silvae Genetica* **33**, 84–90. [18.1, 18.4, 20.6.1]

Cotterill, P.P. (1986a). Genetic gains expected from alternative breeding strategies including simple low cost options. *Silvae Genetica* **35**, 212–223. [18.2.3]

Cotterill, P.P. (1986b). Breeding strategy: don't underestimate simplicity. In *Proceedings IUFRO conference on breeding theory, progeny testing and seed orchards*, Williamsburg, Virginia, October 1986, pp. 8–23. North Carolina State University, Raleigh. [18.2.2, 20.2, 20.4]

Cotterill, P.P. (1990). Numbers of families and progeny required for provenance testing. *Silvae Genetica* **39**, 82–83. [3.3]

Cotterill, P.P. and Dean, C.A. (1990). *Successful tree breeding with index selection.* CSIRO, Melbourne. [18.3, 18.3.2, 20.2]

Cotterill, P.P. and James, J.W. (1984). Number of offspring and plot sizes required for progeny testing. *Silvae Genetica* **33**, 203–209. [21.2.2]

Cotterill, P.P., Moran, G.F., and Grigg, B.R. (1985). Early growth of 36 species of eucalypts near Mount Gambier, South Australia. *Australian Forest Research* **15**, 409–416. [17.2]

Cotterill, P., Dean, C., Cameron, J., and Brindbergs, M. (1989). Nucleus breeding: a new strategy for rapid improvement under clonal forestry. In *Breeding tropical trees*, Proceedings IUFRO Conference, Pattaya, Thailand, November 1988, (ed. G.L. Gibson, A.R. Griffin, A.C. Matheson) pp. 39–51. Oxford Forestry Institute. [18.2, 18.2.2, 18.2.3]

Cremer, K.W. (1965a). Effects of fire on seed shed from *Eucalyptus regnans. Australian Forestry* **29**, 252–262. [19.4.4, 19.5]

Cremer, K.W. (1972). Morphology and development of the primary and accessory buds of *Eucalyptus regnans. Australian Journal of Botany* **20**, 175–195. [2.3.2, 22.3.1]

Cremer, K.W. (1977). Distance of seed dispersal in eucalypts estimated from seed weights. *Australian Forest Research* **7**, 225–228. [19.6.1]

Cremer, K.W., Cromer, R.N., and Florence, R.G. (1984). Stand establishment. In *Eucalypts for wood production* (2nd edn) (ed. E.W. Hillis and A.G. Brown) pp. 81–135. [5.6.3]

Cunningham, T.M. (1960). The natural regeneration of *Eucalyptus regnans. Bulletin* 1. School of Forestry, University of Melbourne. [19.4.2, 19.4.4]

Darrow, W.K. (1983a). Provenance studies of *Eucalyptus nitens* (Deane et Maiden) in South Africa. *Silvicultura*, São Paulo **31**, 433–434. [12.4.1, 12.4.4]

Darrow, W.K. (1983b). Provenance-type trials of *Eucalyptus camaldulensis* and *E. tereticornis* in south Africa and South West Africa: eight-year results. *South African Forestry Journal* **124**, 13–22. [6.4.1]

Darrow, W.K. (1983c). Provenance-type trials of *Eucalyptus grandis* and *E. saligna* in South Africa: eight year results. *South African Forestry Journal* **126**, 30–38. [3.4, 11.4.1]

Darrow, W.K. (1984). Provenance studies of frost-resistant eucalypts in South Africa. *South African Forestry Journal* **129**, 31–39. [9.2, 9.3.1, 17.2, 17.4.2]

Darrow, W.K. and Roeder, K.R. (1983). Provenance trials of *Eucalyptus urophylla* and *E. alba* in South Africa: seven-year results. *South African Forestry Journal* **125**, 20–28. [16.4]

Davidson, J. (1972). Variation, association and inheritance of morphological and wood characters in an improvement programme for *Eucalyptus deglupta* Blume. Unpublished Ph.D. thesis, Australian National University, Canberra. [20.1.6, 20.1.7, 22.1; Table 3.1; Fig. 7.2]

Davidson, J. (1973a). Conservation of the gene resources of tropical *Eucalyptus deglupta*. *Tropical Forestry Research Note* SR 12. Department of Forests, Papua New Guinea. [3.5.1]

Davidson, J. (1973b). A description of *Eucalyptus deglupta*. *Tropical Forestry Research Note* SR 7. Department of Forests, Papua New Guinea. [7.1]

Davidson, J. (1973c). Decayed wood in living trees of *Eucalyptus deglupta* Blume. *Proceedings IUFRO Section 5 Meeting, 1973, South Africa* Vol. 2, pp. 208–224. [7.2]

Davidson, J. (1974a). Reproduction of *Eucalyptus deglupta* by cuttings. *New Zealand Journal of Forestry Science* **4**, 191–203. [22.1, 22.3.1, 22.4.3]

Davidson, J. (1974b). Grafting *Eucalyptus deglupta*. *New Zealand Journal of Forestry Science* **4**, 204–210. [22.1, 22.3.4, 22.4.1]

Davidson, J. (1975). The use of principal components and rotated factor analyses to study variation in the wood of *Eucalyptus deglupta*. *Wood Science and Technology* **9**, 275–291. [20.1.6, 20.6.2]

Davidson, J. (1978a). Exploration, collection, evaluation, conservation and utilization of the gene resources of tropical *Eucalyptus deglupta* Bl. In *Documents FAO third world consultation on forest tree breeding* Vol. 1, pp. 75–102. CSIRO, Canberra. [3.5.1, 7.3]

Davidson, J. (1978b). Breeding *Eucalyptus deglupta* — a case study. In *Documents FAO third world consultation on forest tree breeding* Vol. 2, pp. 1187–1203. CSIRO, Canberra. [7.1, 18.2.2, 19.6.1]

Davidson, J. (1981). Status and methods of land classification in Papua New Guinea. In *Assessing tropical forest lands: their suitability for sustainable uses* (ed. R.A. Carpenter), pp. 247–276. Tycooly International, Dublin. [4.2.2]

Davidson, J. (1983a). Provenance trials of *Eucalyptus deglupta* in Papua New Guinea. *Silvicultura*, São Paulo **31**, 434–440. [7.4, 7.6; Tables 3.1, 7.1]

Davidson, J. (1983b). Progress in breeding *Eucalyptus deglupta*. *Silvicultura*, São Paulo **31**, 529–533. [18.2.2, 19.2.2, 19.3.2, 19.6.1]

Davidson, J. (1983c). Forestry in Papua New Guinea: A case study of the Gogol woodchip project near Madang. In *Forest and watershed development and conservation in Asia and Pacific* (ed. L.S. Hamilton) pp. 19–138. Westview, Boulder, Colarado [4.2.1, 4.2.2]

Davidson, J. (1983d). Conservation of tree genetic resources: role of protected areas in Amazonia. *The Environmentalist* 3 (Supplement No. 5), 62–74. [In *Ecological structures and problems of Amazonia*, IUCN Commission on Ecology Papers No. 5]. [3.5.2]

Davidson, J. (1985a). Setting aside the idea that eucalypts are always bad. *Working Paper* 19. UNDP-FAO Project BGD/79/017, Bangladesh. [1.1, 2.3, 23.1]

Davidson, J. (1985b). Plantation forestry in relation to tropical moist forests in south east Asia. *The Environmentalist* 5 (Supplement No. 10), 101–110. [In *The future of tropical rain forests in south east Asia* (ed. J. Davidson, Tho Yow Pong, M. Bijleveld), IUCN Commission on Ecology Papers No. 10.] [1.1]

Davidson, J., (1986a). Underplanting, interplanting and buffer planting for forest plantations. *Working Paper* No. 18, UNDP/FAO Project BGD/79/017. FAO, Rome. [1.1]

Davidson, J. (1986b). *Eucalyptus deglupta* Bl. In *Databook on endangered tree and shrub species and provenances*. FAO Forestry Paper No. 77, 262–270. [7.6]

Davidson, J. (1987a). *Conservation planning in Indonesia's transmigration programme: case studies from Kalimantan*. IUCN, Gland, Switzerland. [4.2.2]

Davidson, J. (1987b). Pre-feasibility study of the potential for development of a transmigration model in south east Irian Jaya to produce wood chips and wood pulp. Unpublished report to the Indonesian Department of Transmigration. Lavalin International Inc., Montreal. [4.2.2]

Davidson, J. (1989). Ethiopia: Eucalyptus tree improvement and breeding. *Field Document* No. 1, UNDP/FAO Project ETH/88/010, Ethiopia. FAO, Rome. [1.2.2, 10.2; Table 1.1]

Davidson, J. and Das, S. (eds). (1985). Eucalypts in Bangladesh. *Silviculture Division Bulletin* No. 6. Bangladesh Forest Research Institute, Chittagong. [3.1, 5.2, 6.4.1, 15.2, 15.3]

Davidson, J. and Khan, M.N. (1986). Computerized matching of tree species and sites. *Working Paper* No. 1 FAO/UNDP Project BGD/79/017, Bangladesh. FAO, Rome. [4.5]

Davidson, N.J. and Reid, J.B. (1985). Frost as a factor influencing the growth and distribution of subalpine eucalypts. *Australian Journal of Botany* **33**, 657–667. [14.3.2]

Davidson, J., Das, S., Khan, S.A., Latif, M.A., and Zashimuddin, M. (1985). Tree volume tables for small eucalypt roundwood in Bangladesh. *Silvicultural Research Division Bulletin* No. 4. Bangladesh Forest Research Institute, Chittagong. [5.6.5]

Davis, G.L. (1969). Floral morphology and the development of gametophytes in *Eucalyptus stellulata* Sieb. *Australian Journal of Botany* **17**, 177–190. [19.6.1]

Davis, J. and O'Connor, A.E. (1988). Evaluation of *E. regnans* seed harvesting techniques and impact of future seed crops. *Gippsland Technical Report* No. 88–18. APM Forests, Melbourne. [21.2.6, 21.4.4]

Dawkins, H.C. (1975). *Statforms: pro-formas for the guidance of statistical calculations*. Edward Arnold, London. [5.6.2]

Day, B.B. and Austin, L. (1939). A three-dimensional lattice design for studies in forest genetics. *Journal of Agricultural Research* **59**, 101–120. [5.1]

DeBell, D.S., Whitesett, C.D., and Schubert, T.H. (1985). Mixed plantations of Eucalyptus and leguminous trees enhance biomass production. *Research Paper* PSW-175. USDA Forest Service, Berkeley. [1.1]

de Little, D.W. and Tibbits, W.N. (1988). Breeding *Eucalyptus nitens* (shining gum) for pulpwood production in cool-temperate climates. In *Proceedings of the International Forestry Conference for the Australian Bicentenary 1988* Vol. 5, pp. 13. Australian Forest Development Institute, Albury. [12.5]

Delwaulle, J.C. (1983). Creation et multiplication végétative par bouturage d'Eucalyptus hybrides en République Populaire du Congo. *Silvicultura*, São Paulo **32**, 775–778. [1.6.1, 22.1, 22.4.3]

Delwaulle, J.C. (1985a). Plantations clonales d'Eucalyptus hybrides au Congo. *Bois et Forêts des Tropiques* **208**, 37–42. [1.5, 1.6.1, 18.2.2, 22.1]

Delwaulle, J.C. (1985b). Production papetiere de peuplements clonaux d'Eucalyptus au Congo. *Revue Bois et Forêts des Tropiques* **208**, 43–48. [22.2.4]

Delwaulle, J.C. and Martin, B. (1984). Stratégie d'amélioration des Eucalyptus au Congo. In *Proceedings IUFRO colloque international sur les eucalyptus resistants au froid*, Bordeaux, September 1983, pp. 491–508. AFOCEL, Paris. [22.1, 22.2.4]

Denison, N.P. and Quaile, D.R. (1987). The applied clonal eucalypt programme in Mondi Forests. *South African Forestry Journal* **142**, 60–67. [1.6.1, 22.1]

Deo, A.D., Nair, J.M., Nair, T.J.C., Kumar, D., and Ramaswamy, V. (1986). Provenance trials of *Eucalyptus grandis* in Kerala. In *Proceedings of the national seminar on eucalypts in Indian forestry* (ed. Sharma, J.K., Nair, C.T.S., Kedharnath, S., and Kondas, S.). Kerala Forest Research Institute, Peechi. [11.4.3]

Destremau, D.X., Jolly, H., and Korba, M. (1973). Les enseignements de l'essai comparative de provenances d' *Eucalyptus camaldulensis* de Sidi Slimane. *Annales de la Recherche Forestière au Maroc* 13, 121–154. [6.5]

Destremau, D.X., Marien, J.N., and Boulay, M. (1983). Selection et multiplication vegetative d'Eucalyptus resistante au froid. *Silvicultura*, São Paulo **32**, 781–785. [22.1]

Diabate, K. (1978). Premiers resultats de l'essai de provenances d'*Eucalyptus urophylla* mis en place en 1974 a San Pedro (Côte d'Ivoire). In *Documents FAO third world consultation on forest tree breeding* Vol. 1, pp. 317–325. CSIRO, Canberra. [16.4]

Diabate, K. (1983). L'amélioration de l'*Eucalyptus deglupta* en Côte d'Ivoire. *Silvicultura*, São Paulo 31, 534–536. [7.4]

Dianese, J.C., Haridasan, M., and Moraes, T.S. de A. (1984a). Tolerance to 'Mal do Rio Doce', a major disease of *Eucalyptus* in Brazil. *Tropical Pest Management* **30**, 247–252. [11.4.3]

Dianese, J.C., Moraes, T.S. de A., and Silva, R.A. (1984b). Response of *Eucalyptus* to field infection by *Puccinia psidii*. *Plant Disease* **68**, 314–316. [11.4.3]

Diels, B.Z. (1922). Die Myrtaceen von Papusien. In *Botanische Jahrbucher* (ed. A. Engler) **57**, 423. [7.3]

Dillner, B., Ljunger, A., Herud, O.A., and Thune-Larsen, E. (1971). The breeding of *Eucalyptus globulus* on the basis of wood density, chemical composition and growth rate. Paper to *Symposium on the Production and Industrial Utilization of Eucalyptus, Lisbon*. Supplement 5 to Vol. 23 of *Timber Bulletin for Europe*, pp. 120–151. United Nations Economic Commission for Europe, Geneva. [18.2.2]

Doran, J.C. (1974). Genetic variation in wood density of *Eucalyptus regnans*. In *Proceedings fourth meeting (Melbourne), Research Working Group No. 1, Forest Genetics*, pp. 99–101. Australian Forestry Council, Canberra. [20.1.6]

Doran, J.C. (1975). Occurrence of kino veins in two provenance trials of *Eucalyptus regnans*. *Australian Forest Research* **7**, 21–27. [14.3.1]

Doran, J.C. (1978). Collections of *Eucalyptus grandis* Hill ex Maiden seed in north Queensland, Australia in 1977. *Forest Genetic Resources Information*, FAO Rome 8, 22–26. [11.5]

Doran, J.C. (1983). Recent collections of *Eucalyptus* in Australia and Indonesia and availability of seed for provenance research. *Silvicultura*, São Paulo **31**, 443–449. [15.3, 15.4]

Doran, J.C. (1984). Collection of *Eucalyptus camaldulensis* seed from 178 open-pollinated families in the Petford region of North Queensland. Unpublished report, Tree Seed Centre, CSIRO Division of Forest Research, Canberra. [3.1]

Doran, J.C. (1986). Seed, nursery practice and establishment. In *Multipurpose Australian trees and shrubs: lesser-known species for fuelwood and agroforestry* (ed. J.W. Turnbull), pp. 59–80. Australian Centre for International Agricultural Research, Canberra. [5.4, 5.6.3]

Doran, J.C. and Boland, D.J. (1978). Collections of *Eucalyptus camaldulensis* Dehnh. and *E. tereticornis* Sm. seed from northern Australia in 1977 for *ex situ* conservation-selection stands. *Forest Genetic Resources Information*, FAO Rome **8**, 12–21. [3.5.3, 6.5]

Doran, J.C. and Brophy, J.J. (1991). Tropical red gums — a source of 1,8-cineole-rich *Eucalyptus* oil. *New Forests*, **4**, 157–178. [6.4.3, 18.3]

Doran, J.C. and Burgess, I.P. (1992). Variation in floral bud morphology in the intergrading zones of *Eucalyptus camaldulensis* and *E. tereticornis* in northern Queensland. *Commonwealth Forestry Review* (in press). [15.1, Fig. 15.3]

Doran, J.C. and Kleinig, D.A. (1979). *Eucalyptus urophylla* seed collections 1979. *Division of Forest Research Seeds Section Report 2*, pp. 29. CSIRO, Canberra. [16.4, 16.5]

Doran, J.C., Boland, D.J., Turnbull, J.W., and Gunn, B.W. (1983). *Handbook on seeds of dry-zone acacias. A guide for collecting, extracting, cleaning, storing the seed and for treatment to promote germination of dry-zone acacias*. FAO, Rome. [3.5.3]

Doran, J.C., Turnbull, J.W., and Kariuki, E.M. (1987). Effects of storage conditions on germination of five tropical tree species. In *Proceedings IUFRO symposium on forest seed problems, Harare, August 1987* (ed. S.K. Kamra and R.D. Ayling), pp. 84–94. Swedish University of Agricultural Sciences, Umea. [7.6.1]

Drake, D.W. (1974). Fungal and insect attack of seeds in unopened *Eucalyptus* capsules. *Search* **5**, 444. [19.5]

Drake, D.W. (1980). Contrasting success of natural hybridization in two *Eucalyptus* species pairs. *Australian Journal of Botany* **28**, 167–191. [19.6.3]

Drake, D.W. (1981a). Reproductive success of two *Eucalyptus* hybrid populations. I. Generalised seed output model and comparison of fruit parameters. *Australian Journal of Botany* **29**, 25–35. [19.6.3]

Drake, D.W. (1981b). Reproductive success of two *Eucalyptus* hybrid populations. II. Comparison of predispersal seed parameters. *Australian Journal of Botany* **29**, 37–48. [19.6.3]

Drinnan, A.N. and Ladiges, P.Y. (1991). Floral development in the 'Symphyomyrtus Group' of the eucalypts (*Eucalyptus*: *Myrtaceae*). *Australian Systematic Botany* **4**, 553–562. [19.1]

Dvorak, W.S., Franklin, E.C., and Meskimen, G. (1981). Breeding strategy for *E. robusta* in southern Florida. *Proceedings 16th Southern Forest Tree Improvement Conference*, pp. 116–122. US Forest Service, Athens, Georgia. [18.2.2, 21.2.6]

Dyson, W.G. (1975). *Eucalyptus regnans* at South Kinangop, Kenya. *Research report*, East Africa Forest Research Organization, Muguga, Kenya. [14.2]

Edwards, M.U. (1963). The use of exotic trees in increasing production with particular reference to north-western Europe. In *Proceedings of the world consultation on forest genetics and tree improvment*, August 1963, Stockholm, FAO/Forgen 63–4/2. FAO, Rome. [3.2]

Eldridge, K.G. (1963a). Effect of honey bees on seed production of *Eucalyptus regnans*. *Forestry Technical Papers*, Forests Commission Victoria **13**, 6–13. [19.3.2]

Eldridge, K.G. (1963b). *Eucalyptus sieberiana* × *regnans* — an artificial hybrid. *Australian Forestry* **27**, 128–129. [20.5.2]

Eldridge, K.G. (1964). Breeding *Eucalyptus regnans* — a five year progress report. *Australian Forestry* **28**, 34–50. [14.3.1]

Eldridge, K.G. (1968). Physiological studies of altitudinal variation in *Eucalyptus regnans*. *Proceedings of the Ecological Society of Australia* **3**, 70–76. [4.3.4, 14.3.1, 14.3.2; Table 3.1]

Eldridge, K.G. (1969). Altitudinal variation in *Eucalyptus regnans*. Unpublished PhD thesis, Australian National University. [2.2.2]

Eldridge, K.G. (1970). Breeding system of *Eucalyptus regnans*. *Proceedings IUFRO Section 22 Meeting* Vol. 1. Varparanta, Finland. [19.2.2, 19.6.1]

Eldridge, K.G. (1971). Genetically improved eucalypt seed for Australian pulpwood forests. *Appita* **25**, 105–109. [14.3.1, 18.2.2, 21.2.6]

Eldridge, K.G. (1972). Genetic variation in growth of *Eucalyptus regnans*. *Bulletin* 46. Forestry and Timber Bureau, Canberra. [14.3.1, 18.3.7; Tables 3.1, 18.3]

Eldridge, K.G. (1975a). An annotated bibliography of genetic variation in *Eucalyptus camaldulensis*. *Tropical Forestry Papers* 8. Commonwealth Forestry Institute, Oxford. [6.3, 6.4.1]

Eldridge, K.G. (1975b). Eucalyptus species. In *Seed orchards* (ed. R. Faulkner), *Forestry Commission Bulletin* No. 54, pp. 134–139. Her Majesty's Stationery Office, London. [18.2.2]

Eldridge, K.G. (1976). Breeding systems, variation and genetic improvement of tropical eucalypts. In *Tropical trees, variation, breeding and conservation* (ed. J. Burley and B.T. Styles), pp. 101–108. Academic Press, London. [19.2.2, 20.4.1]

Eldridge, K.G. (1978). Genetic improvement of eucalypts. *Silvae Genetica* **27**, 205–209. [5.8]

Eldridge, K.G. (1981). Eucalyptus seed supply for Minas Gerais, Brazil. UNDP, FAO, FO:DP/BRA/78/003 *Field Document* 1. IBDF, Brasilia. [21.5; Fig. 21.1]

Eldridge, K.G. (1982). Genetic improvements from a radiata pine seed orchard. *New Zealand Journal of Forestry Science* **12**, 404–411. [18.5]

Eldridge, K.G. (1984). Breeding trees for fuelwood. In *Genetics: new frontiers, Proceedings XV International Congress of Genetics* (New Delhi 1983), (ed. V.L. Chopra, B.C. Hoshi, R.P. Sharma, and H.C. Bansal), Vol. 4. pp. 339–349. Oxford & IBH, Bombay. [18.1, 22.5; Table 18.2]

Eldridge, K.G. (1990). Conservation of forest genetic resources with particular reference to Eucalyptus species. *Commonwealth Forestry Review* **69**, 45–53. [14.4, 23.2]

Eldridge, K.G. and Cromer, R.N. (1987). Adaptation and physiology of eucalypts in relation to genetic improvement. *Proceedings CIEF Simposio sobre silvicultura y mejoramiento genetico de especies forestales* Vol. 1, pp. 85–100. Centro de Investigaciones y Experiencias Forestales, Buenos Aires. [2.3]

Eldridge, K.G. and Griffin, A.R. (1983). Selfing effects in *Eucalyptus regnans*. *Silvae Genetica* **32**, 216–221. [19.6.1, 20.4.1]

El-Lakany, M.H. (1983). Provenance-site interaction in irrigated plantations of *Eucalyptus camaldulensis*. *Silvicultura*, São Paulo **31**, 450–451. [6.4.1]

El-Lakany, M.H., El-Osta, M.L., and Badran, A.O. (1980). Evaluation of newly introduced *Eucalyptus camaldulensis* provenances in Egypt. *Alexandria Journal of Agricultural Research* **28**, 309–319. [6.4.2]

El-Lakany, M.H., Badran, O.A., Omran, T.A., and Youness, M.I. (1984). Intra specific variability in some leaf characteristics of *Eucalyptus camaldulensis* Dehnh. as related to seed source. *Alexandria Science Exchange* **5**, 305–321. [6.4.1]

Elliott, H.J. and de Little, D.W. (1984). *Insect pests of trees and timber in Tasmania*. Forestry Commission, Hobart. [12.2]

Elmer, A.D.E. (1914). Myrtaceae from Mt. Urdaneta. *Leaflets of Philippine Botany* **7**, 2351. [7.3]

Elmer, A.D.E. (1915). *Myrtaceae*. *Leaflets of Philippine Botany* **8**, 2776. [7.3]

Emery, B.M. and Ledig, F.T. (1987). Provenance variation in *Eucalyptus camaldulensis* Dehnh. in California. *Silvae Genetica* **36**, 172–180. [6.4.1]

Evans, G.C. (1976). A sack of uncut diamonds: the study of ecosystems and the future resources of mankind. *Journal of Ecology* **64**, 1–39. [18.2]

Evans, J. (1982). *Plantation forestry in the tropics*. Clarendon Press, Oxford. [5.6.3, 11.2]

Falconer, D.S. (1983). *Introduction to quantitative genetics* (2nd edn). Longman, London. [18.3, 18.3.2, 18.3.3, 18.3.6, 18.3.7]

FAO (Food and Agriculture Organization of the United Nations) (1969). *Report of the first session of the FAO panel of experts on forest gene resources*, Rome, October 1968, FO: FGR/1/Rep. FAO, Rome. [3.3]

FAO (1971–1982). *FAO/UNESCO soil map of the world 1 : 5 000 000* [10 volumes]. UNESCO, Paris. [4.2.3, 4.4]

FAO (1975). Proposals for a global programme for improved use of forest genetic resources. *Forest Genetic Resources Information* No. 4, pp. 2–54. FAO, Rome. [3.5.3]

FAO (1977). *Report of the fourth session of the FAO panel of experts on forest gene resources*, Canberra, 9–11 March 1977. FAO, Rome. [3.5.3]

FAO (1984). *Agroclimatological data for Africa* [2 volumes]. *Plant Production and Protection Series* No. 22. FAO, Rome. [4.3.2]

FAO (1985). *Agroclimatological data for Latin America and the Caribbean*. *Plant Production and Protection Series* No. 24. FAO, Rome. [4.3.2]

FAO (1987). *Agriclimatological data for Asia* [2 volumes]. *Plant Production and Protection Series* No. 25. FAO, Rome. [4.2.3]

FAO (1988a). The *eucalypt dilemma*. Food and Agriculture Organization of the United Nations, Rome. [2.3]

FAO (1988b). *Case studies of farm forestry and wasteland development in Gujurat, India* (based on the work of Dr Shobhita Jain). FAO Regional Office for Asia and the Pacific, Bangkok. [2.3.1]

Faulkner, R. (ed.) (1975). *Seed orchards*. *Forestry Commission Bulletin* No. 55. Her Majesty's Stationery Office, London. [1.4.5, 21.4]

Ferguson, K.V.M. (1948). Some statistics of timber yields from virgin stands of white mountain ash (*Eucalyptus regnans*). *Australian Forestry* **12**, 13–15. [14.1]

Fielding, J.M. (1948). Eucalyptus cuttings. *Wildlife* **10**, 153. [22.1, 22.3.1]

Figueroa, J.C. and Whitmore, J.L. (1980). Three species of *Eucalyptus* tested in Puerto Rico: five years after outplanting. *Southern Journal of Applied Forestry* **4**, 169–174. [7.4]

Florence, R.G. (1964). A comparative study of flowering and seed production in six blackbutt (*Eucalyptus pilularis* Sm.) forest stands. *Australian Forestry* **28**, 23–33. [19.2.2, 19.6.1]

Florence, R.G. (1969). Variation in blackbutt. *Australian Forestry* **33**, 83–93. [Table 3.1]

Florence, R.G. (1981). The biology of the eucalypt forest. In *The biology of Australian plants* (ed. J.S. Pate and A.J. McComb) pp. 147–180. University of Western Australia Press, Nedlands. [1.2.1]

Florence, R.G. (1986). Cultural problems of eucalypts as exotics. *Commonwealth Forestry Review* **65**, 141–163. [1.1, 2.3, 2.3.2]

Fonseca, S.M. da, Ferreira, M., and Kageyama, P.Y. (1979). Resultados e perspectivas do programade melhoramento genético com eucalyptos, conduzidos pelo IPEF, na região sul do Brasil. *Boletim informativo* No. 7. IPEF, Piracicaba. [17.2, 17.4.1, 17.5]

Ford, H.A., Paton, D.C., and Forde, N. (1979). Birds as pollinators of Australian plants. *New Zealand Journal of Botany* **17**, 509–519. [17.3.2]

Fortech (1989). Fuelwood use and supply in Australia. *Forestry Technical Services Pty Ltd Consultancy Study Report* No. 28. Department of Primary Industries and Energy, Canberra. [1.2.2]

Franclet, A. (1956). Premiers travaux d'amélioration génétique des Eucalyptus. *Annales de la Recherche Forestiére au Maroc* **4**, 63–85. [6.4, 6.5, 22.1, 22.3.1, 22.3.2, 22.3.3; Tables 22.1, 22.2]

Franclet, A. (1963). Amélioration des reboisements d'Eucalyptus par multiplication végétative. In *Proceedings FAO first world consultation on forest genetics and tree improvement*, FAO/FORGEN-63, Vol. 2, 5/5, pp. 1–8. [22.1, 22.3.1]

Franclet, A. (1969). Propagation by cuttings, resinous trees and other difficult species. In *Documents FAO second world consultation on forest tree breeding* Vol. 2, pp. 1304–1313. Washington. [22.1]

Franclet, A. (1970). *Techniques de bouturage des Eucalyptus camaldulensis. Note technique* 12. FAO/IRT, Institut de Reboisement de Tunisie. [22.1, 22.3.4]

Franclet, A. (1984). Perspective pour la micropropagation des eucalyptus resistants au froid. In *Proceedings IUFRO colloque international sur les eucalyptus resistants au froid*, Bordeaux, September 1983, pp. 625–635. AFOCEL, Paris. [22.3.5, 22.4.1, 22.4.2]

Frankel, O.H. (1972). Genetic conservation — a parable of the scientists' social responsibility. *Search* **3**, 193–210. [3.5]

Frankel, O.H. (1974). Genetic conservation: our evolutionary responsibility. *Genetics* **78**, 53–65. [3.5]

Frankel, O.H. (1978). Philosophy and strategy of genetic conservation in plants. In *Documents FAO third world consultation on forest tree breeding* Vol. 1, pp. 6–11. CSIRO, Canberra. [3.5]

Franklin, E.C. (1971). Estimates of frequency of natural selfing and of inbreeding coefficients in loblolly pine. *Silvae Genetica* **20**, 194–195. [19.6.1]

Franklin, E.C. (1986). Estimation of genetic parameters through four generations of selection in *Eucalyptus grandis*. In *Proceedings IUFRO Conference on breeding theory, progeny testing and seed orchards*, Williamsburg, Virginia, USA October 1986, pp. 200–209. North Carolina State University, Raleigh. [18.2.2, 18.2.3]

Franklin, E.C. and Meskimen, G.F. (1973). Genetic improvement of *Eucalyptus robusta* Sm. in southern Florida. In *Tropical provenance research and international co-operation* (ed. J. Burley and D.G. Nikles), pp. 421–424. Commonwealth Forestry Institute, Oxford. [1.4.5]

Franklin, E.C. and Meskimen, G.F. (1974). Progeny testing in a *Eucalyptus* breeding programme. In *Proceedings IUFRO Working Party S2.04.3*, pp. 2–4. Knoxville, Tennessee. [1.4.5]

Franklin, E.C. and Meskimen, G. (1984). Choice of species and provenances in cold summer rainfall climates. In *Proceedings IUFRO Colloque international sur les eucalyptus resistants au froid*, Bordeaux, September 1983, pp. 341–357. AFOCEL, Paris. [1.4.5, 3.4, 11.4.2, 11.5, 18.1, 18.2, 18.2.3, 21.2.6]

Free, J.B. (1970). *Insect pollination in crops*. Academic Press, London. [21.3.5]

Freese, F. (1980). *Elementary statistical methods for foresters*. Agriculture Handbook 317. USDA Forest Service, Washington. [5.6.2]

Freitas, M., Silva, A.P., Diniz, A.S., Kageyama, P.Y., and Ferreira, M. (1983). The research programme with *Eucalyptus grandis* Hill ex Maiden at Champion Papel e Celulose S.A. *Silvicultura*, São Paulo **31**, 537–539. [21.1]

Fripp, Y.J. (1982). Allozyme variation and mating system in two populations of *Eucalyptus kitsoniana* (Luehm.) Maiden. *Australian Forest Research* **13**, 1–10. [Table 19.2]

Galvão, A.P.M. and Couto, H.T.Z. (1984). Man-made industrial forests Brazil: prospects and constraints. In *Strategies and designs for afforestation, reforestation and tree planting* (ed. K.F. Wiersum), pp. 129–136. Pudoc, Wageningen. [1.2.2, 21.5; Table 1.1]

Gardiner, C.A. and Crawford, D.F. (1987). 1987 seed collections of *Eucalyptus globulus* subsp. *globulus* for tree improvement purposes. Unpublished report Australian Tree Seed Centre CSIRO, Canberra. [10.3, 10.5]

Gardiner, C.A. and Crawford, D.F. (1988). 1988 seed collections of *Eucalyptus globulus* subsp. *globulus* Labill. for tree improvement purposes. Unpublished report Australian Tree Seed Centre CSIRO, Canberra. [10.3, 10.5]

Garner, R.J. (1968). *The grafter's handbook*. Faber, London. [21.3.3]

Geary, T.F., Meskimen, G.F., and Franklin, E.C. (1983). Growing eucalypts in Florida for industrial wood production. *General Technical Report* SE-23. USDA Forest Service, Asheville. [21.2.6]

Gentilli, J. (1961). The resistance of eucalypts to low temperatures in the growing season. In *Report and documents FAO second world Eucalyptus conference*, São Paulo, Vol. 1, pp. 495–511. FAO, Rome. [4.3.4]

George, W. (1981). Wallace and his line. In *Wallace's line and plate tectonics* (ed. T.C. Whitmore) pp. 3–8. Clarendon Press, Oxford. [Fig. 1.1]

Ghosh, R.C., Gupta, B.N., and Shedha, M.D. (1977). Provenance trial of *Eucalyptus camaldulensis* Dehn. *Indian Forester* **103**, 441–449. [6.4.2]

Giordano, E. (1956). Osservazioni preliminari sulla moltiplicazione vegetative negli eucalitti. *Publicazioni del Centro di Sperimentazione Agricola e Forestale* **1**, 131–152. [22.1, 22.3.1]

Giordano, E. (1961). Mist propagation of eucalypt cuttings. In *Report and documents FAO second world Eucalyptus conference*, São Paulo, Vol.1, pp.375–379. FAO, Rome. [6.4]

Goes, E. (1977). *Os Eucaliptos — ecologia, cultura, producoes e rentabilidade*. Portucel, Lisbon. [10.2]

Goes, E. (1984). *Arvores monumentais de Portugal*. Portucel, Lisbon. [10.2]

Goes, E. (1985). *Os eucaliptos. Identificação e monographia de 121 espécies em Portugal*. Serviços Florestais e Aquicolas, Lisboa. [1.2.1, 2.1.1]

Golfari, L. (1985). Distribución regional y condiciones ecológicas de los eucaliptos cultivados en la Argentina. Problemas inherentes. *CIEF Publicacion Técnica* No. 1–1985. Centro de Investigaciones y Experiencias Forestales, Buenos Aires. [Table 1.1]

Golfari, L., Caser, R.L., and Moura, V.P.G. (1978). Zoneamento ecologica esquematico para reflorestamento no Brasil. PRODEPEF, PNUD/FAO/IBDF/BRA-45, *Serie Tecnica* **11**, 66 pp. [4.3]

Goor, A.Y. and Barney, C.W. (1976). *Forest tree planting in arid zones* (2nd edn). Ronald Press, New York. [6.2]

Gore, P.L., Potts, B.M., Volker, P.W., and Megalos, J. (1990). Unilateral cross-incompatibility in *Eucalyptus*: the case of hybridisation between *E. globulus* and *E. nitens*. *Australian Journal of Botany* **38**, 383–394. [20.5.2]

Graca, M.E.C. (1987). Avaliação do florescimento e do potencial de produção de sementes de *Eucalyptus dunnii* Maid. no Brasil. *Boletim de Pesquisa Florestal*, Curitiba **14**, 1–12. [19.2.1]

Gray, A. (1854). United States Exploring Expedition during the years 1838–42 under the command of Charles Wilkes, U.S.N. *Phanerogamia* **15**, 554–555. [7.3]

Green, J.W. (1969). Taxonomic problems associated with continuous variation in *Eucalyptus pauciflora* (snowgum) (*Myrtaceae*). *Taxon* **18**, 269–276. [Table 3.1]

Green, J.W. (1971). Variation in *Eucalyptus obliqua* l'Hérit. *New Phytologist* **70**, 897–909. [3.2, 13.1, 13.3, 13.4.1; Table 3.1; Fig. 13.4]

Griffin, A.R. (1980). Floral phenology of a stand of mountain ash (*Eucalyptus regnans* F. Muell.) in Gippsland, Victoria. *Australian Journal of Botany* **28**, 393–404. [19.2.2, 19.2.3, 19.3.2, 19.6.1]

Griffin, A.R. (1982). Pollination ecology of eucalypts — a framework for study. In *Pollination '82* (ed. E.G. Williams, R.B. Knox, J.H. Gilbert, and P. Bernhardt) pp. 42–56. School of Botany, University of Melbourne. [19.3.2]

Griffin, A.R. (1983). A provenance testing strategy for *Eucalyptus regnans* in Australia. *Silvicultura* São Paulo, **31**, 452–453. [13.6, 14.3.1]

Griffin, A.R. (1985). The relevance of eucalypt breeding system research. *Australian Forest Grower* **8**, 22–23. [3.4]

Griffin, A.R. (1989a). Strategies for the genetic improvement of yield in *Eucalyptus*. In *Biomass production by fast-growing trees* (ed. J.S. Pereira and J.J. Landsberg), pp. 247–265. Kluwer, Dordrecht. [14.3.1]

Griffin, A.R. (1989b). Sexual reproduction and tree improvement strategy — with particular reference to *Eucalyptus*. In *Breeding tropical trees*, Proceedings IUFRO Conference, Pattaya, Thailand, November 1988 (ed. G.L. Gibson, A.R. Griffin, and A.C. Matheson) pp.52–67. Oxford Forestry Institute. [21.4.2]

Griffin, A.R. (1990). Effects of inbreeding on growth of forest trees and implications for management of seed supplies for plantation programs. In *Reproductive ecology of tropical forest plants* (ed. K.S. Bawa and M. Hadley), pp.363–382. UNESCO, Paris, and Parthenon, Carnforth. [19.6.1]

Griffin, A.R. and Cotterill, P.P. (1988). Genetic variation in growth of outcrossed, selfed and open-pollinated progenies of *Eucalyptus regnans* and some implications for breeding strategy. *Silvae Genetica* **37**, 124–131. [18.3.2, 19.3.2, 19.6.1, 19.6.2, 20.4.1; Table 18.4]

Griffin, A.R. and Eldridge, K.G. (1980). A field trial of progeny of trees intermediate between *Eucalyptus regnans* and *E. obliqua*. *Australian Forest Research* **10**, 1–8. [13.5, 19.6.3]

Griffin, A.R. and Hand, F.C. (1979). Post-anthesis development of flowers of *Eucalyptus regnans* F. Muell. and the timing of artificial pollination. *Australian Forest Research* **9**, 9–15. [19.3.1, 19.3.4]

Griffin, A.R. and Ohmart, C.P. (1986). Pollination ecology of *Eucalyptus*. *Biennial Report 1983–1985*, pp.26–28. CSIRO Division of Forest Research, Canberra. [19.3.2; Table 19.1]

Griffin, A.R., Williams, E.R., and Johnson, K.W. (1982a). Early height growth and frost hardiness of *Eucalyptus regnans* provenances in twelve field trials in south- east Australia. *Australian Forest Research* **12**, 263–279. [3.1, 14.1, 14.3.1, 14.3.2, 21.2.6; Table 14.1; Fig. 14.4]

Griffin, A.R., Ching, K.K., Johnson, K.W., Hand, F.C., and Burgess, I.P. (1982b). Processing *Eucalyptus* pollen for use in controlled pollination. *Silvae Genetica* **31**, 198–203. [5.3.3]

Griffin, A.R., Moran, G.F., and Fripp, Y.T. (1987). Preferential outcrossing in *Eucalyptus regnans* F. Muell. *Australian Journal of Botany* **35**, 465–475. [19.3.4, 19.3.5]

Griffin, A.R., Burgess, I.P. and Wolf, L. (1988). Patterns of natural and manipulated hybridisation in the genus *Eucalyptus* L'Hérit. — a review. *Australian Journal of Botany* **36**, 41–66. [2.1.2, 6.3, 6.5, 11.1, 17.1, 19.6.3, 20.5.2]

Griffin, A.R., Whiteman, P., Rudge, T., Burgess, I.P., and Moncur, M.W. (1992). Enhancement of flower initiation in two species of *Eucalyptus* with paclobutrazol. *Canadian Journal of Forest Research* (in press). [20.5.2, 21.4.3]

Grose, R.J. (1957). A study of some factors associated with the natural regeneration of alpine ash *Eucalyptus delegatensis* R.T. Baker syn. *E. gigantea* Hook f. *Bulletin* **4**. Forests Commission Victoria, Melbourne. [19.5]

Grose, R.J. (1960). The silviculture of *Eucalyptus delegatensis*. Unpublished Ph.D. thesis, University of Melbourne. [8.4.2, 8.4.3]

Grose, R.J. (1961). Some silvicultural characteristics and notes on the silviculture of alpine ash *Eucalyptus delegatensis* R.T. Baker. In *Report and documents FAO second world Eucalyptus conference*, São Paulo, Vol. 1, pp. 704–714. [19.4.4]

Grunwald, C. and Karschon, R. (1977a). Further observations on the effect of seed origin on coppice regeneration in *Eucalyptus camaldulensis* Dehn. *Leaflet* **61**. Division of Forestry, Agricultural Research Organisation, Israel. [6.4.4]

Grunwald, C. and Karschon, R. (1977b). Leaf characters and early growth of two provenances of *Eucalyptus camaldulensis* Dehn. from NW Victoria. *Leaflet* **62.**

Division of Forestry, Agricultural Research Organisation, Israel. [6.4.1]

Grunwald, C. and Karschon, R. (1977c). Clinal variation in leaf dry-matter content in *Eucalyptus camaldulensis* Dehn. and its relation to frost resistance. *La-Yaaran* **27**(1–4), 48–51. [6.4.6]

Grunwald, C. and Karschon, R. (1982). Leaf xylem water saturation deficits as related to seed origin of *Eucalyptus camaldulensis* Dehn. *Australian Forest Research* **12**, 175–181. [6.3, 6.4.5]

Grunwald, C. and Karschon, R. (1983). Variation of *Eucalyptus camaldulensis* from north Australia grown in Israel. *Silvae Genetica* **32**, 165–173. [6.5; Table 3.1]

Grunwald, C. and Karschon, R. (1984). Juvenile-mature correlations and seed tree selection in two provenances of *Eucalyptus camaldulensis* Dehn. from NW Victoria. *La-Yaaran* **34**(1–4), 58–60. [6.4.1]

Guinness Book of Records (1975). Twenty-second edition 1975. Guinness Superlatives, Enfield, Middlesex. [7.2, 14.2]

Guldager, P. (1975). *Ex situ* conservation stands in the tropics. In *The methodology of conservation of forest genetic resources*, FO. MISC/75/8, pp. 85–92. FAO, Rome. [3.5.3]

Gunn, B.V. and McDonald, M.W. (1991). *Eucalyptus urophylla* seed collections. *Forest Genetic Resources Information*, FAO Rome 19, 34–37. [16.3, 16.4; Fig. 16.1, 16.5]

Gurgel, J.T.A. and Cavalcanti, G.R.A. (1978). Behaviour and survival of *Eucalyptus* species in the state of São Paulo, Brasil. In *Documents FAO third world consultation on forest tree breeding*, Vol.1, pp. 181–189. CSIRO, Canberra. [5.2, 16.2, 16.3, 16.5, 19.2.2, 21.1]

Gutierrez, G. de la (1977). [see Lama Gutierrez, G. de la (1977)].

Hackett, C. (1988). Matching plants and land : development of a general broadscale system from a crop project for Papua New Guinea. *Natural Resources Series* No. 11. CSIRO Division of Water and Land Resources, Canberra. [4.5]

Hagedorn, S.F. (1983). *Eucalyptus nitens*. In *Report for 1982–1983, Wattle Research Institute* pp. 20–21. University of Natal, Pietermaritzburg. [12.4.2, 12.4.3]

Hagedorn, S.F. (1984). A history of seed transactions by the Directorate of Forestry and the origin of plantations of *Eucalyptus elata, E. fastigata, E. fraxinoides, E. macarthurii* and *E. nitens* in the Piet Retief area of the southeastern Transvaal. In *Report for 1983–1984, Wattle Research Institute* pp. 138–143. University of Natal, Pietermaritzburg. [9.4]

Hall, N., Johnston, R.D., and Marryatt, R. (1963). The natural occurrence of the eucalypts. *Leaflet* 65. Forestry and Timber Bureau, Canberra. [4.3]

Hall, N. *et al.* (1972). *The use of trees and shrubs in the dry country of Australia.* Australian Government Publishing Service, Canberra. [6.2]

Hall, N., Johnston, R.D., and Chippendale, G.M. (1975). *Forest trees of Australia* (3rd edn). Australian Government Publishing Service, Canberra. [2.2.4]

Hall, N., Wainwright, R.W., and Wolf, L.J. (1981). Summary of meteorological data in Australia. *Divisional Report* 6. Division of Forest Research, CSIRO, Canberra. [1.2.1, 4.3.2, 6.4.6, 9.1, 12.4.1]

Hallam, P.M. and Reid, J.B. (1989). Seasonal and genetic variation in frost hardiness in *Eucalyptus delegatensis*. *Canadian Journal of Forest Research* **19**, 480–488. [8.4.2]

Hallam, P.M., Reid, J.B., and Beadle, C.L. (1989). Frost hardiness of commercial *Eucalyptus* species at different altitudes. *Canadian Journal of Forest Research* **19**, 1235–1239. [8.4.2]

Hallauer, A.R. and Miranda, J.B.F. (1981). *Quantitative genetics in maize breeding*. Iowa State University Press, Ames. [18.3]

Hamilton, D.I. and Cowley, N.B. (1987). Southern Tablelands eucalypts of New South Wales. *Proceedings 1987 Conference of the Institute of Foresters of Australia* pp. 233–247. [8.2]

Hand, F.C., Raymond, C.A., and Eldridge, K.G. (1989). Establishment report 1985 *Eucalyptus regnans* field trials in northern Tasmania. *DFFP User Series* No. 3. CSIRO Division of Forestry and Forest Products, Canberra. [14.4]

Hanson,W.D. (1963). Heritability. In *Statistical genetics and plant breeding* (ed. W.D. Hanson and H.F. Robinson), National Academy of Sciences Publication No. 982, pp. 125–140. National Research Council, Washington. [18.3.2, 20.4]

Harker, A.P., Sandels, A., and Burley, J. (1982). Calorific values for wood and bark and a bibliography for fuelwood. *Report* G162. Tropical Products Institute, London. [20.1.2]

Harries, E.D., McKenzie, B.C., and Nightingale, J.L. (1973). A quantitative assessment of defective wood in plantation grown *Eucalyptus deglupta* Bl. In *Proceedings IUFRO Section 5 Meeting*, October 1973, South Africa, Vol.2, pp. 428–437. [7.2]

Harris, A.C. (1966). Karri blossom and seeding: pollination to stimulate seed production. In *Report on the operations of the Forests Department Western Australia for the year ended 30th June, 1966.* Forests Department, Perth. [19.3.2]

Hartney, V.J. (1980). Vegetative propagation of the eucalypts. *Australian Forest Research* **10**, 191–211. [3.5.3, 22.3.5, 22.4.1]

Hartney, V.J. (1982). Tissue culture of eucalypts. *Proceedings International Plant Propagators' Society* **32**, 98–109. [6.4.7]

Hartney, V.J. and Barker, P.K. (1983). The vegetative propagation of *Eucalyptus* by tissue culture. *Silvicultura*, São Paulo **32**, 791–793. [3.5.3]

Hartney, V.J. and Kabay, E.D. (1984). From tissue culture to forest trees. *International Plant Propagators' Society*

Combined Proceedings for 1984 **34**, 93–99. [1.6.2, 6.4.7]

Hartney, V.J. and Svensson, J.G.P. (1992). The role of micropropagation for Australian tree species. In *Rapid propagation of fast-growing woody species* (ed. F.W.G. Baker) pp. 7–28. C.A.B. International, Wallingford. [22.3.5, 22.4.2]

Harwood, C.E. (1980). Frost resistance of subalpine *Eucalyptus* species. I. Experiments using a radiation frost room. *Australian Journal of Botany* **28**, 587–599. [Table 3.1]

Harwood, C.E. (1981). Frost resistance of subalpine *Eucalyptus* species. II. Experiments using the resistance index method of damage assessment. *Australian Journal of Botany* **29**, 209–218. [4.3.4, 5.1; Table 3.1]

Harwood, C.E. (1984). Studies on eucalyptus frost resistance in a sub-alpine frost hollow. *Proceedings IUFRO colloque international sur les eucalyptus resistants au froid*, Bordeaux, September 1983, pp. 126–144. [Table 3.1]

Haslett, A.N. (1988). A guide to handling and grade-sawing plantation-grown eucalypts. *FRI Bulletin* No. 142. Forest Research Institute, Rotorua. [20.1.4]

Hayward, W.J. (1987). Volume and taper of *Eucalyptus regnans* grown in the Central North Island of New Zealand. *New Zealand Journal of Forestry Science* **17**, 109–120. [14.2]

Heather, W.A. (1955). The kamarere forests of New Britain. *Empire Forestry Review* **34**, 255–278. [7.1]

Herdt, R.W. (1988). *Equity considerations in setting priorities for third world rice biotechnology research.* Agricultural Sciences Section, Rockefeller Foundation, New York. [18.2].

Heslop-Harrison, J. (1964). Forty years of genecology. *Advances in Ecological Research* **2**, 159–247. [3.1]

Heth, D., Fanger-Vexler, L., and Reuveni, O. (1986). Mass production of cuttings of *Eucalyptus camaldulensis* Dehn. *Commonwealth Forestry Review* **65**, 215–225. [22.4.1]

Heybroek, H.M. (1984). Clones in forestry and in nature. *Arboricultural Journal* **8**, 275–286. [1.6.1]

Higgins, H.G. (1984). Pulp and paper. In *Eucalypts for wood production* (2nd edn.) (ed. W.E. Hillis and A.G. Brown), pp. 290–316. CSIRO/Academic Press, Melbourne. [4.2.4, 9.2, 14.1, 20.1.1, 20.1.7]

Higgs, M.L. (1969). Genetic and environmental factors influencing commercially important wood properties of *Eucalyptus regnans* F. v. M. Unpublished Ph.D. thesis. Australian National University, Canberra. [20.1.6, 20.1.7]

Hillis, W.E. (1972). Properties of eucalypt woods of importance to the pulp and paper industry. *Appita* **26**, 113–122. [20.1.1]

Hillis, W.E. (1984). Wood quality and utilization. In *Eucalypts for wood production* (2nd edn.) (ed. W.E. Hillis and A.G. Brown) pp. 259–289. CSIRO/Academic, Melbourne. [4.2.4, 20.1.1, 20.1.4]

Hillis, W.E. and Brown, A.G. (ed.) (1984). *Eucalypts for wood production* (2nd edn). CSIRO/Academic Press, Melbourne. [1.2.1, 1.2.2, 15.2, 16.2, 17.2]

Hodges, C.S., Geary, T.F., and Cordell, C.E. (1979). The occurrence of *Diaporthe cubensis* on *Eucalyptus* in Florida, Hawaii, and Puerto Rico. *Plant Disease Reporter* **63**(3): 216–226. [7.2]

Hodgson, L.M. (1975). Some aspects of reproductive behaviour in *Eucalyptus grandis* (Hill) Maiden. D.Sc. thesis, University of Pretoria, Republic of South Africa. [19.2.3]

Hodgson, L.M. (1976a). Some aspects of flowering and reproductive behaviour in *Eucalyptus grandis* (Hill) Maiden at J.D.M. Keet Forest Research Station. 1. Flowering, controlled pollination methods, pollination and receptivity. *South African Forestry Journal* **97**, 18–28. [19.2.3, 19.3.1, 19.3.2, 19.3.4, 19.6.1]

Hodgson, L.M. (1976b). Some aspects of flowering and reproductive behaviour in *Eucalyptus grandis* (Hill) Maiden at J.D.M. Keet Forest Research Station. 2. The fruit, seed, seedlings, self-fertility, selfing and inbreeding effects. *South African Forestry Journal* **98**, 32–43. [19.6.1]

Hodgson, L.M. (1977a). Methods of seed orchard management for seed production and ease of reaping in *Eucalyptus grandis*. *South African Forestry Journal* **100**, 38–42. [21.3.5, 21.4.3]

Hodgson, L.M. (1977b). *In situ* grafting in *Eucalyptus grandis* seed orchards. *South African Forestry Journal* **102**, 58–60. [21.3.3]

Hopper, S.D. (1980). Pollen and nectar feeding by purple-crowned lorikeets on *Eucalyptus occidentalis*. *Emu* **80**, 239–240. [19.3.2]

Hopper, S.D. and Moran, G.F. (1981). Bird pollination and the mating system of *E. stoatei*. *Australian Journal of Botany* **29**, 625–638. [19.3.2, 19.6.1; Table 19.2]

Hunt, R. and Zobel, B. (1978). Frost hardy *Eucalyptus* grow well in the southeast. *Southern Journal of Applied Forestry* **2**, 6–10. [17.2, 17.4.2]

Hutchinson, M.F. (1984). A summary of some surface and contouring programs for noisy data. CSIRO Division of Mathematics and Statistics, Consulting Report 84–6. CSIRO, Canberra. [4.2.3, 4.3.3]

Ireland, J.C. and Griffin, A.R. (1984). Observations on pollination ecology of *Eucalyptus muellerana* Howitt in East Gippsland. *Victorian Naturalist* **101**, 207– 211. [19.3.2]

IUCN (International Union of Conservation for Nature and Natural Resources) (1980). *World conservation strategy*. IUCN, Morges, Switzerland. [3.5, 3.5.2, 3.5.4; Fig. 3.2]

Jackson, J.K. and Ojo, G.O.A. (1973). Provenance trials of *Eucalyptus camaldulensis* in the Savanna region of Nigeria. *Savanna Forestry Research Station Research Paper* 14. Federal Department of Forest Research, Ibadan. [5.8, 6.4.5]

Jackson, N. and Turner, H.N. (1972). Optimal structure for a co-operative nucleus breeding system. *Proceedings of the Australian Society for Animal Production* **9**, 55–64. [18.2.3]

Jacobs, M.R. (1955). *Growth habits of the eucalypts.* Forestry and Timber Bureau, Canberra. [1.2.1, 2.2.2, 2.3.2, 20.1.4, 22.3]

Jacobs, M.R. (undated). *The forests we care for.* Associated Pulp and Paper Mills Limited, Melbourne. [This undated leaflet of 34 pages was published about 1975]. [14.2]

Jacobs, M.R. (1981). *Eucalypts for planting. FAO Forestry Series* No. 11. FAO, Rome. [1.2.1, 1.2.2, 1.2.3, 4.3, 5.4, 5.6.3, 6.2, 6.4.7, 10.2, 11.2, 14.2, 15.2, 16.1, 17.2, 19.1]

Jahnke, R., Carr, D.J., and Carr, S.G.M. (1983). Lignotuber development and growth parameters in *Eucalyptus camaldulensis* (Dehnh.): effects of phosphorus and nitrogen levels. *Australian Journal of Botany* **31**, 283–292. [6.4.3]

Jahromi, S.T. (1982). Variation in cold resistance and growth in *Eucalyptus viminalis. Southern Journal of Applied Forestry* **6**, 221–225. [17.4.2]

Janaki Ammal, E.K. and Khosla, S.N. (1969). Breaking the barrier to polyploidy in the genus *Eucalyptus. Proceedings of the Indian Academy of Science B* **70**(8), 248–249. [2.2.2]

Janzen, D.H. (1977). Why don't ants visit flowers? *Biotropica* **9**, 252. [19.3.2]

Jean, B. and Delwaulle, J.C. (1981). Le reboisement en République Populaire du Congo. *Revue Forestière Française* **33**, 157–169. [22.1]

Jennings, J.N. and Mabbutt, J.A. (1977). Physiographic outlines and regions. In *Australia — a geography* (ed. D.N. Jeans), pp. 80–96. Sydney University Press. [2.2.1; Fig. 2.1]

John, J.A. (1987). *Cyclic designs.* Chapman and Hall, London. [5.6.2]

Johnson, L.A.S. (1976). Problems of species and genera in *Eucalyptus* (*Myrtaceae*). *Plant Systematics and Evolution* **125**, 155–167. [1.2.1, 2.1.2, 2.2.2, 19.6.3]

Johnson, L.A.S. and Briggs, B.G. (1983). *Myrtaceae.* In *Flowering plants of Australia* (ed. B.D. Morely and H.R. Toelken) pp. 175–180. Rigby, Adelaide. [1.2.1, 2.1.2; Table 2.1]

Jones, N., and Burley, J. (1973). Seed certification, provenance, nomenclature and genetic history in forestry. *Silvae Genetica* **22**, 53–58. [3.1]

Kapoor, M.L. and Kedharnath, S. (1974). Induced tetraploidy in sandalwood and eucalyptus. *Indian Journal of Genetics and Plant Breeding* **34A**, 408–412. [2.2.2]

Kapoor, M.L. and Sharma, V.K. (1984). Hybrids between *Eucalyptus citriodora* Hook. and *E. torelliana* F.v.Muell. in India. *Silvae Genetica* **33**, 42–46. [2.2.2]

Kapoor, M.L. and Sharma, V.K. (1985). Experimentally synthesized allotetraploids in *Eucalyptus. Silvae Genetica* **34**, 19–22. [2.2.2]

Kardell, L., Steen, E., and Fabiao, A. (1986). Eucalyptus in Portugal — a threat or a promise? *Ambio* **15**, 6–13. [1.1, 1.2.2]

Karschon, R. (1960). Lime tolerance and seed provenance in eucalypts. *Document, 4th Session, Working Party on Eucalyptus*, FAO Joint Sub-commission on Mediterranean Forestry Problems, Lisbon FAO/SCM/EU60-9B. [6.4]

Karschon, R. (1966). Inter-provenance variation in *Eucalyptus camaldulensis* Dehn. (preliminary results). *Document, Mediterranean Forestry Research Committee*, FAO Joint Sub-Commission on Mediterranean Forestry Problems, Madrid. [6.4.3]

Karschon, R. (1967). Ecotypic variation in *Eucalyptus camaldulensis* Dehn. *National and University Institute of Agriculture, Ilanot; Contributions on Eucalypts in Israel* III, 35–53. [6.3, 6.4.3]

Karschon, R. (1968). Variation in frost resistance in ecotypes of *Eucalyptus camaldulensis* Dehn. *Document, Mediterranean Forestry Research Committee*, FAO Joint Sub-Commission on Mediterranean Forestry Problems, 3rd Session, Rome 1968. FO: SCM/FR/68-8B. [6.4.6]

Karschon, R. (1974). The relation of seed origin to growth of *Eucalyptus camaldulensis* Dehn. in Israel. *Israel Journal of Agricultural Research* **23**(3–4), 159–173. [6.3, 6.5]

Karschon, R. and Zohar, Y. (1975). Effects of flooding and of irrigation water salinity on *Eucalyptus camaldulensis* Dehn. from three seed sources. *Leaflet* 54. Division of Forestry, Agricultural Research Organisation, Israel. [6.4.7]

Kaul, R.N. (ed.) (1970). *Afforestation in arid zones.* Dr W. Junk, The Hague. [6.2]

Keast, A. (ed.) (1981). *Ecological biogeography of Australia.* Volumes 1–3. Dr W. Junk, The Hague. [2.2]

Kedharnath, S. and Vakshasya, R.K. (1978). Estimates of components of variance, heritability and correlations among some growth parameters in *Eucalyptus tereticornis.* In *Documents FAO third world consultation on forest tree breeding* Vol. 2, pp. 667–676, CSIRO, Canberra. [Table 18.4]

Khoshoo, T.N. (1986). Conservation and production forestry. In *Eucalypts in India — past, present and future* (ed. J.K. Sharma, C.T.S. Nair, S. Kedharnath, and S. Kondas), pp. 7–13. Kerala Forest Research Institute, Peechi. [1.1]

King, J.N. and Wilcox M.D. (1988). Family tests as a basis for the genetic improvement of *Eucalyptus nitens* in New Zealand. *New Zealand Journal of Forestry Science* **18**, 253–266. [12.4.1, 18.3.3]

King, J.N. and Young, G.D. (1992). Provenance variation in growth rates and internal wood checking for *Eucalyptus delegatensis* grown in New Zealand. *New Zealand Journal of Forestry Science* (in press). [8.4.1]

Kirkpatrick, J.B. (1973). Geographic variation in *Eucalyptus globulus* Labill. Unpublished Ph.D. thesis, University of Melbourne. [Table 3.1]

Kirkpatrick, J.B. (1974). The numerical intraspecific taxonomy of *Eucalyptus globulus* Labill. *Botanical Journal of the Linnean Society* **69**, 89–104. [10.3, Fig. 10.4]

Kirkpatrick, J.B. (1975). Geographical variation in *Eucalyptus globulus*. *Bulletin* 47. Forestry and Timber Bureau, Canberra. [10.3; Table 3.1]

Knockaert, C. and Ait Hamou, B. (1979). Tarifs de cubage pour l'*Eucalyptus camaldulensis* — origine Lake Albacutya. *Annales de la Recherche Forestière au Maroc* **19**, 3–14. [Table 6.2]

Krenke, N.P. (1940). The theory of the cycle of senescence and rejuvenation of plants and its practical application. *Seljhozgiz* 1940, 1–135, from *Plant Breeding Abstracts* **15**, 181. [22.3]

Krug, C.A. and Alves, A.S. (1949). Eucalyptus improvement. Part 1. *Journal of Heredity* **40**, 133–149. [19.3.2, 19.6.1]

Labillardière, J.J.H. de (1799). *Relation du voyage a la recherche de la Pérouse* Vol. 1, pp. 151–153 and plate 13. Jansen, Paris. [10.1]

Lacaze, J.F. (1970). Studies on the ecological adaptation of *Eucalyptus*. Operational report for 1970 on project No. 6. *Document, 4th Session Committee on the Coordination of Mediterranean Forestry Problems*, Ankara, No. FO: SCM/FR 70/2/10. FAO, Rome. [6.4, 6.4.1]

Lacaze, J.F. (1978). Etude de l'adaptation écologique des eucalyptus, étude de provenances *Eucalyptus camaldulensis*, project FAO no. 6. In *Documents FAO third world consultation on forest tree breeding* Vol. 1, pp. 393–409. CSIRO, Canberra. [6.4, 6.4.1]

Lacey, C.J. (1974). Rhizomes in tropical eucalypts and their role in recovery from fire damage. *Australian Journal of Botany* **22**, 29–38. [2.3.2]

Ladiges, P.Y. (1974a). Differentiation in some populations of *Eucalyptus viminalis* Labill. in relation to factors affecting seedling establishment. *Australian Journal of Botany* **22**, 471–487. [17.3]

Ladiges, P.Y. (1974b). Variation in drought tolerance in *Eucalyptus viminalis* Labill. *Australian Journal of Botany* **22**, 489–500. [17.3]

Ladiges, P.Y. and Ashton, D.H. (1974). Variation in some central Victorian populations of *Eucalyptus viminalis* Labill. *Australian Journal of Botany* **22**, 81–102. [17.1, 17.3; Table 3.1; Fig. 17.3]

Ladiges, P.Y. and Ashton, D.H. (1977). A comparison of some populations of *Eucalyptus viminalis* Labill. growing on calcareous and acid soils in Victoria, Australia. *Australian Journal of Ecology* **2**, 161–178. [17.1, 17.3]

Ladiges, P.Y. and Humphreys, C.J. (1983). A cladistic study of *Arillastrum, Angophora* and *Eucalyptus* (*Myrtaceae*). *Botanical Journal of the Linnean Society* **87**, 105–134. [17.6.3]

Ladiges, P.Y. and Kelso, A. (1977). The comparative effects of waterlogging on two populations of *Eucalyptus viminalis* Labill. and one population of *E. ovata* Labill. *Australian Journal of Botany* **25**, 159–169. [17.3]

Lakshmi Sita, G. (1979). Morphogenesis and plant regeneration from cotyledonary cultures of *Eucalyptus*. *Plant Science Letters* **14**, 63–68. [22.3.5]

Lama Gutierrez, G. de la (1977). *Atlas del Eucalypto*, Vol. 1–4. INIA, Sevilla. [1.2.1, 2.1.1]

Larsen, C.S. (1934). *Forest tree breeding*. Royal Veterinary and Agricultural College, Copenhagen, pp. 111–112. [18.1]

Larsen, C.S. (1956). *Genetics in silviculture* (trans. M.L. Anderson). Oliver and Boyd, London. [18.1]

Larsen, E. (1965). A study of the variability of *Eucalyptus maculata* Hook. and *Eucalyptus citriodora* Hook. *Leaflet* 95. Forestry and Timber Bureau, Canberra. [Table 3.1]

Larsen, E. (1967). Geographic variation in *Eucalyptus camaldulensis* Dehn. ANZAAS 39th Congress, Melbourne. [6.4.3, 6.4.6]

Laurie, M.V. (1974). *Tree planting practices in African savannas. Forestry Development Paper* 19. FAO Rome. [6.2]

Ledig, F.T. (1983). Eucalypt improvement for California: progress and plans. In *Eucalyptus in California* (ed. R.B. Staniford and F.T. Ledig), *General Technical Report* PSW-69, pp. 115–120. USDA Forest Service, Berkeley. [12.4.1]

Leeper, G.W. (1970). *The Australian environment* (4th edn.). CSIRO, Melbourne. [2.2]

Lembke, C.A. (1977). N.Z. Forest Products Limited turns to eucalypt. *Australian Forest Industries Journal* **43**(7), 26–31. [4.2.3]

L'Héritier de Brutelle, C.L. (1788). *Sertum anglicum*. Facsimile edn 1963. Hunt Botanical Library, Pittsburg. [1.2.1; Fig. 1.2]

Libby, W.J. (1973). Domestication strategies for forest trees. *Canadian Journal of Forest Research* **3**, 265–276. [1.3, 3.5, 3.5.3, 18.1; Fig. 18.2]

Libby, W.J. (1982). What is a safe number of clones per plantation? In *Resistance to pests and diseases*. Proceedings of the third international workshop on the genetics of host-parasite interactions in forestry, Wageningen, September 1980 (ed. H.M. Heybroek, B.R. Stephan, and K. von Weissenberg) pp. 342–360. Pudoc, Wageningen. [1.6.1]

Libby, W.J. (1985). Potential of clonal forestry. In *Clonal forestry: its impact on tree improvement and our future forests*, Symposium Proceedings of the nineteenth meeting of the Canadian Tree Improvement Association, Part 2, (ed. L. Zsuffa, R.M. Rauter, and C.W. Yeatman) pp. 1–11. Toronto, Ontario. [22.2]

Lindgren, D., Libby, W.J., and Bondesson, F.L. (1989). Deployment to plantations of numbers and proportions of clones with special emphasis on maximizing gain at a constant diversity. *Theoretical and Applied Genetics* **77**, 825–831. [1.6.1]

Linnard, W. (1969). Cultivation of eucalypts in the U.S.S.R. *Forestry Abstracts*, **30**, 199–207. [4.3.4, 17.2]

Little, T.M. and Hills, F.J. (1978). *Agricultural experimentation, design and analysis*. Wiley, New York. [5.6.2]

Logan, A.F., Phillips, F.H., and E.D. Harries (1987). The effect of growth conditions, bark and defective wood on sulphate pulps from plantation grown *Eucalyptus deglupta* Bl. *Appita* **40**, 176–180. [7.2]

Loneragan, O.W. (1979). Karri (*Eucalyptus diversicolor* F.Muell.) phenological studies in relation to reforestation. *Bulletin* 90. Forests Department of Western Australia, Perth. [12.1, 19.3.2]

Long, A.J. and Dykstra, G.F. (1983). Examination of *Eucalyptus* provenances in East Kalimantan. *Silvicultura*, São Paulo **31**, 458–462. [7.4]

Lovejoy, T.E. (1985). Rehabilitation of degraded tropical forests lands. *The Environmentalist* **5**, 1–8. [22.2.2]

Lundquist, J.E. (1983). Assessment methods for measuring intensity of leafspot disease of *Eucalyptus nitens*. In *Research results April 1982 to April 1983*, pp. 99–114. Plant Protection Research Institute, Pretoria. [12.4.4]

McComb, J.A. and Bennett, I.J. (1985). Micropropagation in *Eucalyptus*. In *Biotechnology in agriculture and forestry* Vol.1 Trees, (ed. Y.P.S. Bajaj), pp. 340–362. Springer-Verlag, Berlin. [1.6.2, 22.4.2]

McKimm, R.J. (1985a). Characteristics of wood of young fast-grown trees of *Eucalyptus nitens* Maiden with special reference to provenance variation. I Variations in growth, strain and density associated with provenance. *Australian Forest Research* **15**, 207–218. [12.4.2]

McKimm, R.J. (1985b). Characteristics of wood of young fast-growing trees of *Eucalyptus nitens* Maiden with special reference to provenance variation. II Strength, dimensional stability and preservation characteristics. *Australian Forest Research* **15**, 219–234. [12.4.2]

McKimm, R.J. and Ilic, Y. (1987). Characteristics of wood of young fast-growing trees of *Eucalyptus nitens* Maiden with special reference to provenance variation. III Anatomical and physical characteristics. *Australian Forest Research* **17**, 19–28. [12.4.2]

Machado da Fonseca, S. *et al.* (1979). [see Fonseca, S.M. da *et al.* (1979)].

Maiden, J.H. (1903–1933). *A critical revision of the genus Eucalyptus*. 8 volumes. Government Printer, Sydney. [16.3]

Malan, F.S. (1988). Genetic variation in some growth and wood properties among 18 full-sib families of South African grown *Eucalyptus grandis* : a preliminary investigation. *South African Forestry Journal* **146**, 38–43. [18.3.7, 20.1.6, 20.1.7; Table 18.4]

Mann, M.J. (1988). Yield curves for *Eucalyptus regnans* plantations. *APM Forests Technical Report* No. 88–23. APM Forests Pty Ltd, Morwell. [14.2]

Marcar, N.E. (1989). Salt tolerance of frost-resistant eucalypts. *New Forests* **3**(2), 141–149. [5.1]

Marien, J.N. (1979). La sélection juvénile des eucalyptus pour leur résistance au froid. *Annales de Recherches Sylvicoles AFOCEL* 1979, 224–253. [8.4.2; Fig. 8.5, 8.6]

Marien, J.N. (1984). Difficultés dans l'introduction de *E. delegatensis*. In *Proceedings IUFRO Colloque international sur les eucalyptus resistants au froid*, Bordeaux, September 1983, pp. 281–291. AFOCEL, Paris. [8.2]

Martin, B. (1971). Amélioration génétique des espèces exotiques introduites en République Populaire du Congo. *Bois et Forêts des Tropiques* **138**, 3–26. [1.3, 15.2, 22.2.4]

Martin, B. (1978). L'*Eucalyptus* comme exotique: recents progres dans le choix des espèces et des provenances. In *Documents FAO third world consultation on forest tree breeding* Vol. 1, pp. 155–172. CSIRO, Canberra. [15.3]

Martin, B. (1983). Les essais internationaux de provenances d'*Eucalyptus*: état actuel et developpement souhaitable. *Silvicultura*, São Paulo **29**, 21–27. [16.5]

Martin, B. (1987). Amélioration génétique des eucalypts tropicaux. Contribution majeure á la forestière clonale. Thèse de doctorat. Université de Paris-Sud, Centre d'Orsay. ENGREF, Nancy, 2 volumes. [1.5, 2.1.2, 19.6.3, 20.5.2, 21.4.3]

Martin, B. (1989). The benefits of hybridization. How do you breed for it? In *Breeding tropical trees*, Proceedings IUFRO Conference, Pattaya, Thailand, November 1988 (ed. G.L. Gibson, A.R. Griffin, A.C. Matheson) pp. 79–92. Oxford Forestry Institute. [1.5, 18.2.2, 18.2.3, 20.5.2]

Martin, B. and Cossalter, C. (1975–1976). Les eucalyptus des iles de la Sonde. *Bois et Forêts des Tropiques* **163**, 3–25; **164**, 3–14; **165**, 3–20; **166**, 3–22; **167**, 3–24; **168**, 3–17; **169**, 3–13. [16.1, 16.3, 16.4; Table 3.1; Fig. 16.1]

Martin, B. and Quillet, G. (1974). Bouturage des arbres forestiers au Congo: résultats des essais effectués à Pointe Noire de 1969 a 1973. *Bois et Forêts des Tropiques* **154**, 41–57. [22.1]

Matheson, A.C. (1990). Breeding strategies for MPTs. In *Tree improvement of multipurpose species* (ed. N. Glover and N. Adams), *Multipurpose Tree Species Series Network Technical Series*, Vol. 2, pp. 67–99. Winrock, Arlington. [18.2.2, 20.5.2]

Matheson, A.C. and Cotterill, P.P. (1990). Utility of genotype xenvironment interactions. *Forest Ecology and Management* **30**, 159–174. [18.3.5]

Matheson, A.C. and Lindgren, D. (1985). Gains from the clonal and clonal seed-orchard options compared for tree breeding programs. *Theoretical and Applied Genetics* **71**, 242–249. [22.2.3]

Matheson, A.C. and Mullin, L.J. (1987). Variation among neighbouring and distant provenances of *Eucalyptus grandis* and *E. tereticornis* in Zimbabwean field trials. *Australian Forest Research* **17**, 233–250. [11.4.1, 15.3]

Matheson, A.C. and Raymond, C.A. (1984). The impact of genotype x environment interactions on Australian *Pinus radiata* breeding programs. *Australian Forest Research* **14**, 11–25. [18.3.5]

Matheson, A.C., Turner, C.H., and Dean, G.H. (1986). Genetic variation in the pulp qualities of *Eucalyptus obliqua* L'Hérit. *Appita* **39**, 205–212. [13.4.1, 13.4.2; Table 18.3]

May, A.F. (1969). *Beekeeping in Southern Africa.* Haun, Pretoria. [19.3.2]

Mayo, O. (1987). *The theory of plant breeding* (2nd edn). Clarendon Press, Oxford. [18.3]

Mead, R. and Curnow, R.N. (1983). *Statistical methods in agriculture and experimental biology.* Chapman Hall, London. [5.6.2]

Mendonza, L.A. (1968). Morphological characters and cold resistance in some natural *Eucalyptus* hybrids. Unpublished M.Sc. thesis, Australian National University. [17.1]

Mendonza, L.A. (1974). Ensayos de procedencias de *Eucalyptus camaldulensis* Dehn. y *E. viminalis* Labill., en el norte de la provincia de Buenos Aires. *IDIA Suplemento Forestal* 1973–74, 53–60. [6.4.1, 17.4.1; Table 17.1]

Meskimen, G. (1965). A tool for emasculating eucalypts. *Silvae Genetica* **14**, 57–58. [20.5.1]

Meskimen, G. (1983). Realized gain from breeding *Eucalyptus grandis* in Florida. In *Eucalyptus in California* (ed. R.B. Standiford and F.T. Ledig) *General Technical Report* PSW-69, pp. 121–128. USDA Forest Service, Berkeley. [1.1, 3.4, 18.2.2, 18.2.3, 18.5, 21.1, 21.2.6]

Meskimen, G.F., Rockwood, D.L., and Reddy K.V. (1987). Development of *Eucalyptus* clones for a summer rainfall environment with periodic severe frosts. *New Forests* **1**, 197–205. [18.5, 21.2.6, 22.4.1]

Metro, A. (1955). *Eucalypts for planting. FAO Forestry and Forest Products Studies* II. FAO, Rome. [4.3]

Midgley, S.J., Eldridge, K.G., and Doran, J.C. (1989). Genetic resources of *Eucalyptus camaldulensis. Commonwealth Forestry Review* **68**, 295–308. [6.2, 6.3; Table 6.1]

Mohrdiek, O. (1983). Discussion: future possibilities for poplar breeding. *Canadian Journal of Forest Research* **13**, 465–471. [18.1, 22.2.2]

Molleda, J. (1984). Eucalyptus pulp production and trade. *Paper* **202**(8), 30–34. [1.2.2]

Moncur, M.W. and Boland, D.J. (1989). Floral morphology of *Eucalyptus melliodora* A. Cunn. ex Schau. and comparisons with other eucalypt species. *Australian Journal of Botany* **37**, 125–135. [19.1, 19.2.3]

Money, D.C. (1988). *Climate and environmental systems* [Chapter 8 Soil groups and their distribution]. Unwin Hyman, London. [4.4]

Moore, R.M. and Williams, J.D. (1976). A study of a subalpine- woodland-grassland boundary. *Australian Journal of Ecology* **1**, 145–155. [4.3.4]

Moran, G.F., Bell, J.C., and Griffin, A.R. (1989). Reduction in levels of inbreeding in a seed orchard of *Eucalyptus regnans* F. Muell. compared with natural populations. *Silvae Genetica* **38**, 32–36. [20.4.1; Table 19.2]

Moran, G.F. and Brown, A.H.D. (1980). Temporal heterogeneity of outcrossing rates in alpine ash (*Eucalyptus delegatensis* R.T. Bak.). *Theoretical and Applied Genetics* **57**, 101–105. [8.3, 19.4.2, 19.6.1; Table 19.2]

Moran, G.F., Forrester, R.I., and Rout, A.F. (1990). Growth of *Eucalyptus delegatensis* provenances in four field trials in south-eastern Australia. *New Zealand Journal of Forestry Science* **20**, 148–161. [8.4.1, 8.4.2, 8.5; Table 8.2; Fig. 8.4]

Moran, G.F. and Griffin, A.R. (1983). Recent advances in the study of eucalypt breeding systems. *Silvicultura*, São Paulo **31**, 552–555. [19.6.1]

Moran, G.F. and Hopper, S.D. (1983). Genetic diversity and the insular population structure of the rare granite rock species, *Eucalyptus caesia* Benth. *Australian Journal of Botany* **31**, 161–172. [Table 19.2]

Moran, G.F. and Hopper, S.D. (1987). Conservation of the genetic resources of rare and widespread eucalypts in remnant vegetation. In *Nature conservation* (ed. D.A. Saunders, G.W. Arnold, A.A. Burbidge, A.J.M. Hopkins) pp. 151–162. Surrey Beatty and CSIRO, Perth. [3.5.4]

Moreshet, S. (1981). Physiological activity, in a semi-arid environment, of *Eucalyptus camaldulensis* Dehn. from two provenances. *Australian Journal of Botany* **29**, 97–110. [6.4.5]

Moura, V.P.G. (1977). Altitudinal variation in *Eucalyptus urophylla* S.T. Blake. Unpublished M.Sc. thesis, University of Melbourne. [16.3, 16.4]

Moura, V.P.G. (1983). Resultados de pesquisa com várias procêdencias de *Eucalyptus urophylla* S.T. Blake, no centro-leste do Brasil. *Silvicultura*, São Paulo **31**, 474–480. [16.4]

Moura, V.P.G. (1986). Provenance variation of *Eucalyptus camaldulensis* Dehnh. in Brazil. Unpublished D.Phil. thesis, University of Oxford. [6.4.1, 6.4.2]

Moura, V.P.G., Barnes, R.D., and Birks, J.S. (1987). A comparison of three methods of assessing wood density in provenances of *Eucalyptus camaldulensis* Dehnh. and other *Eucalyptus* species in Brazil. *Australian Forest Research* **17**, 83–90. [6.4.2]

Mullin, L.J. and Barnes, R.D. (1977). Frost kill in a six-month-old provenance trial of *Eucalyptus grandis* Hill ex Maiden and *E. saligna* Sm. in Rhodesia. *South African Forestry Journal* **102**, 88–89. [11.4.2]

Mullin, L.J. and Pswarayi, I. (1990). Flowering periodicity in provenances of *Eucalyptus camaldulensis* in Zimbabwe. *Commonwealth Forestry Review* **69**, 69–77. [19.2.2]

Mullin, L.J., Barnes, R.D., and Barrett, R.L. (1981). The improvement of *Eucalyptus* in Zimbabwe. *South African Forestry Journal* **118**, 20–25. [21.2.6]

Namkoong, G. (1969). Non-optimality of local races. In *Proceedings Tenth Southern Conference on Forest Tree*

Breeding, pp. 194–197. Texas A & M University Press, College Station. [3.2]

Namkoong, G. (1979). *Introduction to quantitative genetics in forestry. USDA Forest Service Technical Bulletin* 1588, Washington. [3.5.4, 18.3]

Namkoong, G., Barnes, R.D., and Burley, J. (1980). A philosophy of breeding strategy for tropical forest trees. *Tropical Forestry Paper* 16. Commonwealth Forestry Institute, Oxford. [1.4.3, 21.2.6]

Namkoong, G., Kang, H.C., and Brouard, J.S. (1988). *Tree breeding : principles and strategies. Monographs on Theoretical and Applied Genetics* Vol. 11. Springer-Verlag, New York. [18.2.3, 18.3, 20.4; Fig. 18.1]

NAS (National Academy of Sciences) (1980). *Firewood crops : shrub and tree species for energy production.* NAS, Washington. [2.3.1]

Navarro de Andrade, E. (1961) [see Andrade, E.N. de (1961)].

Nel, P.M. (1965). The breeding of eucalypts in South Africa. *South African Forestry Journal* **54**, 17–21. [1.1, 20.1.3, 20.1.4]

Nelson, N.D. and Heather, W.A. (1972). Wood colour, basic density and decay resistance in heartwood of fast-grown *Eucalyptus grandis* Hill ex Maiden. *Holzforschung* **26**, 291–301. [20.1.6]

New South Wales Forestry Commission (1974). Preserving forest areas. *Forestry and Timber* **10**(1), 4–15. [3.5.2]

Ngulube, M.R. (1989). Provenance variation in *Eucalyptus urophylla* in Malawi. *Forest Ecology and Management* **26**, 265–273. [16.4]

Nicholls, J.W.P. and Griffin, A.R. (1978). Variation in wood characteristics in a field trial of *Eucalyptus obliqua*, *Eucalyptus regnans* and some intermediate forms. *Australian Forest Research* **8**, 93–102. [13.5]

Nicholls, J.W.P. and Matheson, A.C. (1980). Variation in wood characteristics in thinnings from a field trial of *Eucalyptus obliqua*. *Australian Forest Research* **10**, 239–247. [13.4.2]

Nicholls, J.W.P. and Pederick, L.A. (1979). Variation in some wood characteristics of *Eucalyptus nitens*. *Australian Forest Research* **9**, 309–321. [12.4.2]

Nicholson, J.E. (1971). A rapid method for estimating longitudinal growth stresses in logs. *Wood Science and Technology* **5**, 40–48. [20.1.3]

Nicholson, J.E. (1973). Growth stress differences in eucalypts. *Forest Science* **19**, 169–174. [20.1.4]

Nikles, D.G. (1987). A breeding strategy for genetic improvement of selected tropical eucalypts at Dongmen State Forest Farm, People's Republic of China. *China-Australia Afforestation Project at Dongmen State Forest Farm, Technical Communication* No. 30. Queensland Department of Forestry, Brisbane. [1.5, 6.5, 18.2.2]

Nikles, D.G. and Newton, R.S. (1983). Inventory and use of 'Provenance Resource stands' of *Pinus caribaea* Mor. var. *hondurensis* Barr. and Golf. in Queensland. *Silvicultura*, São Paulo **29**, 122–125. [1.3]

Nix, H.A. (1986). A biogeographic analysis of Australian Elapid snakes. In *Atlas of Australian Elapid snakes* (ed. R. Longmore), pp. 4–15. Bureau of Flora and Fauna, Canberra. [4.3.3]

Nixon, K.M. (1984). A test of fifteen *Eucalyptus* species and provenances for frost resistance in Natal. In *Proceedings IUFRO Colloque international sur les eucalyptus resistants au froid*, Bordeaux, September 1983, pp. 367–376. AFOCEL, Paris. [9.3.1, 12.4.3, 17.4.2]

Nixon, K.M. and Hagedorn, S.F. (1983). A five-year old provenance trial of *Eucalyptus nitens* in South Africa. *Silvicultura,* São Paulo **31**, 481–483. [12.4.1, 12.4.4]

O'Brien, T.P. (1989). The impact of severe frost. In *Mediterranean landscapes in Australia, mallee ecosystems and their management* (ed. J.C. Noble and R.A. Bradstock) pp. 181–188. CSIRO, Melbourne. [6.4.6]

Ohmart, C.P. (1984). Is insect defoliation in eucalypt forests greater than that in other temperate forests? *Australian Journal of Ecology* **9**, 413–418. [2.2.3]

Ohmart, C.P., Thomas, J.R., and Stewart, L.G. (1984). Differential defoliation by insects among provenances of *Eucalyptus delegatensis. Journal of the Australian Entomological Society* **23**, 105–111. [8.3]

Old, K.M. (ed.) (1979). *Phytophythora and forest management in Australia.* CSIRO, Melbourne. [3.5.1]

Old, K.M., Kile, G.A., and Ohmart, C.P. (ed.) (1981). *Eucalypt dieback in forests and woodlands.* CSIRO, Melbourne. [3.5.1]

Olesen, P.O. (1978). On cyclophysis and topophysis. *Silvae Genetica* **27**, 173–178. [22.3]

Orme, R.K. (1978). *Eucalyptus globulus* provenances. In *Documents FAO third world consultation on forest tree breeding* Vol. 1, pp. 207–221. CSIRO, Canberra. [10.1, 10.3, 10.4, 10.4.2]

Orme, R.K. (1983). Progress with *E. globulus* provenance research. *Silvicultura*, São Paulo **31**, 483–488. [10.4, 10.4.1]

Orme, R.K. (1988). Genetic improvement of *E. globulus*. In *Proceedings of the International Forestry Conference for the Australian Bicentenary 1988* Vol. 5, pp. 14. Australian Forest Development Institute, Albury. [10.4.1; Table 10.1]

Otegbeye, G.O. (1985). Provenance productivity in *Eucalyptus camaldulensis* Dehnh. and its implications to genetic improvement in the Savanna Region of Nigeria. *Silvae Genetica* **34**, 121–126. [5.8]

Otegbeye, G.O. (1990). Provenance variatlon in *Eucalyptus tereticornis* in a field trial within the Northern Guinea savannah zone of Nigeria. *Silvae Genetica* **39**, 103–107. [15.3]

Ou, Y.Q., Wei, J., and Wu, F.Q. (1989). The ecological and economic benefits of eucalypt plantations. Paper to Fourth Technical Exchange Seminar, October 1989, *China-Australia Afforestation Project at Dongmen State*

Forest Farm Technical Communication No. 40. Queensland Department of Forestry, Brisbane. [23.1]

Ovington, J.D. and Pryor, L.D. (1983). Temperate broad-leaved evergreen forests of Australia. In *Ecosystems of the world* Vol. 10 (ed. J.D. Ovington), pp. 73–101. Elsevier, Amsterdam. [1.2.1, 1.4.1]

Pacheco, I.A., Kageyama, P.Y., Wiendl, F.M., and Berti, E. (1986). Estudo da dispersão de pólen de *Eucalyptus saligna* Smith por abelhas *Apis mellifera* L. utilizando-se o radiofósforo ^{32}P. *IPEF* **34**, 47–52. [19.3.3, 21.4.2]

Passecker, F. (1947). Entwicklungsphasen und vegetative Vermehrung holziger Gewäschse. *Zentralblatt fuer das Gasamte Forstwesen* **70**, 270. [22.3]

Pasztor, Y.P. de C. (1962). Produção de sementes de eucaliptos. *Silvicultura em São Paulo* **1**, 181–187. [18.2.2, 21.2.6]

Pasztor, Y.P. de C. (1978). Growth and morphological characteristics of some Timor eucalypt provenances compared with the eucalypt known as "Brazilian alba". In *Documents FAO third world consultation on forest tree breeding* Vol. 1, pp. 453–462. CSIRO, Canberra. [16.4]

Paton, D.M. (1972). Frost resistance in *Eucalyptus*: a new method for assessment of frost injury in altitudinal populations of *E. viminalis*. *Australian Journal of Botany* **20**, 127–139. [17.4.2]

Paton, D.M. (1984). Vegetative propagation of adult *Eucalyptus*. In *Proceedings IUFRO colloque international sur les eucalyptus resistants au froid*, Bordeaux, September 1983, pp. 570–586. AFOCEL, Paris. [22.3.4]

Paton, D.M. and Willing, R.R. (1974). Inhibitor transport and ontogenetic age in *Eucalyptus grandis*. In *Plant growth substances* 1973, pp. 126–132. Hirokawa, Tokyo. [22.3.4]

Paton, D.M., Willing, R.R., Nicholls, W., and Pryor, L.D. (1970). Rooting of stem cuttings of *Eucalyptus*: a rooting inhibitor in adult tissue. *Australian Journal of Botany* **18**, 175–183. [22.3.1, 22.3.4, 22.3.5]

Paton, D.M., Willing, R.R., and Pryor, L.D, (1981). Root-shoot gradients in *Eucalyptus* ontogeny. *Annals of Botany* **47**, 835–838. [22.3.1, 22.3.4; Tables 22.3, 22.4, 22.5]

Pederick, L.A. (1960). Some seed collection data from adjacent trees of *Eucalyptus obliqua* l'Hérit. *Forestry Technical Paper* 5. Forests Commission of Victoria, Melbourne. [19.4.2, 19.5]

Pederick, L.A. (1974). Genetic variation in *Eucalyptus obliqua* with special reference to Otway messmate. *Research Report* No. 53. Forests Commission, Victoria, Melbourne. [13.4.1, 13.5; Fig. 13.6]

Pederick, L.A. (1976a). The genetic resources of the Victorian eucalypts. *Bulletin* 22, Forests Commission Victoria, Melbourne. [12.3, 13.4.1, 13.5, 19.2.1; Table 3.1]

Pederick, L.A. (1976b). Conservation of gene resources for the improvement of native species in Australia. *Australian Forestry* **39**, 113–120. [3.5.1, 3.5.2]

Pederick, L.A. (1977). Natural variation in shining gum. 1. Growth rate and foliage characteristics to age 3 years. *Research Branch Report* 88. Forests Commission Victoria, Melbourne. [12.3, 12.4.1; Table 12.1]

Pederick, L.A. (1979). Natural variation in shining gum (*Eucalyptus nitens*). *Australian Forest Research* **9**, 41–63. [12.1, 12.3, 12.4.1, 12.4.3, 12.5; Tables 3.1, 12.1; Fig. 12.2, 12.4, 12.5]

Pederick, L.A. (1983). IUFRO international provenance trials with *Eucalyptus grandis* and *E. tereticornis*. *Silvicultura*, São Paulo **31**, 415–418. [11.4.1, 11.4.2, 11.5, 15.3]

Pederick, L.A. (1985). Natural variation in shining gum, *Eucalyptus nitens*. II. Second progress report. *Research Branch Report* 277. State Forests and Lands Service, Melbourne. [12.4.1]

Pederick, L.A. (1986). *Eucalyptus delegatensis* provenance research in Victoria. In *Proceedings of the ninth meeting (Burnie) Research Working Group No. 1, Forest Genetics* pp. 107–109. Australian Forestry Council, Canberra. [8.5]

Pederick, L.A. (1988). Growth and form of *Eucalyptus regnans* families derived from plus trees and other seed sources to ten years of age. *Research Report* No. 332. Department of Conservation, Forests and Lands, Melbourne. [14.3.1]

Pederick, L.A. (1990). Genetic variation in the Victorian ash-type eucalypts: a discussion paper on its relevance to regeneration and reforestation practices. *Research Report* No. 344. Department of Conservation, Forests and Lands, Melbourne. [3.5.1, 14.1]

Pederick, L.A. and Lennox, F.G. (1979). Variation in polyphenolic constituents of *Eucalyptus nitens* Maiden. *Australian Journal of Botany* **27**, 217–226. [12.3]

Pegg, R.E. and Li, H.J. (1989). *Eucalyptus urophylla* at Dongmen. Paper to Fourth Technical Exchange Seminar, October 1989, *China-Australia Afforestation Project at Dongmen State Farm, Technical Communication* No. 40. Queensland Department of Forestry, Brisbane. [16.5]

Penfold, A.R. and Willis, J.L. (1961). *The Eucalypts*. World Crops Books. Inter-science, New York. [1.2.1, 1.2.2, 22.3]

Pereira, H. and Pereira, J.S. (1988). Short rotation forest biomass plantations in Portugal. In *Biomass forestry in Europe : a strategy for the future* (ed. F.C. Hummel, W. Palz, and G. Grassi) pp. 509–539. Elsevier, London. [Table 1.1]

Peters, G.B., Lonie, J.S., and Moran, G.F. (1990). The breeding system, genetic diversity and pollen sterility in *Eucalyptus pulverulenta*, a rare species with small disjunct populations. *Australian Journal of Botany* **38**, 559–570. [Table 19.2]

Petroff, G. and Tissot, M. (1983). Variabilité, caractéristiques papetières et selection clonale de deux hybrides d'eucalyptus plantes au Congo *Eucalyptus tereticornis* 12 ABL × *E. saligna*, *Eucalyptus* PF1 =

E. platyphylla × *E. urophylla*. *Bois et Forêts des Tropiques* **199**, 57–77. [22.2.4]

Phillips, M.A. and Brown, A.H.D. (1977). Mating system and hybridity in *Eucalyptus pauciflora*. *Australian Journal of Biological Sciences* **30**, 337–344. [19.6.1; Table 19.2]

Phillips, R.L. and Reid, J.B. (1980). Clinal variation between *Eucalyptus viminalis* Labill. and *E. dalrympleana* Maiden. *Australian Journal of Botany* **28**, 329–342. [17.1, 17.4.2; Table 3.1]

Pires, C.L.S., Garrido, L.M.A.G., Garrido, M.A.O., Buzatto, O., Fernandes, P.S., and Souza, W.J.M. (1983). Teste de origens, procedências e progenies de *Eucalyptus grandis* no Estado de São Paulo. *Silvicultura*, São Paulo **31**, 418–420. [11.4.1]

Pohjonen, V. and Pukkala, T. (1990). *Eucalyptus globulus* in Ethiopian forestry. *Forest Ecology and Management* **36**, 19–31. [10.2]

Polunina, N.N. (1959). The embryology of *Eucalyptus*. *Trudy glavnogo botanicheskogo Sada Leningrad* **6**, 191–210. [19.3.4]

Poore, M.E.D. and Fries, C. (1985). *The ecological effects of eucalypts*. *FAO Forestry Paper* 59. FAO, Rome. [1.1, 2.3]

Potts, B.M. (1984). Phenetic variation in the *Eucalyptus gunnii-archeri* complex. In *Proceedings IUFRO Colloque international sur les eucalyptus resistants au froid*, Bordeaux, September 1983, pp. 91–101. AFOCEL, Paris. [Table 3.1]

Potts, B.M. (1985). Variation in the *Eucalyptus gunnii - archeri* complex. III. Reciprocal transplant trials. *Australian Journal of Botany* **33**, 687–704. [Table 3.1]

Potts B.M. and Potts, W.C. (1986). Eucalypt breeding in France. *Australian Forestry* **49**, 210–218. [18.2.2, 22.1]

Potts, B.M. and Reid, J.B. (1983). Hybridization between *Eucalyptus obliqua* l'Herit. and *E. pulchella* Desf. *Australian Journal of Botany* **31**, 211–229. [19.6.3]

Potts, B.M. and Reid, J.B. (1985a). Variation in the *Eucalyptus gunnii - archeri* complex. I. Variation in the adult phenotype. *Australian Journal of Botany* **33**, 337–359. [Table 3.1]

Potts, B.M. and Reid, J.B. (1985b). Variation in the *Eucalyptus gunnii - archeri* complex. II. The origin of variation. *Australian Journal of Botany* **33**, 519–541. [Table 3.1]

Potts, B.M. and Reid, J.B. (1985c). Analysis of a hybrid swarm between *Eucalyptus risdonii* Hook. f. and *E. amygdalina* Labill. *Australian Journal of Botany* **35**, 543–562. [19.6.3]

Potts, B.M., Potts, W.C., and Cauvin, B. (1987). Inbreeding and interspecific hybridization in *Eucalyptus gunnii*. *Silvae Genetica* **35**, 194–199. [19.6.1]

Poynton, R.J. (1971). A silvicultural map of Southern Africa. *South African Journal of Science* **67**, 58–68. [4.3]

Poynton, R.J. (1979). *Tree planting in southern Africa*. Vol. 2. *The eucalypts*. Department of Forestry, Pretoria. [1.2.1, 1.3, 3.4, 9.4, 11.1, 11.2, 21.2.6]

Prado, J.A. and Barros, S. (1989). *Eucalyptus, principios de silvicultura y manejo*. Instituto Forestal, Santiago de Chile. [1.2.1; Table 1.1]

Prescott-Allen, R. and Prescott-Allen, C. (1982). Park your genes : protected areas as *in situ* gene banks for the maintenance of wild genetic resources. Invited paper for World National Park Congress, Bali, October 1982. [3.5.2]

Pryor, L.D. (1950). A hybrid *Eucalyptus*. *Proceedings of the Linnean Society of New South Wales* **75**, 96–98. [19.6.3]

Pryor, L.D. (1951). Controlled pollination of *Eucalyptus*. *Proceedings of the Linnean Society of New South Wales* **76**, 135–139. [19.3.3]

Pryor, L.D. (1956). Variation in snow gum *Eucalyptus pauciflora* Sieb. with altitude. *Proceedings of the Linnaean Society of New South Wales* **81**, 299–305. [Table 3.1]

Pryor, L.D. (1957). Selecting and breeding for cold resistance in *Eucalyptus*. *Silvae Genetica* **6**, 98–109. [19.6.1]

Pryor, L.D. (1963). Provenance in tree improvement with particular reference to *Eucalyptus*. In *Proceedings of the world consultation on forest genetics and tree improvement*, Stockholm, Vol. 1, FAO/FORGEN 63–3/2. FAO, Rome. [3.1]

Pryor, L.D. (1966). A report on past performance and some current aspects of the cultivation of quick-growing species (mainly *Eucalyptus*) in India. *Indian Forester* **92**, 615–622. [1.3, 6.5, 15.2]

Pryor, L.D. (1975). Eucalypts. In *The methodology of conservation of forest genetic resources*. FO.MISC-75–8, pp. 79–84. FAO, Rome. [3.5.1]

Pryor, L.D. (1976). *The biology of eucalypts*. Arnold, London. [1.2.1, 2.2, 2.2.2, 2.2.3, 4.3.4, 19.1, 19.3.2, 19.3.5, 19.6.1, 19.6.3; Fig. 1.1]

Pryor, L.D. (1978). Forest establishment and utilization for local needs, small scale operations in Africa. In *Forest policy and forest products in national development and the role of foreign aid*, Conference Proceedings, 4–7 July 1978. Australian National University, Canberra. [2.3.1]

Pryor, L.D. and Boden, R.W. (1962). Blowflies as pollinators in producing *Eucalyptus* seed. *Australian Journal of Science* **24**, 3267. [19.3.2]

Pryor, L.D. and Briggs, J.D. (1981). *Australian endangered species: eucalypts. Australian National Parks and Wildlife Service Special Publication* No. 5. ANPWS, Canberra. [3.5.1]

Pryor, L.D. and Byrne, O.R. (1969). Variation and taxonomy in *Eucalyptus camaldulensis*. *Silvae Genetica* **18**, 64–71. [6.3, 6.4.6; Table 3.1]

Pryor, L.D. and Johnson, L.A.S. (1971). *A classification of the eucalypts*. Australian National University Press, Canberra. [1.2.1, 2.1.2, 2.2.4, 10.3, 11.1, 12.1, 14.1, 16.1, 16.3, 17.1, 19.6.3, 20.5.2, 21.4.2; Tables 2.1, 5.1]

Pryor, L.D. and Johnson, L.A.S. (1981). *Eucalyptus*, the universal Australian. In *Ecological biogeography of*

Australia, Vol. 1, (ed. A. Keast), pp. 499–536. Dr W. Junk, The Hague. [2.1.2, 20.5.2]

Pryor, L.D. and Willing, R.R. (1963). The vegetative propagation of *Eucalyptus* - an account of progress. *Australian Forestry* **27**, 52–62. [22.1, 22.3.1, 22.4.1]

Pryor, L.D. and Willing, R.R. (1983). *Growing and breeding poplars.* Canberra Publishing and Printing Co. [18.3]

Purnell, R.C. (1986). Early results from provenance/progeny trials of *Eucalyptus nitens* in South Africa. In *Proceedings IUFRO conference on breeding theory, progeny testing and seed orchards*, Williamsburg, Virginia, October 1986, pp. 500–513. North Carolina State University, Raleigh. [12.5; Table 18.4]

Purnell, R.C. and Lundquist, J.E. (1986). Provenance variation of *Eucalyptus nitens* on the Eastern Transvaal Highveld in South Africa. *South African Forestry Journal* **138**, 23–31. [12.4.1, 12.4.4, 12.5]

Quaile, D.R. and Mullin, L.J. (1984). Variation in productivity in provenances of *Eucalyptus nitens* (Dean and Maid.) Maid. and *Eucalyptus regnans* F. Muell. in Zimbabwe. In *Proceedings IUFRO Colloque international sur les eucalypts resistants au froid*, Bordeaux, September 1983, pp. 83–393. AFOCEL, Paris. [12.4.1, 12.4.2]

Rakatomanampison, A. (1983). Teste de procedencia de *Eucalyptus grandis* em madagascar. *Silvicultura,* São Paulo **31**, 420–422. [11.4.1]

Rao, D.V. (1984). Provenance trials of *Eucalyptus*. *Indian Forester* **110**, 28–34. [6.4.1]

Raymond, C.A. (1988a). *Eucalyptus globulus - a breeding plan for China, ACIAR Project 8457.* Australian Centre for International Agricultural Research, Canberra. [18.2.2]

Raymond, C.A. (1988b). *Eucalyptus delegatensis* provenance trials in Tasmania and NSW. In *Proceedings of the tenth meeting of Research Working Group No. 1 Forest Genetics*, Gympie, pp. 222–225. Australian Forestry Council, Canberra [8.4.1, 8.5]

Raymond, C.A. (1989). Developing breeding populations for temperate eucalypts in relation to natural population structure. In *Breeding tropical trees: population structure and genetic improvement strategies in clonal and seedling forestry*, Proceedings IUFRO Conference, Pattaya, November 1988. (ed. G.L. Gibson, A.R. Griffin, and A.C. Matheson), pp. 423–424. Oxford Forestry Institute. [5.8]

Raymond, C.A. and Cotterill, P.P. (1990). Methods of assessing crown form of *Pinus radiata*. *Silvae Genetica* **39**, 67–71. [20.1.5]

Raymond, C.A., Harwood, C.E. and Owen, J.V. (1986). A conductivity method for screening populations of eucalypts for frost damage and frost tolerance. *Australian Journal of Botany* **34**, 377–393. [5.1]

Reddy, K.V. and Rockwood, D.L. (1989). Breeding strategies for coppice production in a *Eucalyptus grandis* base population with four generations of selection. *Silvae Genetica* **38**, 148–151. [Table 18.4]

Restrego, G. and Stohr, G.W.D. (1983). Resultados preliminares de ensaios de procedência de *Eucalyptus* spp. L'Hérit. no sudesde do Paraná — Brasil. *Silvicultura,* São Paulo **31**, 497–502. [17.4.2]

Richardson, K.F. and Meakins, R.H. (1986). Inter- and intra-specific variation in the susceptibility of eucalypts to the snout beetle *Gonipterus scutellatus* Gyll. (*Coleoptera*: *Curculionidae*). *South African Forestry Journal* **139**, 21–31. [17.2, 17.4.3; Figs 17.4, 17.5]

Richmond, K.P. (1971). A seed orchard of superior *Eucalyptus regnans*. *Institute of Foresters of Australia Newsletter* **12**(3), 13–15. [18.2.2, 19.6.2, 21.2.6]

Roberts, E.H. (1975). Problems of long-term storage of seed and pollen for genetic resources conservation. In *Crop genetic resources for today and tomorrow* (ed. O.H. Frankel and J.G. Hawkes) pp. 269–296. Cambridge University Press. [3.5.3]

Robinson, D.L. (1987). *REML user manual: Estimation of variance components in non-orthogonal data by residual maximum likelihood.* Scottish Agricultural Statistics Service, Edinburgh. [5.6.3]

Robinson, L.W. and Wareing, P.F. (1969). Experiments on the juvenile-adult phase change in some woody species. *New Phytologist* **68**, 67–78. [22.3]

Roche, L.R. (1975). Guidelines for the methodology of conservation of forest genetic resources. In *The methodology of conservation of forest genetic resources.* FO.MISC-75–8, p. 107. FAO, Rome. [3.5.4]

Rockwood, D.L., Reddy, K.V., Warrag, E.I., and Comer, C.W. (1987). Development of *Eucalyptus amplifolia* for woody biomass production. *Australian Forest Research* **17**, 173–178. [21.2.6]

Roeder, K.R. (1980a). Early provenance performance of *Eucalyptus fastigata* in South Africa. *South African Forestry Journal* **112**, 30–32. [9.3.1]

Roeder, K.R. (1980b). 'Provenance' variation in growth features of *Eucalyptus grandis* and *Eucalyptus saligna* in South Africa. *South African Forestry Journal* **112**, 36–41. [11.4.1]

Roger, J.H. and Rao, M.R. (1990). Agroforestry field experiments: discovering the hard facts. Part One: Statistical considerations. *Agroforestry Today* **2**(1), 4–7. [5.5, 5.6.2]

Rook, D.A., Wilcox, M.D., Holden, D.G., and Warrington, I.J. (1980). Provenance variation in frost tolerance of *Eucalyptus regnans* F. Muell. *Australian Forest Research* **10**, 213–238. [5.1, 14.3.2; Fig. 14.5]

Rudman, P. (1970). The influence of genotype and environment on wood properties of juvenile *Eucalyptus camaldulensis* Dehn. *Silvae Genetica* **19**, 49–54. [6.4.2, 20.1.7]

Ruggeri, C. (1961). Contributo alla cariologia del genere *Eucalyptus* (*Myrtaceae*). *Caryologia* **14**, 111–120. [2.2.2]

Rumpf, G.E. (1743). Arbor versicolor. *Ay Alla. Herbarii Amboinensis* **3**, 122. [7.3]

Rye, B.L. (1980). Chromosome numbers, reproductive biology and evolution in the *Myrtaceae*. Unpublished Ph.D. thesis. University of Western Australia, Perth. [2.2.2]

SAA (Standards Association of Australia) (1983). *Nomenclature of Australian timbers*. *Australian Standard* 2543. SAA, Sydney. [2.1.2; Table 2.1]

Sampaio, A. and Suiter Filho, W. (1983). Sementes florestais melhoradas. *Brasil Florestal* **54**, 71–73. [21.5]

Sampson, J.F., Hopper, S.D., and James, S.H. (1989). The mating system and population genetic structure in a bird-pollinated mallee, *Eucalyptus rhodantha*. *Heredity* **63**, 383–393. [Table 19.2]

Sanchez, P.A. and Isbell, R.F. (1979). A comparison of the soils of tropical Latin America and tropical Australia. In *Pasture production in acid soils of the tropics*, Proceedings of a seminar held at CIAT, Columbia, April 1978 (ed. P.A. Sanchez and L.E. Tergus), pp. 25–52. CIAT, Columbia. [4.4]

Sands, R. (1981). Salt resistance in *Eucalyptus camaldulensis* Dehn. from three different seed sources. *Australian Forest Research* **11**, 93–100. [6.4.7]

San Miguel, A. (1988). Short rotation forest biomass plantations in Spain. In *Biomass forestry in Europe: a strategy for the future* (ed. F.C. Hummel, W. Palz, and G. Grassi) pp. 540–579. Elsevier, London. [Table 1.1]

Saxena, N.C. (1991). Marketing constraints for Eucalyptus from farm lands in India. *Agroforestry Systems* **13**, 73–85. [2.3.1, 4.2.4]

Schaffalitzky de Muckadell, M. (1959). Investigations of ageing of apical meristems in woody plants and its importance in silviculture. *Forstlige Forsogsvaesen i Danmark* **25**, 310–455. [22.3]

Schönau, A.P.G. (1984). Basic silviculture for the establishment of eucalyptus plantations with special reference to *Eucalyptus grandis*. In *Proceedings IUFRO Colloque international sur les eucalyptus resistants au froid*, September 1983, pp. 469–479. AFOCEL, Paris. [11.4.2]

Schönau, A.P.G. (ed.) (1991). *Symposium on intensive forestry: the role of eucalypts*. Proceedings IUFRO Symposium, Durban, September 1991, 2 volumes. South African Institute of Forestry, Pretoria. [1.2.1]

Schönau, A.P.G. and Schulze, R.E. (1984). Climatic and altitudinal criteria for commercial afforestation with special reference to Natal. In *Proceedings IUFRO Symposium on site and productivity of fast growing plantations* (ed. D.C. Grey et al.). Vol. 2, pp. 551–563. South African Forestry Research Institute, Pretoria. [4.3.3]

Schreiner, H.G. (1989). Culturas intercalares de soja em reflorestamento de eucaliptos no sul-sudeste do Brasil. *Boletim de Pesquisa Florestal*, Curitiba **18–19**, 1–10. [23.1]

Sedgley, M. (1989). Ovule and seed development in *Eucalyptus woodwardii* Maiden (*Symphyomyrtus*). *Botanical Gazette* **150**, 271–280. [19.3.4]

Sedgley, M. and Griffin, A.R. (1989). *Sexual reproduction of tree crops*. Academic Press, London. [18.3.6, 19.1, 19.3.2, 19.3.3, 19.3.5, 19.6.1, 19.6.2]

Sedgley, M. and Smith, R.M. (1989). Pistil receptivity and pollen tube growth in relation to the breeding system of *Eucalyptus woodwardii* (*Symphyomyrtus*: Myrtaceae). *Australian Journal of Botany* **64**, 21–31. [19.3.1, 19.3.4, 19.3.5]

Sedgley, M., Hand, F.C., Smith, R.M., and Griffin, A.R. (1989). Pollen tube growth and early seed development in *Eucalyptus regnans* F. Muell. (*Myrtaceae*) in relation to ovule structure and preferential outcrossing. *Australian Journal of Botany* **37**, 397–411. [19.3.4]

Sesbou, A. and Nepveu, G. (1978). Variabilité infraspecifique du retrait avec collapse et de la densité du bois chez *Eucalyptus camaldulensis*. *Annales des Sciences Forestière* **35**, 237–263. [6.4.1, 6.4.2, 6.4.7, 6.5; Table 6.2]

Shah, S.A. (1985). Eucalyptus - friend or foe? *International Tree Crops Journal* **3**, 191–195. [1.1]

Shaw, M.J., Potts, B.M., and Reid, J.B. (1984). Variation within and between *Eucalyptus nitida* Hook. f. and *E. coccifera* Hook. f. *Australian Journal of Botany* **32**, 641–654. [Table 3.1]

Shelbourne, C.J.A. (1969). Tree breeding methods. A review of tree breeding strategies in relation to classical plant breeding methods with quantitative genetic expectations of gain, and predictions using data from *Pinus radiata*. *New Zealand Forest Service Technical Paper* No. 55. [18.2.3, 20.4, 20.4.1]

Shelbourne, C.J.A. (1991). Genetic gains from different kinds of breeding populations and seed or plant production methods. In *Intensive forestry : the role of eucalypts* (ed. A.P.G. Schönau), Proceedings IUFRO Symposium, Durban, September 1991 Vol. 1, pp. 300–317. South African Institute of Forestry, Pretoria. [18.2.3, 18.3.7, 20.4, 21.1]

Shelbourne, C.J.A., Burdon, R.D., Carson, S.D., Firth, A., and Vincent, T.G. (1986). *Development plan for radiata pine breeding*. Forest Research Institute, Rotorua. [18.1, 18.4]

Shelbourne, C.J.A., Carson, J.J., and Wilcox, M.D. (1989). New techniques in the genetic improvement of radiata pine. *Commonwealth Forestry Review* **68**, 191–201. [20.3]

Shepherd, K.R. (1974). Conservation of forest gene resources — Australia's responsibilities. *Australian Forestry* **37**, 70–76. [3.5.2]

Shepherd, K.R., Banks, J.C.G., and Atyeo, W.J. (1976). Variation in *Eucalyptus nitens* Maiden in response to temperature and seed source. *Australian Journal of Botany* **24**, 167–176. [12.3; Table 3.1]

Sherry, S.P. and Pryor, L.D. (1967). Growth and differential frost-resistance of topoclinal forms of *Eucalyptus fastigata* D. & M. planted in South Africa. *Australian Forestry* **31**, 33–44. [9.3.1]

Shiva, V. and Bandyopadhyay, J. (1983). Eucalyptus — a disastrous tree for India. *The Ecologist* **13**, 184–187. [1.1]

Sidaway, S. (1988). The availability and use of eucalyptus pulps. *Tappi Journal* **71**(12), 47–51. [1.2.2]

Siddiqui, K.M., Khan, M., and Akhtar, S. (1979). Results of 10-year old *Eucalyptus camaldulensis* Dehn. provenance study at Peshawar. *Silvae Genetica* **28**, 24–26. [6.4.2, 6.5]

Simmonds, N.W. (1979). *Principles of crop improvement.* Longman, London. [1.5, 18.2, 18.3]

Skolmen, R.G. (1986). Performance of Australian provenances of *Eucalyptus grandis* and *Eucalyptus saligna* in Hawaii. *Research Paper* PSW-181. USDA Forest Service, Berkeley. [11.4.1]

Slatyer, R.O. (1975). Ecological reserves: size, structure and management. In *A national system of ecological reserves in Australia* (ed. F. Fenner) Report 19. Australian Academy of Science, Canberra. [3.5.4]

Slee, M.U. (1985). Genetic Resources — the issues. In *Research for forest management* (ed. J.J. Landsberg and W. Parsons), pp. 140–155. CSIRO, Melbourne. [18.1]

Snedecor, G.W. and Cochran, W.C. (1980). *Statistical methods* (7th edn.). Iowa State University Press. [3.3]

Snyder E.B. (1972). *Glossary for forest tree improvement workers.* Southern Forest Experiment Station, Forest Service, USDA. [Glossary]

Specht, R.L., Roe, E.M., and Boughton, V.A. (1974). Conservation of major plant communities in Australia and Papua New Guinea. *Australian Journal of Botany Supplementary Series* **7**, 1–666. [3.5.1, 3.5.2]

Squillace, A.E. (1974). Average genetic correlations among offspring from open-pollinated forest trees. *Silvae Genetica* **23**, 149–156. [18.3.2]

Stace, H.C.T. et al. (1972). *A handbook of Australian soils.* Rellum, Adelaide. [4.4]

Stebbins, G.L. (1971). *Processes of organic evolution* (2nd edn). Prentice-Hall, New Jersey. [2.2.2]

Streets, R.J. (1962). *Exotic forest trees in the British Commonwealth.* Clarendon Press, Oxford. [4.3, 14.2]

Suseno, O.H., Na'iem, M., and Danarto, S. (1985). 28-month performance of *Eucalyptus urophylla* S.T. Blake in provenance test at Wanagama I, Yogyakarta. In *Proceedings ASEAN-Australia forest tree improvement program, Bangkok meeting,* August 1985. [16.3, 16.4]

Talbert, J.T. (1979). An advanced-generation breeding plan for the N. C. State University — Industry pine tree improvement cooperative. *Silvae Genetica* **28**, 72–75. [18.4]

Thibout, A. (1984). Contribution a la connaissance de l'*Eucalyptus dalrympleana.* In *Proceedings IUFRO Colloque international sur les eucalyptus resistants au froid,* Bordeaux, September 1983, pp. 196–212. AFOCEL, Paris. [Table 3.1]

Thomas, D.A. and Barber, H.N. (1974). Studies in leaf characteristics of a cline of *Eucalyptus urnigera* from Mount Wellington, Tasmania. I Water repellency and the freezing of leaves. *Australian Journal of Botany* **22**, 501–512. [Table 3.1]

Thomson, L.A.J. and Merwin, M.L. (1987). Seed collections of *Eucalyptus camaldulensis* Dehnh. in Western Victoria, May-June 1985. *Forest Genetic Resources Information,* FAO Rome **15**, 56–60. [6.4.1, 6.5; Fig. 6.8]

Thomson, L.A.J., Morris, J.D., and Halloran, G.M. (1987). Salt tolerance in eucalypts. In *Proceedings International symposium on afforestation of salt-affected soils,* February 1987, Vol. 3, pp. 1–12. Central Soil Salinity Research Institute, Karnal, India. [4.4, 5.1, 6.4.7; Table 6.3]

Thornthwaite, C.W. (1948). An approach towards a rational classification of climate. *Geographic Review* **38**, 55–94. [4.3]

Thornthwaite, C.W. and Mather, J.R. (1956). Instructions and tables for computing potential evapotranspiration and the water balance. *Publications in Climatology* **10**(3) pp. 185–311. Certenton, New Jersey. [4.3]

Tibbits, W.N. (1986). Eucalypt plantations in Tasmania. *Australian Forestry* **49**, 216–225. [12.2]

Tibbits, W.N. (1988). Germination and morphology of progeny from controlled pollination of *Eucalyptus nitens* (Deane & Maiden) Maiden. *Australian Journal of Botany* **36**, 677–691. [19.6.1]

Tibbits, W.N. (1989). Controlled pollination studies with shining gum (*Eucalyptus nitens* (Deane & Maiden) Maiden). *Forestry* **62**, 111–126. [19.6.1]

Tibbits, W.N. and Reid, J.B. (1987). Frost resistance in *Eucalyptus nitens* (Deane and Maiden) Maiden : genetic and seasonal aspects of variation. *Australian Forest Research* **17**, 29–47. [4.3.4, 12.4.3]

Tibbits, W.N., de Little, D.W., and Ravenwood, I.C. (1990). Improving short-rotation, fast-growing, cool temperate eucalypts through controlled pollination, hybridisation, and vegetative propagation. In *Proceedings of the Thirteenth Commonwealth Forestry Conference, Rotorua, New Zealand,* September 1989, pp. 19–24. Ministry of Forestry, Wellington. [21.4.3, 22.3.5]

Timoni, J.L., Coelho, L.C.C., Kageyama, P.Y., and da Silva, A.A. (1983). Teste de procedencia de *Eucalyptus* spp. na região de Moji- Guacu (SP). *Silvicultura,* São Paulo **31**, 505–507. [15.3]

Tiwari, K.M. (1983). *Social forestry and rural development.* International Book Distributors, Dehra Dun. [1.1]

Tiwari, K.M., and Mathur, R.S. (1983). Water consumption and nutrient uptake by eucalypts. *Indian Forester* **12**, 851–860. [1.1]

Toda, R. (1964). A brief review and conclusions of the discussion on seed orchards. *Silvae Genetica* **13**, 1–4. [21.2]

Troup, R.S. (1932). *Exotic forest trees of the British Empire.* Clarendon Press, Oxford. [4.3]

Truswell, E.M. (1990). Australian rainforests: the 100 million year record. In *Australian rainforests: science, values, meanings* (ed. L.J. Webb and J. Kikkawa), pp. 7–22. CSIRO, Melbourne. [2.2.1]

Turesson, G. (1923). The scope and import of genecology. *Hereditas* **4**, 171–176. [3.1]

Turnbull, J.W. (1973). The ecology and variation of *Eucalyptus camaldulensis* Dehn. *Forest Genetic Resources Information* 2, FAO Forestry Occasional Paper 1973–2, 32–40. [6.3]

Turnbull, J.W. (1974). *Eucalyptus deglupta* Blume. *Forest Tree Series* 175. Forestry and Timber Bureau, Canberra. [7.1]

Turnbull, J.W. (1975). Seed collection of eucalypts. *Report of FAO/DANIDA training course on forest seed collection and handling, Chiang Mai, Thailand* March 1975, Vol. 2, pp. 337–34. [6.3]

Turnbull, J.W. (1978). Exploration and conservation of eucalypt gene resources. In *Documents FAO third world consultation on forest tree breeding* Vol. 1, pp. 33–44. CSIRO, Canberra. [3.5, 3.5.2, 3.5.3]

Turnbull, J.W. (1979). Geographic variation in *Eucalyptus cloeziana.* Unpublished Ph.D. thesis, Australian National University, Canberra. [Table 3.1]

Turnbull, J.W. (1981). Eucalypts in China. *Australian Forestry* **44**, 222–234. [1.2.1]

Turnbull, J.W. (1983). Provenance variation in *Eucalyptus cloeziana* F. Muell. *Silvicultura*, São Paulo 31, 508–511. [Table 3.1]

Turnbull, J.W. and Brooker, I. (1978). Timor mountain gum, *Eucalyptus urophylla* S.T. Blake. *Forest Tree Leaflet* 214. CSIRO, Melbourne. [16.1]

Turnbull, J.W. and Eldridge, K.G. (1984). The natural environment of *Eucalyptus* as the basis for selecting frost resistant species. In *Proceedings IUFRO Colloque international sur les Eucalyptus resistants au froid*, Bordeaux, September 1983 pp. 43–62. AFOCEL, Paris. [4.3.2, 4.3.4]

Turnbull, J.W. and Griffin, A.R. (1986). The concept of provenance and its relationship to infraspecific classification in forest trees. In *Infra-specific classification of wild and cultivated plants* (ed. B.T. Styles), pp. 157–189. Clarendon Press, Oxford. [3.1]

Turnbull, J.W. and Pryor, L.D. (1984). Choice of species and seed sources. In *Eucalypts for wood production* (ed. W.E. Hillis and A.G. Brown), pp. 6–65. CSIRO/Academic Press, Melbourne. [4.3, 6.4.1]

Turner, C.H., Balodis, V., and Dean, G.H. (1983). Variability in pulping quality of *E. globulus* from Tasmanian provenances. *Appita* **36**, 371–376. [10.4.2]

Van Wyk, G. (1976). Early growth results in a diallel progeny test of *Eucalyptus grandis* (Hill) Maiden. 1. A field study. *Silvae Genetica* **25**, 126–132. [18.3.1, 18.3.6; Table 18.3]

Van Wyk, G. (1977a). Pollen handling, controlled pollination and grafting of *Eucalyptus grandis*. *South African Forestry Journal* **101**, 47–53. [20.5.1, 22.4.1]

Van Wyk, G. (1977b). Graft incompatibility in *Eucalyptus grandis*. *South African Forestry Journal* **103**, 15–17. [21.3.3]

Van Wyk, G. (1981). Pollen management for eucalypts. In *Pollen management handbook* (ed. E.C. Franklin) *Agriculture Handbook* No. 587, pp. 84–88. USDA, Washington. [21.4.2]

Van Wyk, G. (1983). Inbreeding effects on *E. grandis* families with different degrees of relatedness. *Silvicultura,* São Paulo **31**, 568–571. [19.6.1]

Van Wyk, G. (1985a). Tree breeding in support of vegetative propagation of *Eucalyptus grandis* (Hill) Maiden. *South African Forestry Journal* **135**, 33–39. [11.5, 18.2.2, 22.1; Table 18.2]

Van Wyk, G. (1985b). Genetic variation in wood preservation of fast grown *Eucalyptus grandis* families. In *Forest products research international — achievements and the future,* Proceedings IUFRO Symposium Pretoria April 1985 Vol. 1, paper 3–4, pp. 1–16. CSIR, Pretoria. [11.5, 18.3.3]

Van Wyk, G. (1990). Genetic improvement of timber yield and wood quality in *Eucalyptus grandis* (Hill) Maiden. Part I. Genetic parameters of growth characteristics. *South African Forestry Journal* **153**, 1–11. [18.3.1; Tables 18.2, 18.4]

Van Wyk, G. and Hodgson, L.M. (1983). Clonal seed orchards of *Eucalyptus grandis* in South Africa. *Silvicultura,* São Paulo **31**, 566–568. [1.4.5, 20.1.3, 20.6.2, 21.3.6]

Van Wyk, G., and Roeder, K.R. (1978). The status of tree breeding in South Africa. *South African Forestry Journal* **107**, 54–59. [20.1.4]

Van Wyk, G., Schönau, A.P.G., and Schön, P.P. (1989). Growth potential and adaptability of young eucalypt hybrids in South Africa. In *Breeding tropical trees,* Proceedings IUFRO Conference, Pattaya, Thailand, November 1988, (ed. G.L. Gibson, A.R. Griffin, A.C. Matheson) pp. 325–333. Oxford Forestry Institute. [18.2.3]

Venkatesh, C.S. and Sharma, V.K. (1979). Comparison of a *Eucalyptus tereticornis* × *E. grandis* controlled hybrid with a *E. grandis* × *E. tereticornis* putative natural hybrid. *Silvae Genetica* **28**, 127–131. [19.6.1]

Venkatesh, C.S., Arya, R.S., and Sharma, V.K. (1973). Natural selfing in planted eucalyptus and its estimation. *Journal of Plantation Crops* **1**, 23–25. [19.6.1]

Vieira, F.S. (1983). Procedencias de *Eucalyptus urophylla* S.T. Blake. *Silvicultura,* São Paulo **31**, 512–515. [16.4]

Vigneron, P. (1989). Les hybrides artificiels d'Eucalyptus au Congo — création et multiplication. In *Breeding tropical trees*, Proceedings IUFRO Conference, Pattaya, November 1988 (ed. G.L. Gibson, A.R. Griffin, and A.C. Matheson) pp. 425–427. Oxford Forestry Institute. [20.5.2]

Vigneron, P. (1991). Création et amélioration de varietes hybrides d' Eucalyptus au Congo. In *Intensive forestry : the role of eucalypts* (ed. A.P.G. Schönau), IUFRO Symposium, Durban, September 1991, Vol. 1, pp. 345–3 South African Institute of Forestry, Pretoria. [18.2.2, 20.5.2]

Vivekanandan, K. (1978). Variation in density of *Eucalyptus camaldulensis*. *Sri Lanka Forester* **13**, 71–72. [6.4.2]

Volker, P.W. and Orme, P.K. (1988). Provenance trials of *Eucalyptus globulus* and related species in Tasmania. *Australian Forestry* **51**, 257–265. [10.4.1; Tables 10.1, 10.2; Fig. 10.4]

Volker, P.W. and Raymond, C.A. (1989). Potential for breeding eucalypts in Tasmania. *Appita* **42**, 198–200. [Fig. 14.6]

Volker, P.W., Dean, C.A., Tibbits, W.N., and Ravenwood, I.C. (1990). Genetic parameters and gains expected from selection in *Eucalyptus globulus* in Tasmania. *Silvae Genetica* **33**, 18–21. [18.3.3; Tables 18.4, 18.5]

Volker, P.W., Fripp, Y.J., and Griffin, A.R. (1992). Flowering phenology in a multi-provenance seedling seed orchard of *Eucalyptus globulus* subsp. *globulus* in Tasmania. *New Forests* (in press). [19.2.2]

Von Mueller, F. (1886). Description of a hitherto unrecorded species of Eucalyptus from New Britain. *Australian Journal of Pharmacy* **1**, 239–240. [7.3]

Wallenberg (1984). The new eucalypt forest. *Marcus Wallenberg Foundation symposia proceedings* 1. Wallenberg Foundation, Falun. [1.6.1]

Walter, H. and Lieth, H. (1960). *Klimadiagramm-weltatlas*. Fischer, Jena. [4.2.3, 4.3.2]

Waugh, G. and Rozsa, A. (1991). Sawn products from regrowth *Eucalyptus regnans*. In *The young eucalypt report* (ed. C.M. Kerruish and W.H.M. Rawlins) pp. 178–209. CSIRO, Melbourne. [20.1.4]

Webb, D.P., Ellis, R.C., and Hallam, P.M. (1983). Growth check of *Eucalyptus delegatensis* (R.T. Baker) regeneration at high altitudes in northeastern Tasmania. *Information Report* O-X-348, Great Lakes Forest Research Centre, Canadian Forestry Service, Sault Ste Marie. [8.4.2]

Webb, D.B., Wood, P.J., Smith, J.P., and Henman, G.S. (1984). A guide to species selection for tropical and subtropical plantations. *Tropical Forestry Papers* 15 (2nd edn) [1st edn 1980]. Commonwealth Forestry Institute, Oxford. [4.3, 4.3.3, 4.5]

Wencelius, F. (1983). *Eucalyptus urophylla* en Côte d'Ivoire. *Silvicultura*, São Paulo **31**, 515–518. [16.4; Fig. 16.6]

Weston, G.C. (1957). *Exotic forest trees in New Zealand. Bulletin* 13. New Zealand Forest Service, Wellington. [2.1.1, 9.2]

White, B.J. (1971). Karri silvics. *Research Note* 1. Forests Department of Western Australia, Perth. [19.5]

White, K.J. (1985). Intercrop strategies in the Bhabar Terai of central Nepal. *Sagarnath Forest Development Project Manual* No. 1. Ministry of Forests, Kathmandu. [23.1]

White, K.J. (1986). Practical tree breeding strategies and practices in the Bhabar Terai of Central Nepal. *Sagarnath Forest Development Project Manual* No. 5. Ministry of Forests, Kathmandu. [1.6.1, 6.5, 15.4, 22.1]

White, K.J. (1989). *Eucalyptus urophylla* provenance trials Java and Sumatera, Indonesia. Paper to Fourth Technical Exchange Seminar, October 1989, *China-Australia Afforestation Project at Dongmen State Forest Farm, Technical Communication* No. 40. Queensland Department of Forestry, Brisbane. [16.4]

White. K.J. (1990). Eucalyptus research directions in Asia, an assessment of practical management requirements. *Paper presented to International Eucalyptus Symposium, Zhanjiang, China* [23.1]

White, T.L. (1987). A conceptual framework for tree improvement programs. *New Forests* **4**, 325–342. [1.3, 26.1; Fig. 18.3]

White, T.L. and Hodge, G.R. (1988). Best linear prediction of breeding values in forest tree improvement. *Theoretical and Applied Genetics* **76**, 719–727. [18.3, 20.4]

White, T.L. and Hodge, G.R. (1989). *Predicting breeding values with applications in forest tree improvement*. Kluwer Academic Publishers, Dordrecht. [5.6.3]

Whitmore, J.L. (1973). International provenance trial of *Eucalyptus deglupta* in Puerto Rico. In *Tropical provenance and progeny research and international cooperation* (ed. J. Burley and D.G. Nikles) pp. 297–299. Commonwealth Forestry Institute, Oxford. [7.4]

Whitmore, J.L. (1978). Bibliography on *Eucalyptus deglupta* Bl. *Research Note* No. ITF 17, USDA Forest Service Institute of Tropical Forestry, Rio Piedras, Puerto Rico. [7.2]

Whitmore, J.L. (1983). *Eucalyptus deglupta, E. urophylla, E. alba* provenances tested in Puerto Rico. *Silvicultura*, São Paulo **31**, 518–520. [7.4, 7.6, 16.4]

Whitmore, J.L. and Macia, F. (1975). Eucalyptus provenances tested in Puerto Rico: two years after outplanting. *Forest Science* **21**, 410–412. [7.4]

Whitmore, T.C. (ed.) (1981). *Wallace's line and plate tectonics*. Clarendon Press, Oxford. [2.2.1]

Wilcox, M.D. (1980). Genetic improvement of eucalyptus in New Zealand. *New Zealand Journal of Forestry Science* **10**, 343–359. [9.2, 11.4.1, 12.2, 13.6, 14.2]

Wilcox, M.D. (1982a). Preliminary selection of suitable provenances of *Eucalyptus regnans* for New Zealand. *New Zealand Journal of Forestry Science* **12**, 468–479. [14.4]

Wilcox, M.D. (1982b). Selection of genetically superior *Eucalyptus regnans* using family tests. *New Zealand Journal of Forestry Science* **12**, 480–493. [14.4]

Wilcox, M.D. (1982c). Anthocyanin polymorphism in seedlings of *Eucalyptus fastigata* Deane & Maid. *Australian Journal of Botany* **30**, 501–509. [9.3.2, 9.4; Table 9.2]

Wilcox, M.D. (1982d). Genetic variation in frost tolerance, early height growth, and incidence of forking among and within provenances of *Eucalyptus fastigata*. *New Zealand Journal of Forestry Science* **12**, 510–524. [9.3.1; Fig. 9.5]

Wilcox, M.D. (1983). Provenance selection in *Eucalyptus delegatensis. In Report of Forest Research Institute 1982*, pp. 14–15. New Zealand Forest Service, Wellington. [8.3, 8.5; Table 8.1]

Wilcox, M.D., Faulds, T., Vincent, T.G., and Poole, B.R. (1980). Genetic variation in frost tolerance among open-pollinated families of *Eucalyptus regnans* F. Muell. *Australian Forest Research* **10**, 169–184. [14.3.2]

Wilcox, M.D., Rook, D.A., and Holden, D.G. (1983a). Provenance variation in frost resistance of *Eucalyptus fastigata* Deane and Maiden. *Silvicultura,* São Paulo **31**, 521–523. [9.3.1; Fig. 9.5]

Wilcox, M.D., Faulds, T., and Vincent, T.G. (1983b). Genetic improvement of *Eucalyptus saligna* Sm. in New Zealand. *Silvicultura,* São Paulo **31**, 578–583. [11.4.1]

Wilcox, M.D., Miller, J.T., Williams, I.M., and Guild, D.W. (1985). Eucalyptus species trials in Longwood Forest, Southland. *Bulletin* **95**, New Zealand Forest Research Institute, Rotorua. [8.4.1, 8.5]

Willan, R.L. (1979). Eucalypts for South-east Asia. *Tropical Agriculture Research Series* No. 12, pp. 153–161. Tropical Agriculture Research Centre, Tsukuba, Japan. [4.3.3, 5.1]

Wright, J.W. (1976). *Introduction to forest genetics.* Academic Press, New York. [18.3; Table 18.3]

Wright, P.J. and Eldridge, K.G. (1985). Profitability of using seed from the Tallaganda radiata pine seed orchard. *Appita* **38**, 341–344. [18.1]

Yeatman, C.W. (1972). Gene pool conservation for applied breeding and seed production. In *Proceedings IUFRO genetics and SABRAO joint symposium, Tokyo*, paper no. B-8(v). [3.5.2]

Yeh, F.C., Brune, A., Cheliak, W.M., and Chipman, D.C. (1983). Mating system of *Eucalyptus citriodora* in a seed-production area. *Canadian Journal of Forest Research* **13**, 1051–1055. [Table 19.2]

Zacharin, R.F. (1978). *Emigrant eucalypts, gum trees as exotics.* Melbourne University Press. [1.2.1, 4.2.3, 15.2]

Zani, J. and Kageyama, P.Y. (1984). A produção de sementes melhoradas de especies florestais, com enfase en Eucalyptus. *IPEF,* Piracicaba **27**, 49–52. [21.1]

Zhou, W.L. and Bai, J.Y. (1989). Tropical eucalypt trials on Hainan Island, People's Republic of China. In *Trees for the tropics* (ed. D.J. Boland) pp. 79–87. Australian Centre for International Agricultural Research, Canberra. [15.3]

Zobel, B. (1981). Vegetative propagation in forest management operations. In *Proceedings 16th southern forest tree improvement meeting,* Blacksburg, Virginia pp. 149–159. [22.1]

Zobel, B., Campinhos, E., and Ikemori, Y. (1983). Selecting and breeding for desirable wood. *Tappi* **66**, 70–74. [22.2.4]

Zobel, B.J. and Talbert, J. (1984). *Applied forest tree improvement.* Wiley, New York. [1.4.5, 3.4, 18.3, 18.3.1, 20.4, 20.4.1, 21.2]

Zobel, B.J., Van Wyk, G. and Stahl, P. (1987). *Growing exotic forest trees.* Wiley, New York. [5.6.3]

Zohar, Y. (1989). Biomass production of short rotation *Eucalyptus camaldulensis* Dehn. stands growing on peat soil under a high water table in Israel. *South African Forestry Journal* **149**, 54–57. [6.2]

Species Index

Subject Index

THE PHILIPPINES
Manila
Mindanao
BRUNEI
SABAH
SARAWAK
Singapore
Kalimantan
Sulawesi
Moluccas Is.
Ceram
I N D O N E S I A
Java
Bali
Flores
Wetar
Timor
Timor Sea

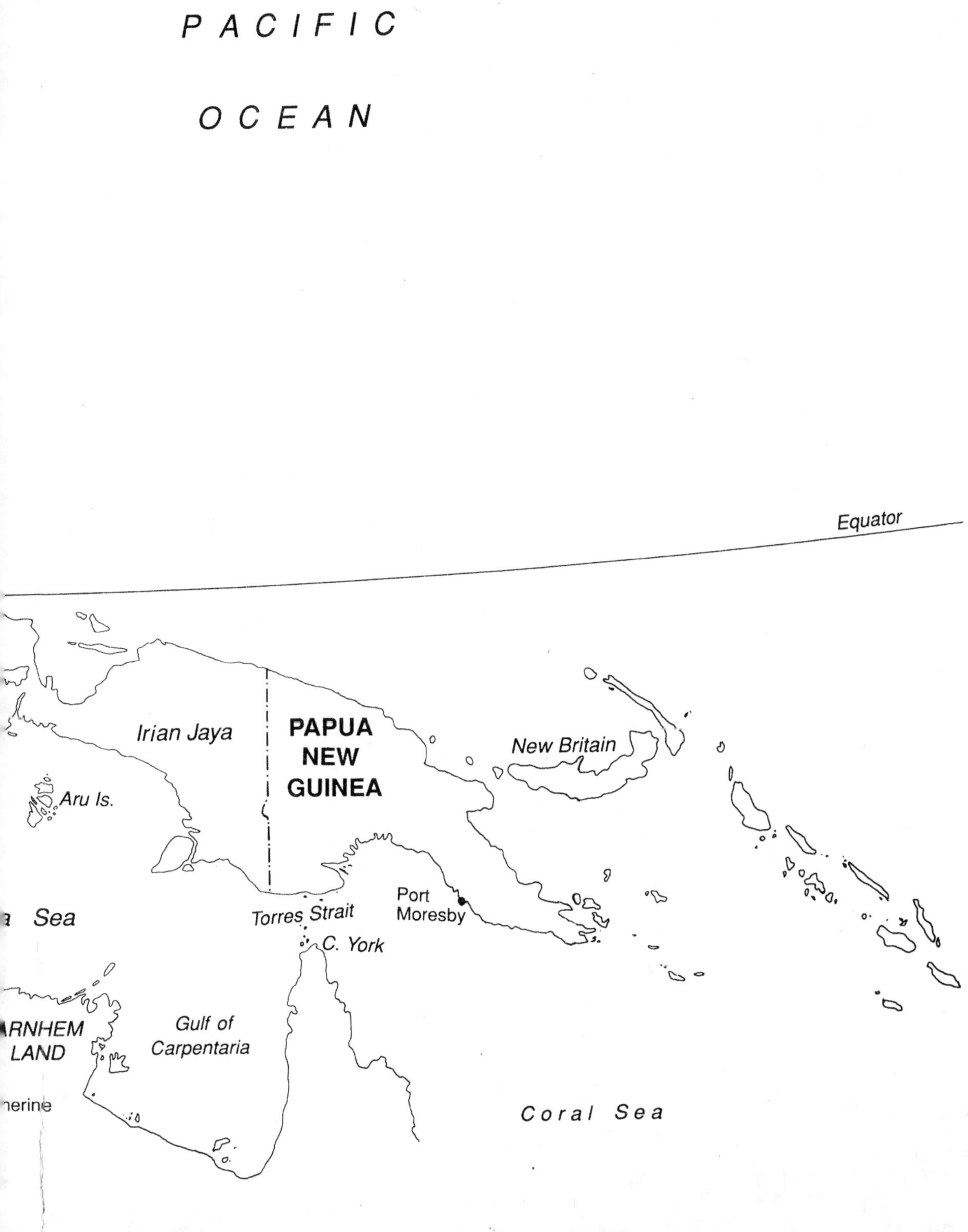
PACIFIC
OCEAN
Equator
Irian Jaya
PAPUA
NEW
GUINEA
New Britain
Aru Is.
Sea
Torres Strait
Port
Moresby
C. York
LAND
Gulf of
Carpentaria
Coral Sea

The manufacturer's authorised representative in the EU for product safety is Oxford University Press España S.A. of el Parque Empresarial San Fernando de Henares, Avenida de Castilla, 2 – 28830 Madrid (www.oup.es/en or product.safety@oup.com). OUP España S.A. also acts as importer into Spain of products made by the manufacturer.

www.ingramcontent.com/pod-product-compliance
Lightning Source LLC
LaVergne TN
LVHW080526210926
848481LV00022B/3121
9780198548669